Methods in Molecular Biology

Series Editor
John M. Walker
School of Life and Medical Sciences
University of Hertfordshire
Hatfield, Hertfordshire, AL10 9AB, UK

For further volumes:
http://www.springer.com/series/7651

Prions

Methods and Protocols

Edited by

Victoria A. Lawson

Department of Pathology
The University of Melbourne
Parkville, VIC, Australia

Humana Press

Editor
Victoria A. Lawson
Department of Pathology
The University of Melbourne
Parkville, VIC, Australia

ISSN 1064-3745 ISSN 1940-6029 (electronic)
Methods in Molecular Biology
ISBN 978-1-4939-7242-5 ISBN 978-1-4939-7244-9 (eBook)
DOI 10.1007/978-1-4939-7244-9

Library of Congress Control Number: 2017952947

Printed on acid-free paper

This Humana Press imprint is published by Springer Nature
The registered company is Springer Science+Business Media LLC
The registered company address is: 233 Spring Street, New York, NY 10013, U.S.A.

Preface

The transmissible spongiform encephalopathy or prion diseases are a group of invariably fatal neurodegenerative disorders with sporadic, familial and acquired etiologies that affect a variety of animal species including humans.

It is now generally accepted that prion diseases arise through the aberrant misfolding of the host-encoded cellular prion protein. Protein misfolding, although not unique amongst neurodegenerative disorders, can in the case of prion diseases be seeded by the misfolded protein, a property which imparts the transmissible nature of prion diseases in the absence of a conventional infectious agent. It is this feature, in conjunction with difficulties associated with decontamination of infected surgical instruments and material, which can be attributed to the level of community interest and public health concerns despite the diseases' low prevalence.

It has been over 30 years since a novel infectious particle was proposed to be the causative agent of the prototypic prion disease, scrapie. Proving the existence of a transmissible protein or 'prion' and understanding prion disease transmission and pathogenesis have led to the development of many innovative methodologies. Understanding the unique nature of the disease has led to a truly multidisciplinary approach, which unites biochemical and biophysical approaches with in vitro and in vivo models and animal biologists and veterinarians with biomedical researchers and clinicians.

This book brings together protocols from each of these disciplines and highlights the contribution each discipline has made to our understanding of the nature of prion disease.

In addition to contributing to our understanding of prion disease, these methods may also find application to the newly emerging and so-called 'prion-like' properties observed in other protein misfolding neurodegenerative diseases as highlighted in the final method which describes a method for detecting the intercellular prion-like conversion and transmission of Cu/Zn superoxide dismutase (SOD1).

Parkville, VIC, Australia *Victoria A. Lawson*

Contents

Part III Methods for Cell Free Propagation of Prions

Part IV In Vivo Investigation of Prion Disease

Part V Methods for the Detection and Treatment of Prion Diseases

Part VI Methods for Investigation of Prion-Like Disease

Contributors

GARY ADAMSON • *MRC Prion Unit at UCL, UCL Institute of Prion Diseases, London, UK*
OLIVIER ANDRÉOLETTI • *INRA, UMR 1225, IHAP, Toulouse, France; Université de Toulouse, INP, ENVT, UMR1225, IHAP, Toulouse, France*
KELSIE J. ANSON • *Laboratory of Persistent Viral Diseases, Rocky Mountain Laboratories, NIAID, NIH, Hamilton, MT, USA*
ZAIRA E. ARELLANO-ANAYA • *INRA, UMR 1225, IHAP, Toulouse, France; Université de Toulouse, INP, ENVT, UMR1225, IHAP, Toulouse, France*
RYUICHIRO ATARASHI • *Division of Microbiology, Department of Infectious Diseases, Faculty of Medicine, University of Miyazaki, Miyazaki, Japan*
GERALD S. BARON • *Rocky Mountain Laboratories, National Institute of Allergy and Infectious Diseases, NIH, Hamilton, MT, USA*
ILIA V. BASKAKOV • *Department of Anatomy and Neurobiology, Center for Biomedical Engineering and Technology, University of Maryland School of Medicine, Baltimore, MD, USA*
JONATHAN A. BECK • *MRC Prion Unit at UCL, UCL Institute of Prion Diseases, London, UK*
SHAYNE A. BELLINGHAM • *Department of Biochemistry and Molecular Biology, Bio21 Molecular Science and Biotechnology Institute, The University of Melbourne, Parkville, VIC, Australia*
IVANA BILJAN • *Department of Chemistry, Faculty of Science, University of Zagreb, Zagreb, Croatia*
AILEEN BOYLE • *The Roslin Institute & R(D)SVS, University of Edinburgh, Edinburgh, UK*
SEBASTIAN BRANDNER • *MRC Prion Unit at UCL, UCL Institute of Prion Diseases, London, UK*
LARA BROCK • *MRC Prion Unit at UCL, UCL Institute of Prion Diseases, London, UK*
KATRINA J. CAMPBELL • *Laboratory of Persistent Viral Diseases, Rocky Mountain Laboratories, NIAID, NIH, Hamilton, MT, USA*
ROBERTO CAPPAI • *Department of Pathology, The University of Melbourne, Parkville, VIC, Australia*
NEIL R. CASHMAN • *Division of Neurology, Department of Medicine, UBC Hospital, University of British Columbia, Vancouver, BC, Canada*
JOAQUÍN CASTILLA • *CIC bioGUNE, Parque Tecnológico de Bizkaia, Derio, Bizkaia, Spain; IKERBASQUE, Basque Foundation for Science, Bilbao, Bizkaia, Spain*
BYRON CAUGHEY • *Laboratory of Persistent Viral Diseases, Rocky Mountain Laboratories, NIAID, NIH, Hamilton, MT, USA*
GIUSEPPE D. CICCOTOSTO • *Department of Pathology, The University of Melbourne, Parkville, VIC, Australia*
BRADLEY COLEMAN • *Department of Biochemistry and Molecular Biology, Bio21 Molecular Science and Biotechnology Institute, University of Melbourne, Parkville, VIC, Australia*
JOHN COLLINGE • *MRC Prion Unit at UCL, UCL Institute of Prion Diseases, London, UK*
S.J. COLLINS • *Department of Medicine, Royal Melbourne Hospital, The University of Melbourne, Parkville, VIC, Australia*

Abigail B. Diack • *The Roslin Institute & R(D)SVS, University of Edinburgh, Edinburgh, UK*
Simon C. Drew • *Department of Medicine (Royal Melbourne Hospital), The University of Melbourne, Parkville, VIC, Australia*
Saioa R. Elezgarai • *CIC bioGUNE, Parque Tecnológico de Bizkaia, Derio, Bizkaia, Spain*
Laura J. Ellett • *Department of Pathology, The University of Melbourne, Parkville, VIC, Australia*
Hasier Eraña • *CIC bioGUNE, Parque Tecnológico de Bizkaia, Derio, Bizkaia, Spain*
Natalia Fernández-Borges • *CIC bioGUNE, Parque Tecnológico de Bizkaia, Derio, Bizkaia, Spain*
Leslie I. Grad • *Division of Neurology, Department of Medicine, UBC Hospital, University of British Columbia, Vancouver, BC, Canada*
Bradley R. Groveman • *Laboratory of Persistent Viral Diseases, Rocky Mountain Laboratories, NIAID, NIH, Hamilton, MT, USA*
Cathryn L. Haigh • *Department of Medicine, Melbourne Brain Centre, Royal Melbourne Hospital, The University of Melbourne, Parkville, VIC, Australia; Laboratory of Persistent Viral Diseases, Rocky Mountain Laboratories, National Institute of Allergy and Infectious Diseases, National Institutes of Health, Hamilton, MT, USA*
Chafik Harrathi • *CIC bioGUNE, Parque Tecnológico de Bizkaia, Derio, Bizkaia, Spain*
Andrew F. Hill • *Department of Biochemistry and Molecular Biology, Bio21 Molecular Science and Biotechnology Institute, The University of Melbourne, Parkville, VIC, Australia; Department of Biochemistry and Genetics, La Trobe Institute for Molecular Science, La Trobe University, Bundoora, VIC, Australia*
Kris Hogan • *The Roslin Institute & R(D)SVS, University of Edinburgh, Edinburgh, UK*
Andrew G. Hughson • *Laboratory of Persistent Viral Diseases, Rocky Mountain Laboratories, NIAID, NIH, Hamilton, MT, USA*
Alvina Huor • *INRA, UMR 1225, IHAP, Toulouse, France; Université de Toulouse, INP, ENVT, UMR1225, IHAP, Toulouse, France*
Gregor Ilc • *Slovenian NMR Centre, National Institute of Chemistry, Ljubljana, Slovenia; EN-FIST Center of Excellence, Ljubljana, Slovenia*
Metta Jana • *Department of Pathology, The University of Melbourne, Parkville, VIC, Australia*
Vanessa A. Johanssen • *Department of Medicine, The University of Melbourne, Parkville, VIC, Australia*
Susan Joiner • *MRC Prion Unit at UCL, UCL Institute of Prion Diseases, London, UK*
G.M. Klug • *Department of Medicine, Royal Melbourne Hospital, The University of Melbourne, Parkville, VIC, Australia*
Allison Kraus • *Laboratory of Persistent Viral Diseases, Rocky Mountain Laboratories, NIAID, NIH, Hamilton, MT, USA*
Victoria A. Lawson • *Department of Pathology, The University of Melbourne, Parkville, VIC, Australia*
Pascal Leblanc • *CNRS, UMR5239, Laboratoire de Biologie Moléculaire de la Cellule (LBMC), Equipe Différenciation Neuromusculaire, Ecole Normale Supérieure-Lyon, Lyon Cedex, France*
Victoria Lewis • *Department of Medicine, Royal Melbourne Hospital, The University of Melbourne, Parkville, VIC, Australia*
Jacqueline M. Linehan • *MRC Prion Unit at UCL, UCL Institute of Prion Diseases, London, UK*

NATALLIA MAKARAVA • *Department of Anatomy and Neurobiology, Center for Biomedical Engineering and Technology, University of Maryland School of Medicine, Baltimore, MD, USA*
MATTEO MANCA • *Laboratory of Persistent Viral Diseases, Rocky Mountain Laboratories, NIAID, NIH, Hamilton, MT, USA*
JEAN C. MANSON • *The Roslin Institute & R(D)SVS, University of Edinburgh, Edinburgh, UK*
HILARY E. MCMAHON • *UCD School of Biomolecular and Biomedical Science, Conway Institute for Biomolecular and Biomedical Research, University College Dublin, Dublin, Ireland*
SIMON MEAD • *MRC Prion Unit at UCL, UCL Institute of Prion Diseases, London, UK*
JULIE A. MORENO • *Cell and Molecular Biology Graduate Program, Molecular, Cellular and Integrative Neuroscience Graduate Program, Department of Microbiology, Immunology and Pathology, Prion Research Center, Colorado State University, Fort Collins, CO, USA*
NORIYUKI NISHIDA • *Department of Molecular Microbiology and Immunology, Nagasaki University Graduate School of Biomedical Sciences, Nagasaki, Japan*
CHRISTINA D. ORRÙ • *Laboratory of Persistent Viral Diseases, Rocky Mountain Laboratories, NIAID, NIH, Hamilton, MT, USA*
JANEZ PLAVEC • *Slovenian NMR Centre, National Institute of Chemistry, Ljubljana, Slovenia; EN-FIST Center of Excellence, Ljubljana, Slovenia; Faculty of Chemistry and Chemical Technology, University of Ljubljana, Ljubljana, Slovenia*
EDWARD POKRISHEVSKY • *Division of Neurology, Department of Medicine, UBC Hospital, University of British Columbia, Vancouver, BC, Canada*
CAROLINE POWELL • *MRC Prion Unit at UCL, UCL Institute of Prion Diseases, London, UK*
SUZETTE A. PRIOLA • *Laboratory of Persistent Viral Diseases, Rocky Mountain Laboratories, National Institute of Allergy and Infectious Diseases, National Institutes of Health, Hamilton, MT, USA*
GREGORY J. RAYMOND • *Rocky Mountain Laboratories, National Institute of Allergy and Infectious Diseases, NIH, Hamilton, MT, USA*
LYNNE D. RAYMOND • *Laboratory of Persistent Viral Diseases, Rocky Mountain Laboratories, NIAID, NIH, Hamilton, MT, USA*
KATSUYA SATOH • *Department of Locomotive Rehabilitation Science, Nagasaki University Graduate School of Biomedical Sciences, Nagasaki, Japan*
REGINA SAVTCHENKO • *Department of Anatomy and Neurobiology, Center for Biomedical Engineering and Technology, University of Maryland School of Medicine, Baltimore, MD, USA*
GLENN C. TELLING • *Cell and Molecular Biology Graduate Program, Molecular, Cellular and Integrative Neuroscience Graduate Program, Department of Microbiology, Immunology and Pathology, Prion Research Center, Colorado State University, Fort Collins, CO, USA*
CAROLIN TUMPACH • *Florey Institute of Neuroscience and Mental Health, Parkville, VIC, Australia*
LAURA J. VELLA • *The Florey Institute of Neuroscience and Mental Health, University of Melbourne, Parkville, VIC, Australia*
VANESA VENEGAS • *CIC bioGUNE, Parque Tecnológico de Bizkaia, Derio, Bizkaia, Spain*
DIDIER VILETTE • *INRA, UMR 1225, IHAP, Toulouse, France; Université de Toulouse, INP, ENVT, UMR1225, IHAP, Toulouse, France*
JONATHAN D.F. WADSWORTH • *MRC Prion Unit at UCL, UCL Institute of Prion Diseases, London, UK*

Part I

Methods for Biochemical and Biophysical Analysis of Prion Disease

Chapter 1

Purification and Fibrillation of Full-Length Recombinant PrP

Natallia Makarava, Regina Savtchenko, and Ilia V. Baskakov

Abstract

Misfolding and aggregation of prion protein are related to several neurodegenerative diseases in humans such as Creutzfeldt-Jakob disease, fatal familial insomnia, and Gerstmann-Straussler-Scheinker disease. A growing number of applications in the prion field including assays for detection of PrP^{Sc} and methods for production of PrP^{Sc} de novo require recombinant prion protein (PrP) of high purity and quality. Here, we report an experimental procedure for expression and purification of full-length mammalian prion protein. This protocol has been proved to yield PrP of extremely high purity that lacks PrP adducts, oxidative modifications, or truncation, which is typically generated as a result of spontaneous oxidation or degradation. We also describe methods for preparation of amyloid fibrils from recombinant PrP in vitro. Recombinant PrP fibrils can be used as a noninfectious synthetic surrogate of PrP^{Sc} for development of prion diagnostics including generation of PrP^{Sc}-specific antibody.

Key words Recombinant prion protein, Inclusion body, Protein purification, Amyloid fibrils, Conformational transition, Prion diseases, IMAC, HPLC

1 Introduction

Recombinant prion protein (PrP) expressed in *E. coli.* Has been used extensively in prion research for various applications. These applications include modeling of prion conversion in vitro, utilization of PrP as immunogen for generating anti-PrP antibody, development of anti-prion therapeutic strategies that involve active immunization using PrP refolded in β-sheet-rich conformations, screening of anti-prion drugs using in vitro conversion assays, and others. These applications require PrP of very high purity with minimal amounts of chemical modifications or degradation. While several methods for purification and refolding of recombinant PrP have been previously described by different groups [1–6], some of the previously developed protocols required a fusion of PrP to histidine tags or produced PrP of insufficient purity or partially degraded. Inconsistent results in converting of PrP into β-sheet-rich conformations described in the past are attributed, at least in part, to differences in experimental protocols for expression and

Victoria A. Lawson (ed.), *Prions: Methods and Protocols*, Methods in Molecular Biology, vol. 1658,
DOI 10.1007/978-1-4939-7244-9_1, © Springer Science+Business Media LLC 2017

purification of PrP employed by different laboratories. Here, we describe a reliable experimental protocol for expression of tag-free full-length recombinant PrP of high purity and with minimal amount of chemical modifications or degradation. This protocol yields approximately 10 mg of mouse PrP or 6–8 mg of hamster PrP per liter of bacterial culture.

The current chapter also describes experimental protocols for converting full-length recombinant PrP into amyloid fibrils developed by our group in the past [7–10]. While recombinant PrP fibrils were able to induce transmissible prion diseases in wild-type animals, their infectivity was found to be very low [11]. Nevertheless, the immunoconformational assay that utilized conformational, PrP^{Sc}-specific antibodies and a broad panel of non-conformational antibodies revealed that the PrP fibrils produced in vitro acquired a surface structure similar to that of PrP^{Sc} [12]. In this regard, PrP fibrils appear to be a suitable synthetic surrogate of PrP^{Sc} and can be utilized for the development of prion diagnostics, high-throughput screening of anti-prion drugs, development of anti-prion decontamination procedures or an antigen for generating PrP^{Sc}-specific antibody, and other important applications in the field.

2 Materials

All solutions are prepared with deionized water purified using Synergy 185 UV Ultrapure Water System (Millipore, Bedford, MA). Water and solutions for desalting and HPLC are de-gassed under vacuum. HPLC buffers are purged with helium. Shaking procedures at 37 °C were performed in an Innova 4300 incubator (New Brunswick Scientific) set at 200 rpm.

2.1 Protein Expression

1. Plasmid DNA encoding mouse PrP 23–230, Syrian hamster PrP 23–231, or human PrP 23–231 (129M or 129V variants) (*see* **Note 1**) in pET101/D-TOPO (Invitrogen).
2. Competent BL21 Star (DE3) One Shot *E. coli* cells and their SOC medium (Invitrogen).
3. Luria-Bertani (LB) Broth.
4. 100 mg/ml carbenicillin disodium salt in water and stored in aliquots at −20 °C.
5. Two 2800 ml baffled PYREX flasks.
6. Terrific Broth (TB) medium composition for 1200 ml (*see* **Note 2**): 14.4 g Bacto tryptone (BD Biosciences, Sparks, MD), 28.8 g Bacto yeast extract (BD Biosciences), 4.8 ml glycerol, water to adjust 1080 ml. TB medium needs to be autoclaved and then supplemented with 120 ml filter-sterilized

solution of 0.17 M KH_2PO_4 and 0.72 M K_2HPO_4 and 100 μg/ml carbenicillin.

7. 1 M isopropyl-beta-D-thiogalactopyranoside (IPTG) in water and stored in aliquots at −20 °C.

2.2 Isolation of Inclusion Bodies

1. Cell lysis buffer: 50 mM Tris–HCl, 1 mM ethylenediaminetetraacetic acid (EDTA), 100 mM NaCl, pH 8.0.
2. 9 mg/ml phenylmethylsulfonyl fluoride (PMSF) in acetonitrile and stored at −20 °C.
3. Lysozyme solution: prepare at 10 mg/ml in lysis buffer. Store in aliquots at −20 °C.
4. Deoxycholic acid.
5. 2 mg/ml deoxyribonuclease I (DNase I, type II) in water and stored in aliquots at −20 °C.

2.3 Immobilized Metal Ion Affinity Chromatography (IMAC) and Oxidative Refolding

1. 9 M urea. After urea is dissolved in deionized water, 10 g/l, add mixed bed Amberlite (MB-150), and stir further the solution for at least 1 h. Before use, the solution is filtered using disposable filter units with polyethersulfone membrane. 9 M urea can be stored at −20 °C.
2. Chelating Sepharose Fast Flow (GE Healthcare).
3. 0.2 M nickel sulfate ($NiSO_4$).
4. Acidic buffer for elution of loosely bound ions from Sepharose: 0.02 M Na acetate, 0.5 M NaCl, pH 3.0.
5. IMAC buffer A: 8 M urea, 0.1 M Na_2HPO_4, 10 mM Tris–HCl, 10 mM reduced glutathione, pH 8.0.
6. IMAC buffer B: 8 M urea, 0.1 M Na_2HPO_4, 10 mM Tris–HCl, 10 mM reduced glutathione, pH 4.5.
7. 0.5 M ethylene glycol bis(2-aminoethyl ether)-*N*,*N*,*N*′*N*′-tetraacetic acid (EGTA), pH 8.0.
8. Desalting buffer: 6 M urea, 0.1 M Tris–HCl, pH 7.5.
9. 50 mM oxidized glutathione and stored in aliquots at −20 °C.
10. Solutions for Sepharose regeneration and preservation: 2 M NaCl; 1 M NaOH; molecular grade ethanol.
11. XK chromatography column (GE Healthcare).
12. HiPrep 26/10 desalting column (GE Healthcare).
13. FPLC system (ÄKTA prime, GE Healthcare).
14. 13 × 100 mm tubes.

2.4 High-Performance Liquid Chromatography (HPLC)

1. HPLC buffer A: 0.1% trifluoroacetic acid in water.
2. HPLC buffer B: 0.1% trifluoroacetic acid in acetonitrile.
3. Protein C4 HPLC column, particle size 10 μm, inner diameter 22 mm, length 250 mm; column guard, particle size 12 μm, cartridge 10 mm (Vydac).

4. Polyethersulfone (PES) membrane disposable filter units.
5. Shimadzu HPLC system (Columbia, MD) operated with EZStart 7.3 SP1 software.

2.5 Amyloid Fibril Formation (Manual Setup)

1. 0.5 M 2-(N-Morpholino)ethanesulfonic acid (MES) buffer, pH 6.0.
2. 10 mM sodium acetate, pH 5.0.
3. 0.5 M thiourea pH adjusted to 6.0.
4. 6 M guanidine hydrochloride, pH adjusted to 6.0.
5. Dialysis tubing (Spectra/Por, molecular weight cutoff [MWCO] 2000) and clips.
6. 1 mM thioflavin T stock in water, stored in the dark at +4 °C.
7. Delfia plate shaker (Perkin Elmer, Wellmix, or similar) with a microcentrifuge tube rack attached, Clay Adams Nutator Mixer (model 1105, Becton Dickinson & Co.)
8. Bath sonicator (Branson 2510, Bransonic, Danbury, CT).
9. Plastic tubes, 1.5 ml.
10. Recombinant full-length prion protein (PrP) (see below).
11. Dialysis tubing (MWCO 2000).

2.6 Amyloid Fibril Formation (Semiautomated Setup).

In addition to the reagents described above (*see* Subheading 2.5):

1. Teflon spheres (3/32 in diameter, McMaster-Carr).
2. 96-well flat bottom nontreated polystyrene assay plates.
3. Transparent plate sealers.
4. Microplate fluorescence reader equipped with 444 nm excitation and 485 nm emission filters.

2.7 Epifluorescence Microscopy

1. Inverted fluorescent microscope (Nikon Eclipse TE2000-U) equipped with the illumination system, sets of objectives, filters (excitation filter 485DF22, beam splitter 505DRLPO2, and emission filter 510LP (Omega Optical, Inc., Brattleboro, VT) and a charge-coupled device (CCD) camera.
2. Immersion oil type FF.
3. Glass Coplin staining jar.
4. Microscope cover glass no. 1.
5. Isopropanol.
6. Acetone.
7. Sulfuric acid.
8. Hydrogen peroxide.

3 Methods

Researchers might face the following technical challenges during expression and purification of PrP:

1. Difficulties in achieving complete solubilization of PrP inclusion bodies.
2. Precipitation and irreversible binding of PrP to the IMAC matrix.
3. Copper-dependent self-cleavage of PrP.
4. Incomplete oxidative refolding of PrP.
5. Spontaneous formation of oxidative adducts.
6. Spontaneous methionine oxidation.

To minimize these problems and to achieve successful purification of PrP of high purity, the protocol described below needs to be closely followed.

3.1 PrP Production

3.1.1 Transformation of Bacterial Cells

Chemical transformation of BL21 Star (DE3) One Shot *E. coli* is based on the protocol described in the pET Directional TOPO Expression Kit Instruction Manual (Invitrogen).

1. Thaw on ice one vial of cells.
2. Add 1 μl (20 ng) plasmid DNA into the vial of cells, and mix by stirring gently with the pipette tip.
3. Incubate on ice for 30 min.
4. Heat-shock the cells for 30 s at 42 °C.
5. Immediately transfer the tube on ice.
6. Add 250 μl of room temperature SOC medium.
7. Tape the tube on its side to the bottom of incubator, and shake at 37 °C at 200 rpm for 30 min.
8. Add the entire transformation reaction into 50 ml centrifuge tube containing 10 ml of LB supplemented with 100 μg/ml carbenicillin.
9. Shake at 37 °C at 200 rpm for 3–5 h.
10. Add the entire volume into the 500 ml flask containing 90 ml of LB supplemented with 100 μg/ml carbenicillin.
11. Shake overnight at 37 °C at 200 rpm.

3.1.2 Induction of PrP Expression

1. Supplement 1080 ml autoclaved TB media with 120 ml filter-sterilized solution of phosphates (0.17 M KH_2PO_4 and 0.72 M K_2HPO_4) and 100 μg/ml carbenicillin. Mix and save 1 ml of resulting mixture as an absorbance reference for cell growth monitoring.
2. Add 60 ml overnight cell culture, mix, and divide equally between two baffled flasks (use sterile 1 L cylinder).

3. Incubate flasks with cell culture shaking at 37 °C at 200 rpm until the absorbance at 600 nm reached 0.6. Dilute with fresh TB, if overgrown.
4. Induce expression by adding 1 mM IPTG.
5. Continue incubation for 4–5 h.

3.1.3 Cell Harvesting

1. To be able to determine cell pellet mass, weigh empty centrifuge bottles.
2. Divide bacterial culture between four 500 ml centrifuge bottles, and centrifuge at 2975 × *g* maximum RCF in Beckman rotor JLA-10.500 for 10 min at 4 °C.
3. Discard supernatant, and calculate cell pellet mass.
4. At this point, cells can be stored overnight at −20 °C.

3.1.4 Cell Lysis and Isolation of Inclusion Bodies

1. Thoroughly resuspend the pellet in lysis buffer (8.7 ml of buffer per each g of bacterial pellet) by vortex and pipetting up and down with 25 ml pipette.
2. Freeze at −80 °C, and thaw the cells at least one time to ensure cell lysis. At −80 °C, cells freeze in about 10 min. Room temperature water bath is used to thaw the pellet quickly.
3. Pour cell lysate into a beaker, add 2 μl PMSF and 20 μl lysozyme per 1 ml lysis buffer, and stir at room temperature for 20–40 min (*see* **Note 3**).
4. Add deoxycholic acid, 1 mg/ml, and stir for 20–30 min until the liquid becomes viscous.
5. Add DNase I to 5 μg/ml, and stir for additional 30–45 min.
6. Divide the lysate between four 50 ml centrifuge tubes, centrifuge at 12,000 × *g* for 30 min at 4 °C, and decant the supernatant.
7. Thoroughly resuspend the pellet in 15 ml lysis buffer by vortex and pipetting up and down.
8. Repeat DNase I treatment: DNase I to 5 μg/ml to each centrifuge tube, incubate on rotating platform for 20 min.
9. Centrifuge at 12,000 × *g* for 30 min at 4 °C, and decant the supernatant.
10. Dilute lysis buffer with water 1:9. Thoroughly resuspend the pellet in 20 ml of diluted buffer. This step removes the excess of EDTA from the inclusion bodies to allow proper binding to Ni^{2+}-charged chromatography column.
11. Centrifuge at 12,000 × *g* for 20 min at 4 °C, and decant the supernatant. Resulting pellet contains recombinant PrP precipitated in the form of inclusion bodies, which can be stored frozen at −80 °C for at least 1 month.

3.2 PrP Purification

3.2.1 Immobilized Metal Ion Affinity Chromatography (IMAC)

IMAC purification is performed using a XK chromatography column (GE Healthcare), packed with 20 ml of Chelating Sepharose Fast Flow. Loading the Sepharose with Ni ions and protein binding are performed in solution. The same Sepharose can be reused several times for purification of the same PrP variant (*see* **Note 4**). Desalting column is stored at 4 °C; however, it should be equilibrated to room temperature before use.

Preparing Ni-Charged Sepharose

1. Take the required aliquot of the Sepharose into two 50 ml centrifuge tubes.
2. Let the Sepharose settle down by gravity, and remove preservative solution.
3. To wash the Sepharose, add water to the top of the tube, cover the tube, and gently resuspend the Sepharose by inverting the tube several times. Let the Sepharose settle down (it takes about 20 min), and remove water.
4. With the same procedure, wash with water again.
5. Remove water, and charge the Sepharose by adding 2 ml 0.2 M $NiSO_4$.
6. Wash the excess of ions with water twice.
7. Elute the loosely bound ions, washing with acidic buffer, pH 3.0.
8. Wash two times with water.
9. Equilibrate the Sepharose by washing twice with IMAC buffer A. Keep the Sepharose under buffer until the protein is solubilized and ready for binding.

Protein Solubilization and Binding

1. Add 10 ml of IMAC buffer A to each tube of inclusion bodies (*see* **Notes 2** and **5**).
2. Thoroughly resuspend the pellet.
3. Incubate on rotating platform 1–1.5 h at room temperature.
4. Centrifuge at 12,000 × *g* for 15 min at 4 °C.
5. Remove equilibration buffer from the Ni-charged Sepharose.
6. Add the supernatant containing solubilized recombinant PrP to the Sepharose.
7. Gently rotate the mixture of the Sepharose and the protein at room temperature allowing 30–40 min for binding of the protein.

IMAC and Desalting

1. Secure empty chromatography column on a holder nearby chromatographer.
2. Close column outlet, and load the mixture of Sepharose and protein solution.

3. Open the outlet, and drain the excess of liquid from the column, collecting it as IMAC flow-through for the analysis of binding efficiency (Figs. 1 and 2). Make sure not to drain the slurry completely: insert the adaptor, and lock it above the

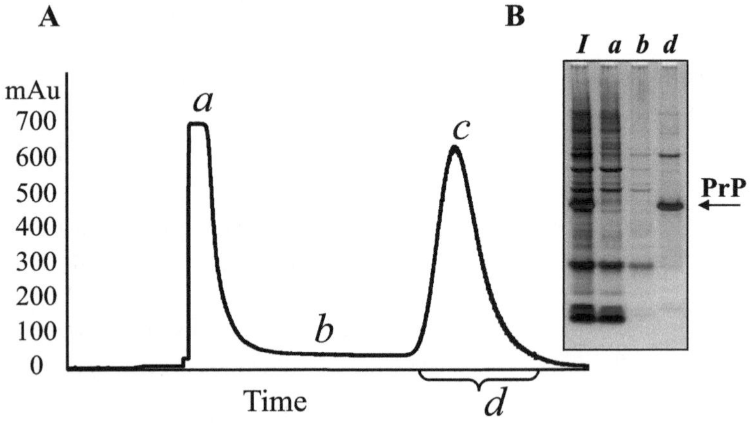

Fig. 1 Typical IMAC profile of mouse recombinant PrP purification. (**a**) IMAC profile. The peak of unbound proteins (stage *a*) typically reaches about 700 mAu. After this peak drops to the baseline (below 50 mAu, stage *b*), the IMAC buffer A (pH 8.0) is changed to the IMAC buffer B (pH 4.5). The PrP peak typically reaches approximately 600 mAu (stage *c*). Fractions with UV values above 50 mAu (stage *d*) are combined for subsequent purification. (**b**) Analysis of IMAC fractions in SDS-PAGE (10% bis-tris) following by silver staining. *I*, solubilized inclusion bodies; *a*, *b*, *d*, fractions collected at the stages *a*, *b*, and *d*, respectively

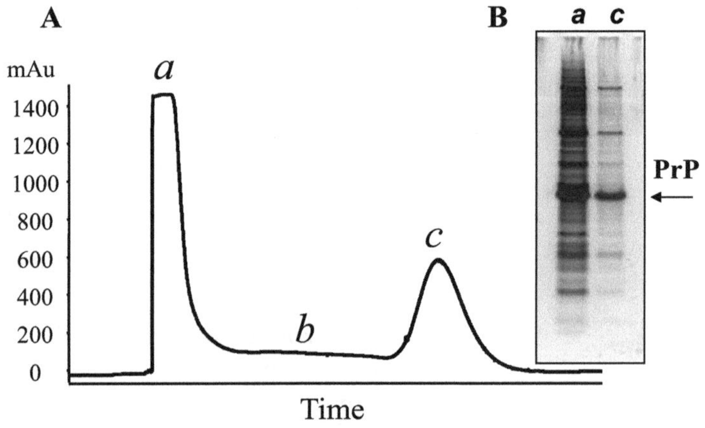

Fig. 2 Leakage of proteins from the IMAC column. (**a**) IMAC profile. Overloading of IMAC column results in higher than normal UV values of unbound proteins (>1400 mAu, stage *a*) and subsequent baseline higher than 100 mAu (stage *b*). The height of the PrP peak remains at typical level of 600 mAu (stage *c*). (**b**) Analysis of IMAC fractions in SDS-PAGE (10% bis-tris) confirms the presence of high amount of PrP in the flow-through (lane *a*)

Sepharose as soon as the liquid front reaches the surface of the Sepharose.

4. Connect the column to the FPLC system (ÄKTA prime, GE Healthcare). Set flow rate to 2 ml/min and fraction size to 5 ml. Wash unbound proteins with IMAC buffer A until the UV readings from the chromatographer reach low plato (Fig. 1; *see* **Notes 4** and **5**).
5. Switch to the IMAC buffer B, pH 4.5, to start elution of PrP. Fractions with protein are collected into borosilicate glass (13 × 100 mm tubes) containing EGTA; the final concentration of EGTA in each tube should be 5 mM after fraction is collected. Typical profile of IMAC purification is shown on Fig. 1.
6. Combine fractions containing PrP (*see* Fig. 1) in a 50 ml centrifuge tube. Typically, we collect 30–35 ml of protein solution and proceed with desalting immediately.
7. To remove the Sepharose from the chromatography column, add water to the column, gently resuspend the slurry with the 25 ml pipette, and transfer the Sepharose with water to a new 50 ml tube for regeneration.
8. Attach HiPrep 26/10 desalting column to the FPLC system, wash out storage solution, and equilibrate with desalting buffer: 6 M urea, 0.1 M Tris–HCl, pH 7.5 (*see* **Note 6**).
9. Desalting step is used to separate the protein from glutathione. Protein solution is loaded through the FPLC super loop. Because the total volume of the protein collected after IMAC exceeds column capacity (14 ml for HiPrep 26/10 desalting column), desalting is performed in two runs. First, 14 ml of the protein solution is loaded and desalted. Then, after the salt is washed out and the column is re-equilibrated once more, the rest of the protein (~11 ml) is run through the column (Fig. 3).

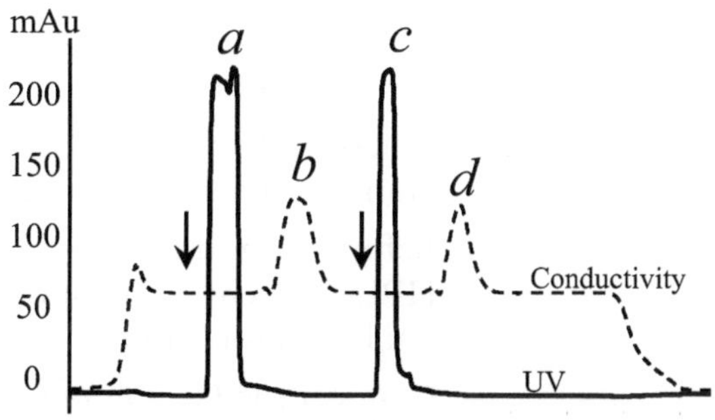

Fig. 3 Desalting profile. PrP collected after IMAC were divided into two parts and desalted using gel filtration chromatography. The *arrows* mark the points of loading of PrP onto the desalting column. During desalting, the protein (peaks *a* and *c*) is separated from glutathione (peaks *b* and *d*)

10. Wash desalting column immediately after the last run. Wash with water until the conductivity is at the baseline level. Disconnect the column from the FPLC system and reconnect it upside down. Wash with 0.2 M NaOH until the conductivity is on high plato. Wash with water again. Finally, fill out the column with 20% ethanol, disconnect, close, and store at 4 °C.
11. Combine the fractions containing recombinant PrP in a new 50 ml tube, and mix. To estimate protein concentration (C), prepare 1:5 dilution of protein solution, measure absorbance at 280 nm, and calculate the concentration using the following equation: $C(\mathrm{mg/ml}) = A280 \times 5 \times 0.37$ (for mouse PrP 23 – 230; 0.37 mg/ml = 1 o.e. at 280 nm). To minimize formation of dimers during oxidative refolding of PrP, dilute PrP solution with the desalting buffer (6 M urea, 0.1 M Tris–HCl, pH 7.5) to such extent that the concentration of PrP does not exceed 0.3 mg/ml.
12. Supplement the PrP solution with 5 mM EGTA and 0.2 mM oxidized glutathione, and gently rotate at room temperature overnight (*see* **Note 3**).

Chelating Sepharose Regeneration

1. After the Sepharose is transferred from the chromatographer column to a 50 ml centrifuge tube, add water to the top of the tube, cover the tube, and gently resuspend the Sepharose by inverting the tube several times. Let the Sepharose settle down, and remove water.
2. Wash with 2 M NaCl using the same procedure.
3. Wash with water three times.
4. Wash with 1 M NaOH.
5. Wash with water two times. When removing water after the last wash, leave 2 ml above the Sepharose, and add 2 ml of pure ethanol for preservation. Keep at 4 °C.

High-Pressure Liquid Chromatography (HPLC)

To perform C4 HPLC, we use Shimadzu HPLC system operated with EZStart 7.3 SP1 software.

1. Prepare HPLC buffers A and B (1 l each).
2. Degas buffers A and B for 15–20 min by stirring under vacuum, and then keep under constant purging of helium.
3. Before connecting the column, wash the tubing and pumps of HPLC system with running buffers A and B consecutively at 5 ml/min for 10 min.
4. Connect the C4 column, and equilibrate with buffer A.
5. Visible precipitation of PrP occurs after overnight oxidation. To remove these precipitates, centrifuge the protein solution at 12,000 × *g* for 30 min, and then filter supernatant using disposable filter units with polyethersulfone (PES) membrane.

6. To reduce urea concentration, dilute the protein solution with HPLC buffer A (1:2 v/v), and load onto C4 column (*see* **Note 7**).
7. Wash unbound proteins from the C4 column with HPLC buffer A at the flow rate of 5 ml/min; monitor UV absorbance at 220 and 280 nm. When the baseline is reached, start the gradient (*see* **Note 8**).
8. Using HPLC profile as guidance, manually collect PrP fractions into borosilicate glass tubes. An α-PrP monomer is eluted in major peak between 52.5 and 54.5 min (Fig. 4a). Slow gradient separates correctly folded α-PrP monomer from PrP adducts.
9. At the end of the run, wash the column with HPLC buffer A until the baseline of UV absorbance is reached (*see* **Note 9**).
10. The quality of the purified protein is checked by SDS-PAGE (Fig. 4b) and by mass spectroscopy (*see* **Note 10**).
11. Freeze collected fractions at −80 °C. Lyophilize (we use Free-Zone 2.5 Plus freeze dry system from Labconco (Kansas City, MO)). The protein is stored as lyophilized powder at −20 °C. The expression/purification yield depends on a species of rPrP. When mouse, Syrian hamster, or human rPrP is expressed, purification typically resulted in 12–15 mg, 7–8 mg, or just 1 mg of protein, respectively, per 1200 ml bacterial culture.
12. In our experience, the longevity of HPLC C4 column is limited to ~25–30 preparations. A change in PrP retention time and significant tailing indicate that the column has reached the end of its lifetime and needs to be replaced. The column longevity depends on a species of PrP it is used to purify. The tendency for aggregation is much higher for human recombinant PrP than that for mouse or hamster PrP. As a result, the longevity of a column used for human PrP is shorter than that used for mouse or hamster PrP.

3.3 Conversion of Full-Length PrP into Amyloid Fibrils in Manual Setup

Full-length PrP is capable of forming a variety of aggregated forms in vitro [7, 9, 13, 14]. Formation of amyloid fibrils is highly dependent on the reaction conditions. Ideal reaction conditions for fibril formation combine neutral or slightly acidic pH (between 5.0 and 7.5) and moderate concentrations of denaturants such as guanidine hydrochloride (up to 2 M) or urea (up to 4 M). Possible complications during fibril formation include inhibition of fibrillization by Cu^{2+} [13], copper-mediated and/or spontaneous N-terminal truncation of the protein [15–17], and side-chain oxidation [14]. In order to minimize these problems, copper ions are removed from the protein during purification, and 10 mM thiourea is added during fibril formation. Additional problem may arise if recombinant PrP used for conversion is of low purity. The

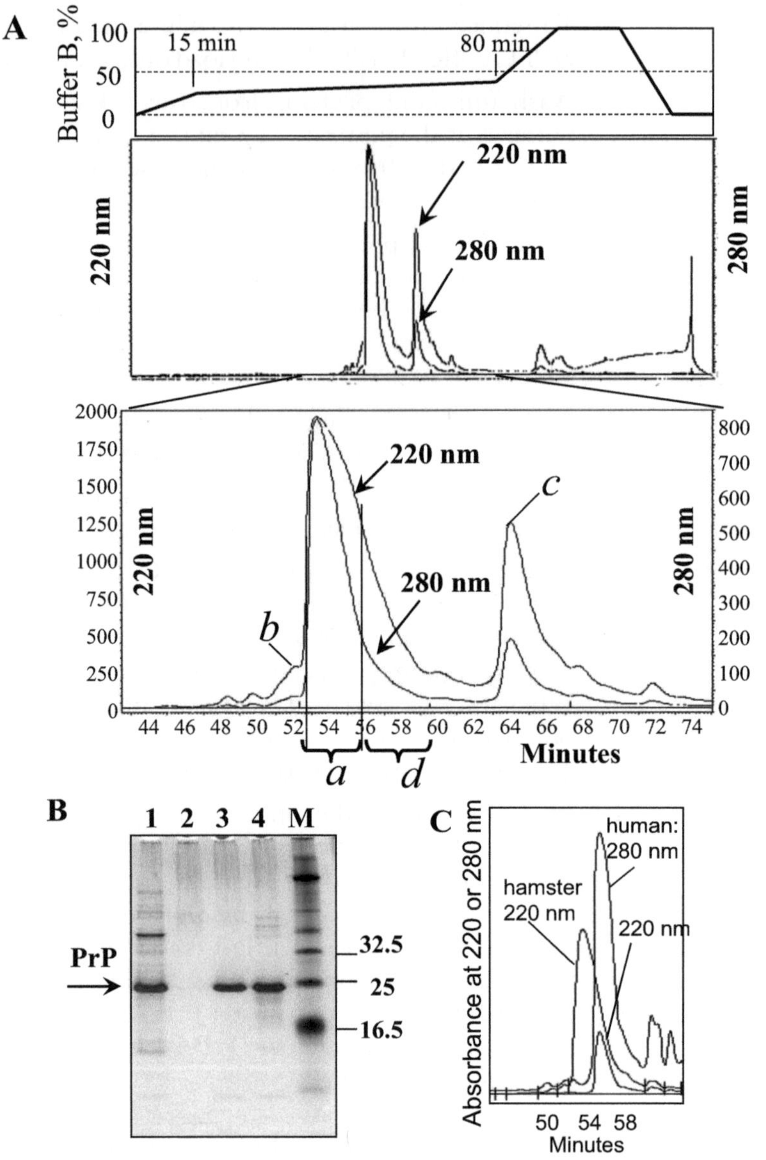

Fig. 4 Purification of PrP on C4 column. (**a**) Typical HPLC profile of elution of mouse recombinant PrP. Major peak contains pure PrP (peak *a*); it is separated from the sub-peak containing PrP with oxidized methionines (peak *b*), from the peaks containing PrP with double glutathione adducts (peak *c*), and from other impurities. The right shoulder of the major HPLC peak (fractions *d* eluted at 56–60 min) is not collected; this shoulder may contain products of PrP degradation (*see* lane *4* in panel **b**). (**b**) Analysis of HPLC fractions in SDS-PAGE followed by silver staining: *lane 1*, PrP collected after IMAC and loaded onto C4 column; *lane 2*, HPLC flow-through; *lane 3*, pure PrP collected from the major HPLC peak *a*; *lane 4*, right shoulder of the major peak (referred to as fraction *d*); and *lane M*, molecular marker. *Lane 4* shows minor amounts of PrP degradation products that occur due to self-cleavage (seen as a smear with Mw < 23 kDa). The extent of PrP degradation may vary from preparation to preparation (*see* **Note 3**). (**c**) Comparative HPLC profile of elution of hamster and human recombinant PrP. The purification yield of human recombinant PrP (detection at 220 and 280 nm) is substantially lower than that of hamster PrP (detection at 220 nm)

conversion conditions described here and referred to as standard are 2 M GdnHCl, pH 6. However, fibrillations can be also performed in the absence of GdnHCl or at concentrations lower than 2 M GdnHCl as previously described [18, 19].

1. Prepare stock solution of recombinant PrP (130 μM, 3 mg/ml) in 6 M GdnHCl, pH 6.0. This solution can be stored at −20 °C for up to 1 week (*see* **Note 11**). Alternatively, the protein can be dissolved directly in the MES buffer (50 mM, pH 6) at lower concentration (0.5–1 mg/ml). This solution, however, cannot be stored and must be used for fibril formation immediately (within several hours).
2. To prepare 500 μl reaction (*see* **Note 12**), mix the following reagents in the conical plastic tube: water (273.3 μl), GdnHCl (6 M, 83.4 μl), MES buffer (0.5 M, pH 6.0, 50 μl), and thiourea (0.5 M, 10 μl). Then add stock solution of PrP in 6 M GdnHCl (3 mg/ml, 83.3 μl). Mix reagents gently, and avoid introducing air bubbles.
3. If you are using previously formed fibrils as seeds, sonicate them for at least 10 s in the bath sonicator, and add to the reaction mixture before adding the PrP stock. Seeding capacity of fibrils decreases upon prolonged storage. Small amounts of seeds (as little as 0.1% of the amount of protein) are sufficient to significantly decrease the lag phase of conversion.
4. Incubate the tube with continuous shaking at 600 rpm using a plate shaker or rotation at 24 rpm using Nutator Mixer at 37 °C (Fig. 5; *see* **Note 13**).
5. Monitor the kinetics of fibril formation using a thioflavin T binding assay. For this assay, withdraw the aliquots (4 μl)

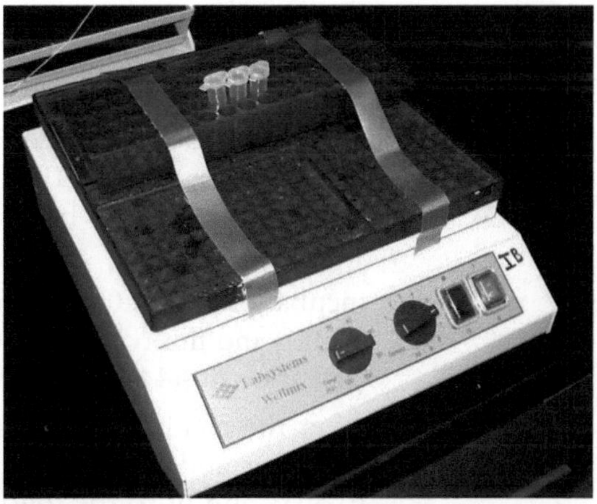

Fig. 5 Plate shaker with a microcentrifuge tube rack used for the conversion reaction. Place plastic tubes next to each other to prevent their rotation inside the rack during shaking

during the incubation. Before taking aliquots, pipette the reaction mixture gently each time (strong pipetting might perturb the kinetics of fibril formation). Dilute each aliquot into 10 mM Na acetate buffer (pH 5.0) to a final concentration of PrP of 0.3 μM, and then add thioflavin T to a final concentration of 10 μM. Record three emission spectra (from 460 to 520 nm) for each time point in 0.4 cm rectangular cuvettes with excitation at 445 nm on a microplate fluorescence reader (we use FluoroMax-3 fluorimeter, Jobin Yvon, Edison, NJ) keeping excitation and emission slits at 4 nm. Average the spectra, and determine the fluorescence intensity at emission maximum (usually around 482 nm). Emission will remain low for a few hours or days, then will start rising, and eventually will rise 10- to 40-fold of the original baseline level (*see* **Note 14**).

6. After the fibril formation has reached a plateau, the fibrils should be dialyzed for prolonged storage. Place the suspension of fibrils in the bag prepared from the dialysis tubing (MWCO 2000), and dialyze against a large volume of 10 mM Na acetate buffer (pH 5.0) with several buffer changes. Fibrils should be stored at +4 °C. Prolonged storage of fibrils at room temperature or at higher pH may lead to their aggregation and copper-mediated protein self-cleavage. Freezing and thawing may cause fragmentation of fibrils into short pieces.

3.4 Conversion of Full-Length PrP into Amyloid Fibrils in Semiautomated Setup

1. Perform the conversion of PrP into amyloid fibrils in semiautomated setup at least in triplicate to ensure reproducibility. Add three Teflon spheres (3/32 in. diameter) per well of 96-well assay plate. Mix the following reagents in the conical plastic tube: water (268.3 μl), GdnHCl (6 M, 163.7 μl), MES buffer (0.5 M, pH 6.0, 50 μl), thioflavin T (1 mM, 5 μl), and thiourea (0.5 M, 10 μl). Add stock solution of PrP in 6 M GdnHCl (3 mg/ml, 5 μl). After thorough mixing, divide the reaction mixture between three wells of the 96-well plate (160 μl per well), and cover the plate with the plate sealer. If previously formed fibrils are used as seeds, they should be sonicated for at least 10 s in the bath sonicator and added to the reaction mixture before addition of the PrP stock.
2. Insert the 96-well plate into the microplate reader. Set up incubation at 37 °C with shaking at 900 rpm (shaking diameter 1 mm) and fluorescence measurements every 5 or 10 min with excitation at 444 nm and emission at 485 nm (Fig. 6).
3. After the completion of the experiment, transfer the data to a graphing and data analysis program (we use Origin (OriginLab)), and fit to the following equation:

$$F = A + (B + ct)/(1 + \exp(k(t_m - t)))$$

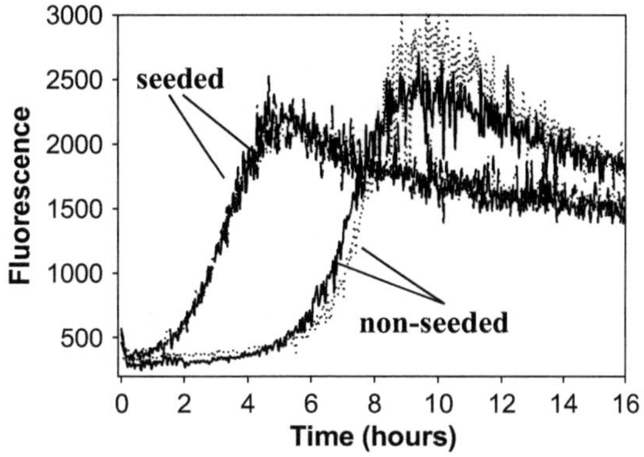

Fig. 6 Kinetics of fibril formation from PrP (1 μM) carried out in semiautomated format under standard solvent conditions. Two repeats for non-seeded and seeded conversion reactions are shown. After reaching a plato, the ThT fluorescence often shows a decline that appears to be due to sorption of fibrils to plate walls and partial aggregation of fibrils into clumps

where A is the initial level of thioflavin T fluorescence, B is the increase in thioflavin T fluorescence during conversion, k is the rate constant of amyloid accumulation (h^{-1}), t_m is the midpoint of conversion of PrP to amyloid, and c is an empirical parameter describing changes in fluorescence after fibril formation. The lag time (t_l) can be calculated as $t_l = t_m - 2/k$.

3.5 Epifluorescence Microscopy

While ThT fluorescence assay is convenient for measuring the kinetics of fibril formation, this assay, however, is not sufficient for providing a definite poof as to whether amyloid fibrils were formed in the reaction mixture. Relatively small (two- to threefold) increase in ThT fluorescence may indicate formation of non-fibrillar PrP isoforms such as soluble β-oligomers, which also bind ThT. Several techniques including electron microscopy and atomic force microscopy have been used for confirming the formation of fibrils; however, their application is laborious and requires costly equipment. In our experience, epifluorescence microscopy in the presence of ThT serves as a rapid and reliable test for the presence of the amyloid fibrils in the sample, for assessing their quality, size, aggregation status, and even growth rate [7, 8, 10, 13, 20]. Here, we describe the experimental procedure for imaging amyloid fibrils using inverted fluorescence microscopy in the presence of ThT.

1. Deposit several cover glasses into a staining jar, and clean them by sonicating it in isopropanol (1 min), then in acetone (1 min), and again in isopropanol (1 min). Wash the cover glass with water, and incubate it in the mixture of sulfuric acid (70%) and hydrogen peroxide (30%) for at least 1 h. Rinse with water several times. Store in water, and dry with compressed air before use.

2. Add 0.5 μl of suspension of PrP fibrils (0.5 mg/ml) in 10 mM acetate buffer (pH 5.0) to the same buffer (99.5 μl) containing 10 μM thioflavin T. Incubate the solution in the plastic tube at 25 °C for 5 min.
3. Place 10–20 μl of the solution on the cover glass. Allow fibrils to sediment on the glass surface for 1–2 min. Examine the sample with an inverted fluorescence microscope (Nikon Eclipse TE2000-U) using 60× or 100× objective. The emission can be isolated from Rayleigh and Raman-shifted light by a combination of filters: an excitation filter 485DF22, a beam splitter 505DRLPO2, and an emission filter 510LP (Omega Optical, Inc.). Digital images can be acquired using a CCD camera.
4. Shape and size of PrP fibrils vary depending on the conditions of their formation including shaking modes and purity of the protein (*see* **Note 14**). Under the standard conditions, PrP yields long (>1 μm) and straight fibrils, if highly pure (Fig. 7a). Fibrils formed at higher pH (pH > 7.0) tend to aggregate into large clusters. Exposure of fibrils formed at pH < 6 to pH > 6.5 or addition of salt induced aggregation into clumps of various size (Fig. 7b). Presence of impurities in the protein (substantial methionine oxidation or N-terminal truncation) can result in formation of shorter fibrils (<1 μm) that often appear bent or very short, dot-like fibrils (Fig. 7c).

4 Notes

1. Plasmids for expression of mouse PrP 23–230, Syrian hamster PrP 23–231, or human PrP 23–231 (129M or 129V) were cloned from mouse, Syrian hamster, or human cDNA, respectively, as described [7, 14].
2. We have found it convenient to grow bacteria in two flasks each containing 600 ml TB media. This typically results in total of 12 g bacterial pellet used for isolation of inclusion bodies. To prevent overloading of HPLC column, only one half of inclusion bodies should be used for IMAC. Another half is stored at −80 °C.
3. Recombinant PrP is prone to copper-dependent self-cleavage [15, 16]. Performing cell lysis steps at lower temperature could slow down degradation; however, this would require prolonged incubation time. Therefore, the cell lysis steps are carried out at room temperature. EDTA and EGTA are added to the lysis buffer and during oxidative refolding, respectively, to minimize protein degradation.
4. Lack of Ni ions on the Sepharose may cause the leakage of protein during wash step. For better performance, recharge the column with Ni ions each time before reuse.

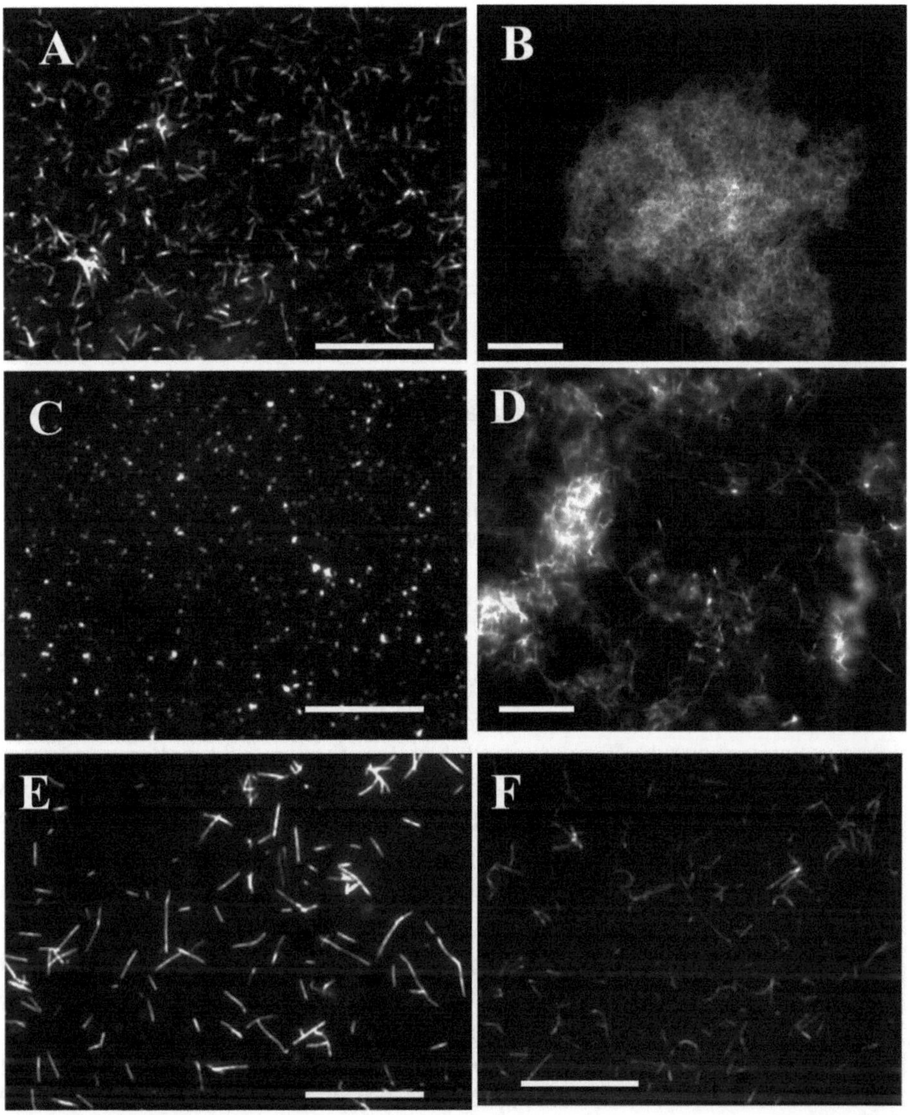

Fig. 7 Epifluorescence microscopy of PrP fibrils stained with thioflavin T. (**a**) Fibrils of mouse PrP prepared under standard conditions (shaking, pH 6). (**b**) Fibrils of mouse PrP prepared under standard conditions and incubated for 24 h in PBS, pH 7.2. (**c**) PrP (mouse) of insufficient purity or with substantial methionine oxidation converts into very short or dot-like fibrils under standard conditions. (**d**) Fibrils of mouse PrP prepared under standard conditions and digested with PK for 1 h at 37 °C. (**e**) Fibrils of mouse PrP prepared under standard solvent conditions upon rotation. (**f**) Fibrils of Syrian hamster PrP prepared under standard solvent conditions upon rotation. When stained with ThT, mouse fibrils show substantially brighter fluorescence than hamster fibrils. Scale bars = 10 μm

5. Loading the column with excessive amounts of protein (Fig. 2) or incomplete solubilization of inclusion bodies (Fig. 8) results in high UV readings during the wash with IMAC buffer A. To achieve proper solubilization of the protein, it is very important to prepare IMAC buffers A and B fresh, on the same day the

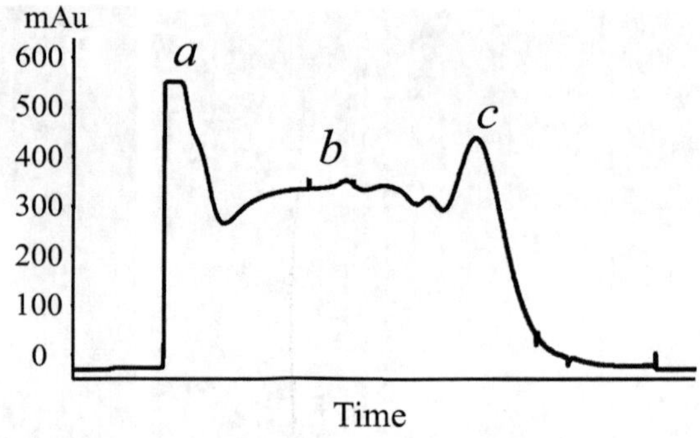

Fig. 8 IMAC profile of incompletely solubilized protein. Reduced glutathione degraded during preparation of stock solution (*see* **Note 5**). As a result, recombinant PrP was incompletely solubilized and leaked from the Sepharose during the wash step. The peak of unbound proteins (stage *a*) was below 600 mAu; the UV values during the wash steps were above 300 mAu (stage *b*), and the yield of eluted PrP was reduced (stage *c*)

IMAC purification is performed. Instead of keeping reduced glutathione in concentrated stock solution, we dissolved it directly in the IMAC buffers to the final concentration of 10 mM.

6. Make sure to follow recommendations on pressure limit and buffer compatibility for the specific column. Our desalting column cannot tolerate urea at the concentrations above 6 M; therefore, desalting buffer contains less urea than IMAC buffers. Depending on solution viscosity, flow rate needs to be adjusted during the run to keep pressure under the limit.
7. Because after overnight oxidation PrP solution contains urea, loading the protein onto C4 column requires pressure above Shimadzu HPLC system limits. To avoid high pressure, we dilute PrP solution threefold with buffer A and use external HPLC pump 2010 from Varian (Walnut Creek, CA) to load the protein. We wash the pump with water and with 20 ml of buffer A before loading PrP. We load protein solution at flow rate 5 ml/min and finish by loading 20 ml of HPLC buffer A, to ensure that no protein is left in the pump. After loading is completed, we disconnect and thoroughly wash the loading pump with water.
8. In our gradient method, the percentage of HPLC buffer B (0.1% trifluoroacetic acid in acetonitrile) grows from 0% to 25% during the first 15 min (at flow rate 5 ml/min). Then, gentle gradient is applied (from 25% to 35% of buffer B in the next 65 min) to ensure efficient separation of PrP adducts (Fig. 4a). Correctly folded, oxidized, α-helical monomeric form of PrP is eluted between 52.5 and 54.5 min. Then, the percentage of buffer B grows from 35% to 100% in 15 min; the column is washed with 100% of buffer B for 15 min, and the gradient drops down to 0% within 10 min.

9. If the same C4 column is used for purification of different PrP variants, running washing program with extended time of 100% acetonitrile will ensure the absence of cross-contamination. In our washing program, buffer B concentration grows from 0% to 100% within 20 min, remains constant 100% for 25 min, and then goes down to 0% during the last 20 min. The run is finished by equilibration of the C4 column with buffer A for 25 min, which prepares the column for loading of new protein.
10. To perform mass spectroscopy, a small amount of recombinant PrP (as a lyophilized powder) is dissolved in 20 μl of 1:1 water/methanol mixture containing 1% acetic acid. The solution is then analyzed on a single quadrupole mass spectrometer operated in positive ion mode.
11. After PrP is purified, it is preferable to store the protein as a lyophilized powder at −20 °C. Storage in 6 M GdnHCl at −20 °C is possible for up to 1 week. However, prolonged storage at 6 M GdnHCl leads to formation of the β-oligomeric non-fibrillar form. Prolonged storage (longer than 24 h) in solution at neutral pH and in the absence of denaturants is not advisable due to self-cleavage and nonspecific aggregation.
12. Fibril formation can proceed sluggishly if the volume of the solution in the tube is too low (<300 μl).
13. Make sure that the plastic tube fits tightly within the rack attached to the platform and the tube does not rotate inside the rack during orbital shaking (Fig. 5). End-over-end rotation on Nutator Mixer or similar device can also be used as an alternative way for fibrillation reactions. However, structurally different types of fibrils are formed under different shaking/rotation modes even when identical solvent conditions are employed for fibrillation reactions [10, 21].
14. Exact value of thioflavin T fluorescence depends on several factors including purity of the protein and conditions of fibril formation. Differences in intensity of ThT fluorescence could be due to several factors including differences in ThT binding affinity for fibrils prepared at different solvent conditions, different yields of fibril formation, or different propensities of fibrils to aggregate into large clumps. The yield of ThT fluorescence of fibrils prepared under nonstandard conditions is often lower than that for standard fibrils. ThT fluorescence of fibrils prepared under different shaking modes or from different PrP variants (e.g., mouse vs. hamster PrP) can also vary [10] (Fig. 7e, f). Values of ThT fluorescence threefold or more above background usually indicate the presence of amyloid fibrils. In addition to ThT fluorescence assay, fluorescence microscopy helps to analyze the yield and quality of fibrils (Fig. 7).

Acknowledgments

This work was supported by the National Institutes of Health grants R01 NS045585 and NS074998.

References

1. Zahn R, von Schroetter C, Wuthrich K (1997) Human prion protein expression in Escherichia coli and purified by high-affinity column refolding. FEBS Lett 417:400–404
2. Mehlhorn I, Groth D, Stöckel J et al (1996) High-level expression and characterization of a purified 142-residue polypeptide of the prion protein. Biochemistry 35:5528–5537
3. Rezaei H, Marc D, Choiset Y et al (2000) High yield purification and physico-chemical properties of full-length recombinant allelic variants of sheep prion protein linked to scrapie susceptibility. Eur J Biochem 267:2833–2839
4. Hornemann S, Korth C, Oesch B et al (1997) Recombinant full-length murine prion protein, *m*PrP(23–231): purification and spectroscopic characterization. FEBS Lett 413:277–281
5. Jackson GS, Hil AF, Joseph C et al (1999) Multiple folding pathways for heterologously expressed human prion protein. Biochim Biophys Acta 1431(1):1–13
6. Yin SM, Zheng Y, Tien P (2003) On-column purification and refolding of recombinant bovine prion protein: using its octarepeat sequences as a natural affinity tag. Protein Expr Purif 32:104–109
7. Bocharova OV, Breydo L, Parfenov AS et al (2005) In vitro conversion of full length mammalian prion protein produces amyloid form with physical property of PrPSc. J Mol Biol 346:645–659
8. Baskakov IV, Bocharova OV (2005) In vitro conversion of mammalian prion protein into amyloid fibrils displays unusual features. Biochemistry 44:2339–2348
9. Bocharova OV, Makarava N, Breydo L et al (2006) Annealing PrP amyloid fibrils at high temperature results in extension of a proteinase K resistant core. J Biol Chem 281:2373–2379
10. Makarava N, Baskakov IV (2008) The same primary structure of the prion protein yields two distinct self-propagating states. J Biol Chem 283:15988–15996
11. Makarava N, Kovacs GG, Bocharova OV et al (2010) Recombinant prion protein induces a new transmissible prion disease in wild type animals. Acta Neuropathol 119:177–187
12. Novitskaya V, Makarava N, Bellon A et al (2006) Probing the conformation of the prion protein within a single amyloid fibril using a novel immunoconformational assay. J Biol Chem 281:15536–15545
13. Bocharova OV, Breydo L, Salnikov VV, Baskakov IV (2005) Cu(II) inhibits in vitro conversion of prion protein into amyloid fibrils. Biochemistry 44:6776–6787
14. Breydo L, Bocharova OV, Makarava N et al (2005) Methionine oxidation interferes with conversion of the prion protein into the fibrillar proteinase K-resistant conformation. Biochemistry 44:15534–15543
15. McMahon HEM, Mange A, Nishida N et al (2001) Cleavage of the amino terminus of the prion protein by reactive oxygen species. J Biol Chem 276:2286–2291
16. Mange A, Beranger F, Peoc'h K et al (2004) Alpha- and beta-cleavages of the amino-terminus of the cellular prion protein. Biol Cell 96:125–132
17. Ostapchenko VG, Makarava N, Savtchenko R, Baskakov IV (2009) The polybasic N-terminal region of the prion protein controls the physical properties of both the cellular and fibrillar forms of PrP. J Mol Biol 383:1210–1224
18. Sun Y, Makarava N, Lee CI et al (2008) Conformational stability of PrP amyloid firbils controls their smallest possible fragment size. J Mol Biol 376:1155–1167
19. Almstedt K, Nyström S, Nilsson KPR, Hammarström P (2009) Amyloid fibrils of human prion protein are spun and woven from morphologically disordered aggregates. Prion 3:224–235
20. Bocharova OV, Breydo L, Salnikov VV et al (2005) Synthetic prions generated *in vitro* are similar to a newly identified subpopulation of PrPSc from sporadic Creutzfeldt-Jakob disease. Protein Sci 14:1222–1232
21. Ostapchenko VG, Sawaya MR, Makarava N et al (2010) Two amyloid states of the prion protein display significantly different folding patterns. J Mol Biol 400:908–921

Chapter 2

Method for Folding of Recombinant Prion Protein to Soluble β-Sheet Secondary Structure

Laura J. Ellett

Abstract

A key event in the pathogenesis of prion diseases is the change in structure of the normal cellular form of the prion protein from a predominantly α-helix form to the β-sheet-rich prion protein found in disease-associated tissue. To allow more detailed structural research into PrP misfolding, it is necessary to have techniques which enable enrichment of the β-sheet content in recombinant PrP.

This method describes the procedure for inducing β-folding of recombinant PrP to resemble a disease-associated structure and ultimately produce soluble β-folded recombinant PrP.

Key words α-helix, β-sheet, β-folding, Misfolded, Protein conformation, Recombinant protein, Secondary structure, Conversion, Buffer exchange, Low pH refolding, Protein structure, Soluble

1 Introduction

From experimental data, it has been inferred that during prion disease the normal cellular prion protein (PrP^C) misfolds into a protease, detergent-resistant isoform [1–6] which has been termed PrP^{Sc}. Structurally PrP^C mainly consists of α-helices, while disease-associated PrP^{Sc} misfolds in such a way that the β-sheet content of the protein increases and the protein aggregates [7–10].

The following method describes the refolding of PrP into the β-sheet isoform based on the protocols published by Jackson et al. [11] and Coleman et al. [12]. Jackson [11] established the conditions required to switch native, α-helical recombinant PrP to a soluble, monomeric, β-sheet-rich form. In this method, following denaturation, the protein is slowly refolded in an acidic, reducing environment, which promotes β-sheet formation. The product formed exhibits partial resistance to PK digestion and can be a precursor to fibrils; both properties are indicative of disease-associated PrP^{Sc}.

The ability to refold recombinant PrP has enabled the investigation of folding pathways [13], prion seeding [14], and the

Victoria A. Lawson (ed.), *Prions: Methods and Protocols*, Methods in Molecular Biology, vol. 1658,
DOI 10.1007/978-1-4939-7244-9_2, © Springer Science+Business Media LLC 2017

thermodynamic stability of disease-associated PrP [15]. Manipulation of the structure of PrP will enable the understanding of how the protein is induced to misfold during disease as well as how the β-sheet structure influences the behavior of PrP during pathogenesis.

2 Materials

1. High-quality recombinant protein (*see* **Note 1**).
2. Serological pipettes and pipette boy.
3. Dialysis tubing and clips.
4. 2× magnetic stirrer and large fleas (*see* **Note 2**).
5. 4 × 5 L beakers.
6. Dithiothreitol (DTT).
7. Guanidine HCl (Gnd-HCl).
8. Tris-acetate.
9. Sodium acetate.
10. Acetic acid (*see* **Note 3**).
11. Ultrapure water (Milli-Q).
12. Unfolding/denaturing buffer: 6 M Gnd-HCl, 100 mM DTT, 10 mM Tris-acetate, 10 mM sodium acetate, pH 8.
13. Folding buffer: 10 mM Tris–acetate (pH 4), 10 mM sodium acetate.
14. pH meter.
15. Vivaspin 20 columns with a 10 kDa molecular weight cut-off (MWCO) (Sartorius).
16. Centrifuge.

3 Methods

1. Dissolve lyophilized recombinant protein to a final concentration of 0.04 mg/mL in unfolding/denaturation buffer. Incubate overnight at room temperature.
2. The following day, take dialysis tubing with the correct MW cutoff, and cut to size (*see* **Note 4**).
3. Fold over one end three times and clip shut.
4. Take a serological pipette, and transfer approximately 25 mL of the protein solution into the dialysis tubing tube (*see* **Note 5**).

5. Fold over the top end several times and clip. You want a small air bubble to remain in the tube which will cause the tube to float in the buffer.
6. Check for leaks before you place the tubing into the dialysis buffer.
7. Dialyze at 4 °C against folding buffer supplemented with 1 mM DTT, twice for 2 h each. 5 L of dialysis buffer is used each time, and the buffer is continually mixed by magnetic stirrer to speed up the diffusion.
8. Continue dialysis in folding buffer (no DTT) at 4 °C for 2 days with a total of four buffer changes, with 5 L of fresh buffer each time.
9. After the dialysis is complete, very carefully take out the tubes, and undo the clips on one end. Insert a serological pipette and carefully recover the sample (*see* **Note 6**).
10. Concentrate the protein using a centrifugal concentrator with the appropriate MWCO if required (*see* **Note 7**).
11. Confirm β-sheet content increase using circular dichroism spectroscopy (CD). It is also worth testing for increases in PK resistance compared to the original native recombinant protein.
12. Store β-folded PrP in folding buffer at 4 °C. Avoid using other buffers or water for storage as the protein will potentially slowly lose its β-sheet content.

4 Notes

1. Full-length and C2 fragments of human and mouse recombinant PrP have been successfully refolded using this procedure.
2. Preferably set up two magnetic stirrers, one at room temperature for the denaturing step and one set up in a cold room (4 °C) for the refolding steps.
3. Acetic acid is used for lowering the pH of the refolding buffers.
4. For example, Thermo Fisher Scientific SnakeSkin Dialysis Tubing with a 10 K MWCO with a 22 mm internal diameter. The tubing will usually come with instructions for converting volume of liquid into length of tubing. Always cut extra to fold over to close the ends.
5. Transfer sample to dialysis tubing very carefully. It is very easy to drop the slippery tubing and spill your sample.
6. Pipette the sample up, and wash the sides of the tubing to help dislodge any PrP remaining bound to the tubing.

7. For example, Vivaspin 20 columns with a 10 kDa MWCO for full-length recombinant protein (23–230).

References

1. Bolton D, McKinley M, Prusiner S (1982) Identification of a protein that purifies with the scrapie prion. Science 218:1309–1311
2. McKinley M, Bolton D, Prusiner S (1983) A protease-resistant protein is a structural component of the scrapie prion. Cell 35:57–62
3. Diringer H, Hilmert H, Simon D et al (1983) Towards purification of the scrapie agent. Eur J Biochem 134:555–560
4. Prusiner S, McKinley M, Bowman K et al (1983) Scrapie prions aggregate to form amyloid-like birefringent rods. Cell 35:349–358
5. De Armond S, McKinley M, Barry R et al (1985) Identification of prion amyloid filaments in scrapie-infected brain. Cell 41:221–235
6. Gabizon R, McKinley M, Groth D et al (1988) Immunoaffinity purification and neutralization of scrapie prion infectivity. Proc Natl Acad Sci U S A 85:667–621
7. Caughey B, Dong A, Bhat K et al (1991) Secondary structure analysis of the scrapie-associated protein PrP 27-30 in water by infrared spectroscopy. Biochemist 30:7672–7680
8. Pan K, Baldwin M, Nguyen J et al (1993) Conversion of alpha-helices into beta-sheets features in the formation of the scrapie prion proteins. Proc Natl Acad Sci U S A 90:10962–10966
9. Safar J, Roller P, Gajdusek D et al (1993) Conformational transitions, dissociation, and unfolding of scrapie amyloid (prion) protein. J Biol Chem 286:20276–20284
10. Safar J, Roller P, Gajdusek D et al (1993) Thermal stability and conformational transitions of scrapie amyloid (prion) protein correlate with infectivity. Protein Sci 2:2206–2216
11. Jackson G, Hosszu L, Power A et al (1999) Reversible conversion of monomeric human prion protein between native and fibrillogenic conformations. Science 283:1935–1937
12. Coleman B, Nisbet R, Han S et al (2009) Conformational detection of prion protein with biarsenical labeling and FlAsH fluorescence. Biochem Biophys Res Commun 380:564–568
13. Jackson G, Hill A, Joseph C et al (1999) Multiple folding pathways for heterologously expressed human prion protein. Biochim Biophys Acta 1431:1–13
14. Legname G, Baskakov I, Nguyen H et al (2004) Synthetic mammalian prions. Science 305:673–676
15. Baskakov I, Legname G, Prusiner S et al (2001) Folding of prion protein to its native alpha-helical conformation is under kinetic control. J Biol Chem 276:19687–19690

Chapter 3

Analysis of Prion Protein Conformation Using Circular Dichroism Spectroscopy

Laura J. Ellett and Vanessa A. Johanssen

Abstract

According to the protein-only hypothesis of prion propagation, the pathogenesis of prion disease is due to the misfolding of cellular PrP (PrP^C) which gives rise to disease-associated PrP^{Sc}. This misfolding results in the predominantly α-helix secondary structure of PrP becoming increasingly β-sheet. Prion protein researchers often employ circular dichroism (CD) spectroscopy to rapidly analyze and identify the degree of α-helix and β-sheet content in their recombinant protein and peptide samples. CD is a nondestructive method of determining protein secondary structure and can be used to monitor the protein structural changes in various environments, e.g., pH and temperature. CD can also be used to investigate kinetic and thermodynamic characteristics of proteins and peptides.

Key words α-Helix, β-Sheet, CD, Circular dichroism, Misfolded, Protein conformation, Recombinant protein, Secondary structure, Spectra, Spectrometer, Spectroscopy

1 Introduction

The secondary structure of proteins can be measured by a number of techniques such as NMR, X-ray crystallography, FTIR, Raman spectroscopy, and circular dichroism (CD) spectroscopy; however, CD is the easiest and most widely used due to rapid accumulation of data which can be obtained with low amounts of protein in aqueous buffers [1]. This low-resolution technique for determining peptide and protein secondary structures exploits the optical activity of peptides and proteins by measuring the difference in absorbance of right and left circularly polarized light [1]. Animations illustrating this phenomenon can be found at http://cddemo.szialab.org/. A comprehensive review on circular dichroism theory and techniques, including an overview on database reference sets, can be read in Whitmore and Wallace [2, 3] and more directly related to the analysis of PrP in Han and Hill [4].

Normal, cellular PrP (PrP^C) mainly consists of α-helices, while disease-associated PrP, PrP^{Sc}, misfolds in such a way that the β-sheet

Victoria A. Lawson (ed.), *Prions: Methods and Protocols*, Methods in Molecular Biology, vol. 1658,
DOI 10.1007/978-1-4939-7244-9_3,

content of the protein increases. Hence, prompt analysis to determine if a purified prion protein or peptide in solution is predominantly α-helical or β-sheet is a valuable tool in prion research. Here, we demonstrate the usefulness and ease at which CD spectroscopy can be used to determine the secondary structure of recombinant prion proteins and peptides in various soluble conditions and manipulations. An example of measuring the secondary structure of full-length mouse recombinant prion protein in a buffer, which induces native α-helix structure, will be used. A protocol describing thermal manipulation of recombinant PrP which can be used to determine the stability of various isoforms of recombinant PrP such as truncations and mutations is also included.

2 Materials

Prepare all solutions using ultrapure water (such as Milli-Q®), and use aseptic techniques (filter barrier tips). Using polypropylene plasticware (tips, tubes, and microtiter plates) is also advantageous to prevent contamination and loss of protein to surface area. Keep reagents and recombinant PrP on ice to minimize aggregation.

2.1 High-Quality Recombinant Prion Protein Reconstitution

1. Recombinant prion protein should be HPLC purified and stored freeze-dried (*see* **Note 1**) at −80 °C until use.
2. Ice-cold, sterile MqH_2O (Milli-Q ultrapure water).
3. 10 × sterile desired buffer such as phosphate buffer, pH 7.4 (*see* **Note 2**).
4. Refrigerated bench top microcentrifuge.

2.2 CD Components

1. Pipette and tips (*see* **Note 3**).
2. CD spectrometer.
3. Quartz CD cuvette (e.g., Starna or Hellma) (*see* **Notes 4** and **5**).
4. Nitrogen gas.
5. Neutral pH detergent (*see* **Note 6**).
6. Mild acid (*see* **Note 7**).
7. For thermal analysis, a water-jacketed cell and a computer-controlled water bath.

3 Methods

3.1 Standard CD

3.1.1 Preparation of Recombinant PrP Solution

1. Prepare fresh recombinant prion protein working stock solutions by dissolving lyophilized/freeze-dried protein in ice-chilled sterile MqH_2O to 1–10 μg/μL, and incubate on ice for 5 min.
2. Unless specified otherwise, centrifuge the prion protein solution for 15–60 min at 13,200 × *g* in a bench top centrifuge at

4 °C. Keep the supernatant on ice until used for experiments (*see* **Note 8**).

3. Verify protein concentration by absorbance spectrophotometry at UV 280 nm using molar extinction coefficient values as calculated by ExPASy [5].
4. Adjust PrP protein solution to the desired final concentration between 0.1–0.3 mg/mL by diluting with ice-cold MqH_2O and 10 × desired buffer (e.g., final concentration of 10 mM phosphate buffer) (*see* **Note 2**).

3.1.2 Acquisition of CD Spectra

1. Turn on the nitrogen tap which is connected to the machine. This displaces the air in the machine and prevents oxygen from damaging the mercury lamp. Make sure the gas flow rate is appropriate for the model of machine, usually around 5 LPM.
2. CD spectra are often recorded at room temperature (24 °C). The temperature may need to be set, be it room temperature or other, before beginning to read samples, therefore allowing time to reach desired temperature.
3. Wash cuvettes before starting as follows: five rinses with neutral pH detergent, ten rinses with ultrapure water, and five rinses in methanol. Have squeeze bottles set up with all three solutions as this is the easiest way of filling up the cuvettes. Be careful when shaking the liquid out of the cuvette – avoid handling the optical surface, and do not drop and damage the cuvette. Dry the inside of the cuvette completely with nitrogen gas and the outside with a lens cleaner before adding the sample. Perform this cleaning process in between measuring each sample (*see* **Notes 6, 7** and **9**).
4. Load blank (solution the protein is solubilized in) into the quartz cuvette (*see* **Note 3**).
5. Place cuvette into machine carefully noting the direction of the light path and positioning the sample in this path.
6. Set software parameters (*see* **Note 10**).
7. Commence measurement (*see* **Notes 11** and **12**).
8. Remove sample and keep for further analysis or discard appropriately.
9. Clean and dry cuvette and then add sample and read (*see* **Note 13**).
10. Once all measurements are complete, clean and dry cuvette again before storage.

3.1.3 CD Spectra Analysis

1. Subtract the baseline acquired in the absence of protein.
2. For publication, CD spectra are most commonly presented in mean residue ellipticity (MRE) with the units **degrees cm^2 dmol^{-1} residue^{-1}**, but these are not the units the machine generally provides. The data obtained from the machine

(in units of θ machine units) is converted to MRE (or [θ]) using the following equations (equation from DICHROWEB http://dichroweb.cryst.bbk.ac.uk/html/userguide.shtml).

$$\Delta\varepsilon = \theta \times \left(\frac{0.1 \times \mathrm{MRW}}{\mathrm{P} \times \mathrm{conc}}\right) \times 3298$$

$$\mathrm{MRE} = \Delta\varepsilon \times 3298$$

where:

P is pathlength in cm (0.1 cm).

Conc is protein concentration in mg/ml.

MRW is mean residue weight.

MRW = proteins weight (Da)/number of residues.

3. CD spectra of proteins and peptides have characteristic minima, depending on their secondary structure. Examples of predominantly α-helical, β-sheet, and unfolded/random coil protein spectra are given in Fig. 1. Qualitative interpretation of CD spectra can be reported by eye with some practice.

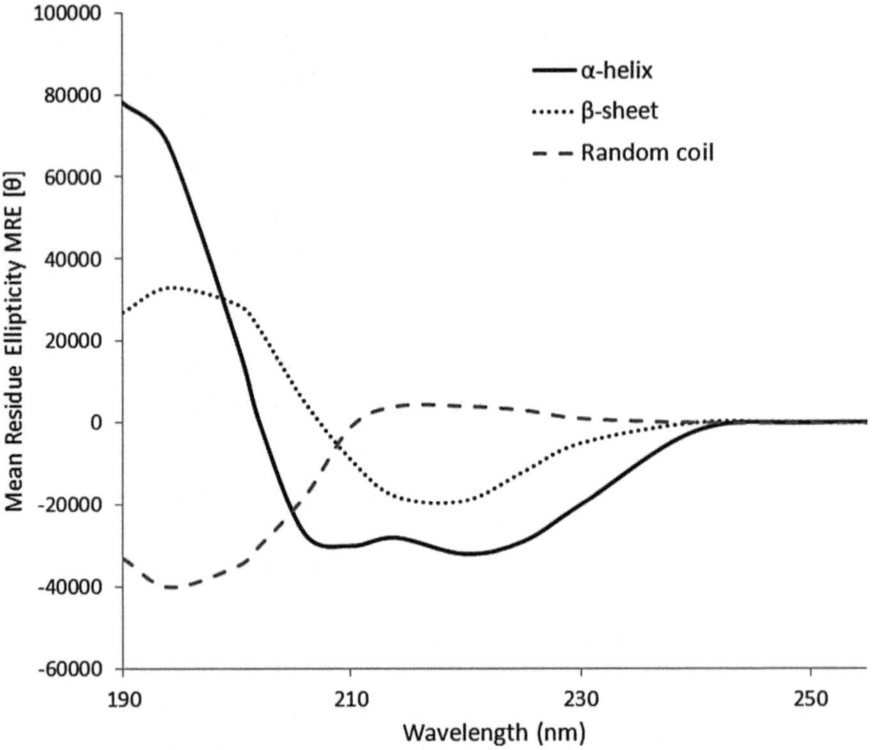

Fig. 1 Examples of expected CD spectra for predominantly α-helical, β-sheet, and unfolded/random coil proteins

4. More quantitative analysis of CD spectra can be performed by using algorithms available through DICHROWEB (http://dichroweb.cryst.bbk.ac.uk/html/home.shtml).

3.2 Thermal Denaturation Experiments

1. Prepare recombinant PrP solution as per Subheading 3.1.1 at a concentration of 10 μM.
2. Follow changes in ellipticity at 222 nm (α-helical fold) or 218 nm (β-sheet fold) as a function of temperature.
3. The measurements to be performed will require a CD spectrometer equipped with a water-jacketed cell and a computer-controlled water bath. We use a Jasco-810 CD spectrometer. Set software parameters to a heating rate of 1 deg/min.
4. Start at room temperature, and then heat sample to 90 °C at 1 degree/min followed by cooling again to room temperature at the same rate to also determine reversibility of protein folding (*see* **Note 14**).
5. Convert machine units to MRE (*see* Subheading 3.1.3) and plot temperature versus MRE.
6. Analyze thermal unfolding curves assuming a two-step process as suggested by previous studies [6, 7], yielding the midpoint denaturation temperature, T*m* (*see* **Note 15**). The thermal stability of human prion protein 90–231 in its pathogenic-associated isoform (predominantly β-sheet) is given in Fig. 2.

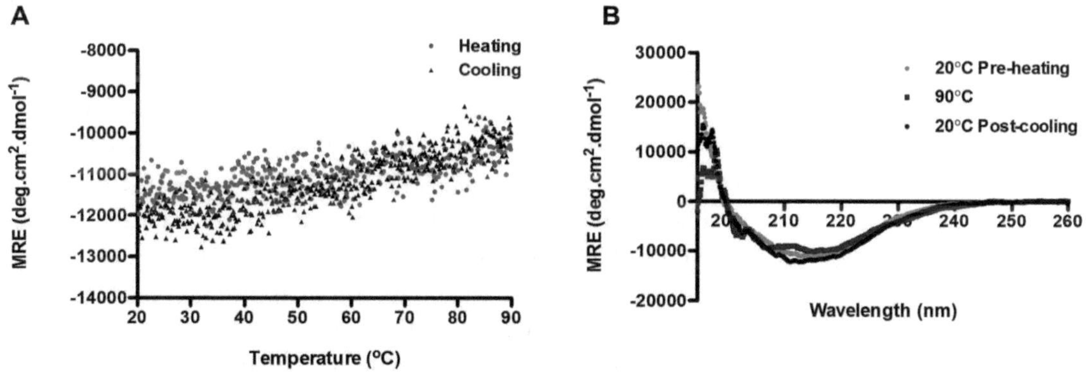

Fig. 2 Thermal stability of human recombinant β-PrP-(90-231). (**a**) β-PrP-(90–231) was diluted to 10 μM. The fraction of molecules folded was determined by CD at 218 nm. The heating (*gray*) profile of β-PrP-(90–231) is superimposed with data for cooling (*black*). (**b**) Secondary structure of β-PrP-(90–231) before thermal melt at 20 °C (*light gray*), at 90 °C (*dark gray*), and post-thermal melt at 20 °C (*black*). Before thermal melt, at 90 °C, and after thermal melt (20 °C), β-PrP-(90–231) displays a predominantly β-sheet structure with a single minimum at approximately 218 nm. This research was originally published in the *Biochemical Journal*. Johanssen et al. (2014) [11]

4 Notes

1. High-purity and high-quality HPLC purified synthetic and recombinant PrP can be produced as per methods described in Karas et al. (2012) and Johanssen et al. (2012), respectively [8, 9]. Freeze-drying is more desirable than lyophilizing purified peptide to avoid aggregate formation which can occur as peptide concentration increases during the lyophilization process.
2. The environment the protein is in is very important as it impacts on the structure. Research the best buffering system for the particular protein within the context of the experiment at hand. For example, ultrapure water will give the least background, but if a certain pH is required, an appropriate buffer will need to be utilized. 10 mM phosphate or Tris buffer is recommended but avoid chloride (this includes using HCl to pH solutions) and imidazole [10]. The pH of Tris is highly dependent on temperature and is therefore unsuitable for thermal melt analysis. Always record a spectra of the buffer alone to determine if it will impact on absorbance readings.
3. Gel loading tips are easiest to use when filling the cuvette as the cuvettes are very narrow.
4. To minimize the volume of sample required, 1 mm cuvettes are often used (requiring only approximately 200 μL of sample). However, the longer the pathlength, the lower the concentration of sample you can use. Yet then, the longer the pathlength, the more noise.
5. For better consistency, do experiments in one cuvette. If you plan on using two cuvettes within an experiment (such as one containing the reference buffer), they should be a matched pair.
6. There are detergents available for washing cuvettes and cells, such as Starna CellClean. Try testing different detergents to see which gives the best result and the least residue.
7. For more thorough cleaning, use a dilute acid such as 2 M nitric acid or NOCHROMIX® (an inorganic oxidizer diluted with sulfuric acid). Dilute acid also helps remove detergent residues.
8. The purpose of centrifuging the sample is to remove any preformed protein aggregates. Alternatively, a syringe filter specialized for small volumes can be used such as Minisart® NML syringe filters (Sartorius); however, this method will incur greater sample loss.
9. Don't let the cuvette dry out before cleaning as this makes it much more difficult to clean protein and salt from the cuvette. Only use lens cleaning wipes when drying the outside of the

cuvette so as not to scratch the optical surface and never handle the optical surface. Any scratches on the optical surface will prevent the machine from taking a quality reading.

10. Software Parameters

    Sensitivity: standard.

    Starting wavelength: 250 nm.

    End wavelength: 185 nm.

    Scanning rate: 50 nm/min.

    Data pitch: 0.1 nm steps.

    Accumulation: 3–5. Start with three and if noisy increase to five.

    If desired, the machine can be set to automatically subtract the background reading.

11. Baseline between −1 and +1 mdeg is good. If over − 5 + 5, wash the cuvette or assess the components of the buffering solution (*see* **Note 9**).

12. Be aware of the following during analysis of your sample:

    Check that the protein solution is not cloudy or contains particles before loading into the machine. This indicates that the protein is coming out of solution ("crashing") and will not be able to be properly detected by the machine.

    If the spectra readings seem low, check the voltage. If the voltage reading plateaus/flatlines, this can indicate that the protein concentration is too high, and this impairs the machine's ability to detect the sample properly. Dilute the sample and try again.

13. You can start analyzing data while still running other samples.

14. Following heat-induced structural changes with CD by application of a time-dependent temperature gradient (20 °C–90 °C) can give insight into protein stability. Often, under the experimental conditions described here, the thermal unfolding of PrP is reversible. However, if the refolding curve (cooling pathway) of PrP finishes at a slightly different point, this suggests a modest change in secondary structure (as seen in Fig. 2a).

15. The T*m* can be calculated using the equation $\alpha = (\theta_{obs} - \theta_U)/(\theta_F - \theta_U)$, where θ_{obs} is the observed ellipticity at any temperature and the ellipticity of the fully folded form and the unfolded form is denoted as θ_F and θ_U, respectively. T*m* is the temperature where $\alpha = 0.5$, in other words, when there is an equal amount of folded and unfolded protein.

Acknowledgment

This work was supported by National Health and Medical Research Council.
(grant numbers 400202 and 400183) and the Bethlehem Griffiths Research Foundation.
(grant number 802270).

References

1. Johnson W (1990) Protein secondary structure and circular dichroism: a practical guide. Proteins 7:205–214
2. Whitmore L, Wallace B (2004) DICHROWEB, an online server for protein secondary structure analyses from circular dichroism spectroscopic data. Nucleic Acids Res. doi:10.1093/nar/gkh371
3. Whitmore L, Wallace B (2008) Protein secondary structure analyses from circular dichroism spectroscopy: methods and reference databases. Biopolymers 89:392–400
4. Han S, Hill A (2008) Analysis of PrP conformation using circular dichroism. Methods Mol Biol 459:145–159
5. Gill S, von Hippel P (1989) Calculation of protein extinction coefficients from amino acid sequence data. Anal Biochem 182:319–326
6. Swietnicki W, Petersen R, Gambetti P et al (1998) Familial mutations and the thermodynamic stability of the recombinant human prion protein. J Biol Chem 273:31048–31052
7. Baskakov I, Legname G, Prusiner S et al (2001) Folding of prion protein to its native alpha-helical conformation is under kinetic control. J Biol Chem 276:19687–19690
8. Karas J, Boland M, Haigh C et al (2012) Microwave synthesis of prion protein fragments up to 111 amino acids in length generates biologically active peptides. In J Pept Res Ther 18:21–29
9. Johanssen V, Barnham K, Masters C (2012) Generating recombinant C-terminal prion protein fragments of exact native sequence. Neurochem Int 60:318–326
10. Kelly S, Jess T, Price N (2005) How to study proteins by circular dichroism. Biochim Biophys Acta 1751:119–139
11. Johanssen V, Johanssen T, Masters C et al (2014) C-terminal peptides modelling constitutive PrP^{C} processing demonstrate ameliorated toxicity predisposition consequent to α-cleavage. Biochem J 459:103–115

Chapter 4

Analysis of Prion Protein Structure Using Nuclear Magnetic Resonance Spectroscopy

Ivana Biljan, Gregor Ilc, and Janez Plavec

Abstract

Nuclear magnetic resonance (NMR) spectroscopy is a powerful experimental tool for obtaining information on three-dimensional (3D) structures of proteins at atomic resolution. In inherited forms of prion diseases, misfolding of cellular prion protein, PrP^C, into its pathological form, PrP^{Sc}, is caused by mutations in the human prion protein gene (*PRNP*). Understanding of the earliest stages of the conformational changes leading to spontaneous generation of prions in inherited forms of prion diseases may benefit from detailed structural analysis of different human (Hu) PrP variants. Here, we describe the protocol for structure determination of HuPrP variants by NMR spectroscopy in solution that consists of preparation of NMR samples, acquisition of NMR data, NMR resonance assignments, and structure calculation.

Key words Protein structure, NMR structure determination, Prion protein, Mutants, Prion diseases

1 Introduction

Detailed atomic-level analysis of the three-dimensional (3D) structural features of cellular prion protein, PrP^C, and its pathological form, PrP^{Sc}, is central for understanding the conversion process between healthy and malicious PrP forms. NMR spectroscopy has proven to be an invaluable tool for gaining structural information on PrPs expressed by various mammalian species [1–9]. Over the last two decades, NMR solution-state structural studies were mainly performed on full-length or truncated recombinant PrPs (recPrPs) lacking posttranslational modifications at mildly acidic pH conditions. The first solution-state NMR structure of the wild-type (WT) human (Hu) PrP was determined in year 2000 by Wütrich [9]. It was found that WT HuPrP consists of two parts with distinct structural and dynamic features. The first part is a highly flexible N-terminal domain containing residues 23–124, whereas the second part is a well-defined C-terminal globular domain comprising residues 125–231. The structured part contains three α-helices, spanning residues 144–154 (α1), 173–194

Victoria A. Lawson (ed.), *Prions: Methods and Protocols*, Methods in Molecular Biology, vol. 1658,
DOI 10.1007/978-1-4939-7244-9_4, © Springer Science+Business Media LLC 2017

(α2), and 200–228 (α3), and a very short antiparallel β-sheet, formed by residues 128–131 (β1) and 161–164 (β2). The globular domain is further stabilized by Cys179-Cys214 disulfide bond bridging helices α2 and α3. 3D structures of other mammalian PrPs display very similar features.

Approximately 10–15% of prion diseases in humans are caused by mutations in the PrP gene (*PRNP*) [10]. The molecular mechanism by which mutations favor spontaneous generation of PrP^{Sc} in specific brain regions leading to neuronal cell degeneration and death is largely still unknown [11–13]. The majority of mutations are positioned in the C-terminal globular domain of HuPrP enabling structural characterization of different HuPrP variants that may be helpful for better understanding of the earliest stages of prion diseases. Our recent NMR studies were focused on 3D structure analysis of various truncated recombinant HuPrP variants. We determined NMR structures of HuPrPs with pathological Q212P and V210I mutations [14–16] causing Gerstmann–Sträussler–Scheinker syndrome (GSS) and familial Creutzfeldt–Jakob disease (fCJD), respectively, and HuPrP carrying naturally occurring E219K polymorphism considered to protect against sporadic CJD (sCJD) [17]. This chapter focuses on the application of high-resolution solution-state NMR spectroscopy in the analysis of 3D structural features of HuPrP variants. We describe the general protocol for solving NMR structures of HuPrP variants including preparation of NMR samples, acquisition of NMR data, NMR resonance assignments, and structure calculation.

2 Materials

2.1 Cloning and Mutagenesis of HuPrP

1. Vent polymerase (New England Biolabs).
2. pET-11a vector (Novagen).
3. QuikChange™ kit (Stratagene).
4. pProExHTa plasmid (Invitrogen).
5. NdeI, BamHI, and EcoRI restriction enzymes.

2.2 Expression of HuPrP

1. *Escherichia coli* BL21 (DE3) strain (Stratagene).
2. M9 minimal medium (MM) plus ampicillin (100 μg/mL).
3. ^{13}C-labeled glucose and ^{15}N-labeled ammonium chloride.
4. Isopropyl-β-D-galactopyranoside (IPTG).

2.3 Isolation of Inclusion Bodies

1. French press (EmulsiFlex-C3).
2. Inclusion body buffer: 25 mM Tris–HCl, 5 mM ethylenediaminetetraacetic acid (EDTA), 0.8% TritonX100, pH 8.

2.4 Purification of HuPrP

1. 6 M GndHCl.
2. 5 mL HisTrap column (GE Healthcare).
3. Binding buffer: 2 M GndHCl, 500 mM NaCl, 20 mM Tris–HCl, 20 mM imidazole, pH 8.
4. 500 mM imidazole.
5. Reverse-phase column (Jupiter C4, 250 mm × 21.2 mm, 300 Å pore size, Phenomenex).
6. 0–95% acetonitrile.
7. 0.1% trifluoroacetic acid.

2.5 Expression and Purification of Tobacco etch virus *(TEV) Protease*

1. Zym-5052 medium for autoinduction plus kanamycin (100 μg/mL).
2. 6 M GndHCl.
3. 5 mL HisTrap column (GE Healthcare).
4. TEV binding buffer: 6 M GndHCl, 500 mM NaCl, 20 mM Tris–HCl, 20 mM imidazole, pH 8.
5. 500 mM imidazole.
6. TEV refolding buffer: 50 mM Tris base and 0.5 mM EDTA, pH 8.0.
7. Dithiothreitol (DTT).
8. Spectra/Por membrane (MW 3000).
9. TEV storage buffer: 50 mM Tris base, 100 mM NaCl, 0.5 mM EDTA, 10% glycerol, pH 7.9.

2.6 His-Tag Cleavage

1. 8 M GndHCl.
2. TEV reaction buffer: 50 mM Tris base, 1 mM EDTA, 5 mM DTT, pH 8.
3. TEV protease.
4. 5 mL HisTrap column (GE Healthcare).
5. Binding buffer: 500 mM NaCl and 20 mM Tris–HCl, pH 8.

2.7 Protein Refolding

1. 8 M GndHCl.
2. Refolding buffer: 20 mM sodium acetate-d_3, 0.005% NaN_3, pH 5.5.
3. Spectra/Por membrane (MW 3000).

2.8 NMR Sample Preparation

1. Amicon centrifugal cells (Millipore).
2. Samples of HuPrP variants of concentration between 0.6 and 1 mM.
3. Deuterium oxide (D_2O).

2.9 NMR Acquisition

1. External chemical shift reference compound: 2,2-dimethyl-2-silapentane-5-sulfonate (DSS).
2. 5 mm NMR tubes (Shigemi Inc.).

3 Methods

3.1 Preparation of NMR Samples

1. Amplify the WT HuPrP(90–231, M129) from genomic human DNA by PCR using Vent polymerase and primers 5′-GGA ATT CCA TAT GAA GAA GCG CCC GAA GCC TGG A-3′ and 5′-CGG GAT CCC TAA TGG TGA TGG TGA TGG TGG CTC GAT CCT CTC TGG TAA TAG GCC TGA-3′. The reverse primer encodes for an uncleavable His_6 tag at the C-terminus. After digestion with the enzymes NdeI and BamHI, ligate the restricted fragment into the pET-11a vector. Verify the cloned DNA sequence by sequencing.
2. For construction of Q212P mutant, use the QuikChange™ kit utilizing primers 5′-CGC GTG GTT GAG CCG ATG TGT ATC ACC C-3′ and 5′- GGG TGA TAC ACA TCG GCT CAA CCA CGC G -3′ and HuPrP(90–231, M129) as template. Insert the DNA product into a pET-11a vector containing a His-tag sequence at the C-terminus of the inserted sequence. Verify the cloned DNA sequence by sequencing.
3. For V210I mutant, use the HuPrP(90–231, M129) gene as template, and clone it into pProExHTa plasmid using BamHI and EcoRI restriction sites. The plasmid contains the TEV protease cleavage site between the His_6 tag and the protein fragment. After cleavage, five extra residues are present at the N-terminus of the sequence (GAMDP). Construct the V210I mutant using the QuikChange™ kit with primers 5′-ATG ATG GAG CGC GTG ATT GAG CAG ATG TGT AT-3′ and 5′-ATA CAC ATC TGC TCA ATC ACG CGC TCC ATC AT-3′ and the previous plasmid as template. Verify the cloned DNA sequence by sequencing.
4. For construction of the E219K polymorphism, use the QuikChange™ kit, primers 5′-GTA TCA CCC AGT ACA AGA GGG AAT CTC AG-3′ and 5′-CTG AGA TTC CCT CTT GTA TGG GTG ATA C-3′, and HuPrP(90–231, M129) as template. Verify the cloned DNA sequence by sequencing.
5. HuPrPs expression: Add freshly transformed overnight culture of *Escherichia coli* BL21 (DE3) cells at 37 °C to 2 L of M9 MM plus ampicillin (100 mg/mL). For isotope labeling, add 4 g/L [$^{13}C_6$] glucose and 1 g/L [^{15}N] ammonium chloride. At an OD600 of 0.8, induce expression with IPTG to a final concentration of 0.8 mM (*see* **Note 1**). Harvest the cells 12 h after induction.

6. Isolation of inclusion bodies: Lyse the cells by a French press (EmulsiFlex-C3), and wash the inclusion bodies in inclusion body buffer and then in bi-distilled water several times.
7. HuPrPs purification: Solubilize pure inclusion bodies in five volumes of 6 M GndHCl, load onto a 5 mL HisTrap column equilibrated in binding buffer, and elute with 500 mM imidazole. Subsequently, purify the protein by reverse-phase (Jupiter C4, 250 mm × 21.2 mm, 300 Å pore size), and separate using a gradient of 0–95% acetonitrile and 0.1% trifluoroacetic acid.
8. TEV protease expression: Add an overnight culture of *Escherichia coli* BL21 (DE3) cells freshly transformed with pET24a::TEV at 37 °C to 2 L of an autoinduction (Zym-5052) medium plus kanamycin (100 μg/mL). Harvest the cells 16 h after induction, and lyse by a French press. Wash the inclusion bodies in inclusion body buffer and then in bi-distilled water several times.
9. TEV protease purification: Solubilize pure inclusion bodies in five volumes of 6 M GndHCl, load onto a 5 mL HisTrap column equilibrated in TEV binding buffer, and elute with 500 mM imidazole.
10. TEV protease refolding: Dilute the protein to 0.1 mg/mL in TEV refolding buffer supplemented with 10 mM DTT, and dialyze against the TEV refolding buffer supplemented with 5 mM DTT using a Spectra/Por membrane (MW 3000). Dialyze the protein against TEV storage buffer and store at −80 °C. Check the purity of the protein with sodium dodecyl sulfate polyacrylamide gel electrophoresis (SDS-PAGE).
11. His-tag cleavage: Lyophilize the purified protein and dissolve it in 8 M GndHCl. Dilute the protein to a final concentration of 0.5 mg/mL in TEV reaction buffer, and add TEV protease to a final concentration of 75 μg/mL. Incubate the reaction at 22 °C overnight. Load the cleaved sample onto a 5 mL HisTrap column (GE Healthcare) equilibrated with binding buffer, and collect the flow-through containing only the cleaved protein.
12. Protein refolding: Lyophilize the purified protein, and dissolve it in 8 M GndHCl. Perform refolding by dialysis against refolding buffer (*see* **Note 2**) using a Spectra/Por membrane (MW 3000). Analyze the purified protein by SDS-PAGE under reducing condition, Western blot, and electrospray mass spectroscopy.
13. NMR sample preparation: Concentrate refolded proteins to 0.6–1 mM using Amicon centrifugal cells. Add D_2O for deuterium lock, transfer to 5 mm NMR tube, and store at 4 °C prior to acquisition of NMR data.

3.2 Acquisition of NMR Data

1. Carry out all NMR experiments on uniformly ^{13}C, ^{15}N doubly labeled HuPrP variants at 298 K. We performed our measurements on Varian VNMRS 800 MHz NMR spectrometer using triple $^{1}H/^{13}C/^{15}N$ resonance cold probe head with inverse detection. NMR samples typically contained 300 μL of 0.6–1 mM doubly labeled HuPrP variants in sodium acetate buffer (same as refolding buffer).
2. Perform NMR experiments with detection of HN in 90%/10% H_2O/D_2O at pH 5.5, whereas for HC-detected NMR data sets, use 100% deuterated buffer.
3. The first NMR experiment to be recorded on HuPrP variants is the ^{1}H-^{15}N heteronuclear single-quantum coherence (HSQC) experiment (*see* **Note 3**). Fig. 1 shows examples of ^{1}H-^{15}N HSQC spectra of HuPrP(Q212P), HuPrP(V210I), and HuPrP(E219K) proteins. Well-resolved signals of backbone amide resonances in the spectra indicate that all three HuPrP variants adopt a globular fold and are amenable to structure determination by NMR spectroscopy.
4. In order to obtain the data for the sequence-specific ^{1}H, ^{15}N, and ^{13}C backbone assignments, record the following set of 3D NMR experiments: CA(CO)NH, HNCA, CBCA(CO)NH, HNCACB, HAHB(CO)NH, ^{15}N-edited nuclear Overhauser enhancement spectroscopy (NOESY)-HSQC, and HNCO [18–20] (*see* **Note 4**).

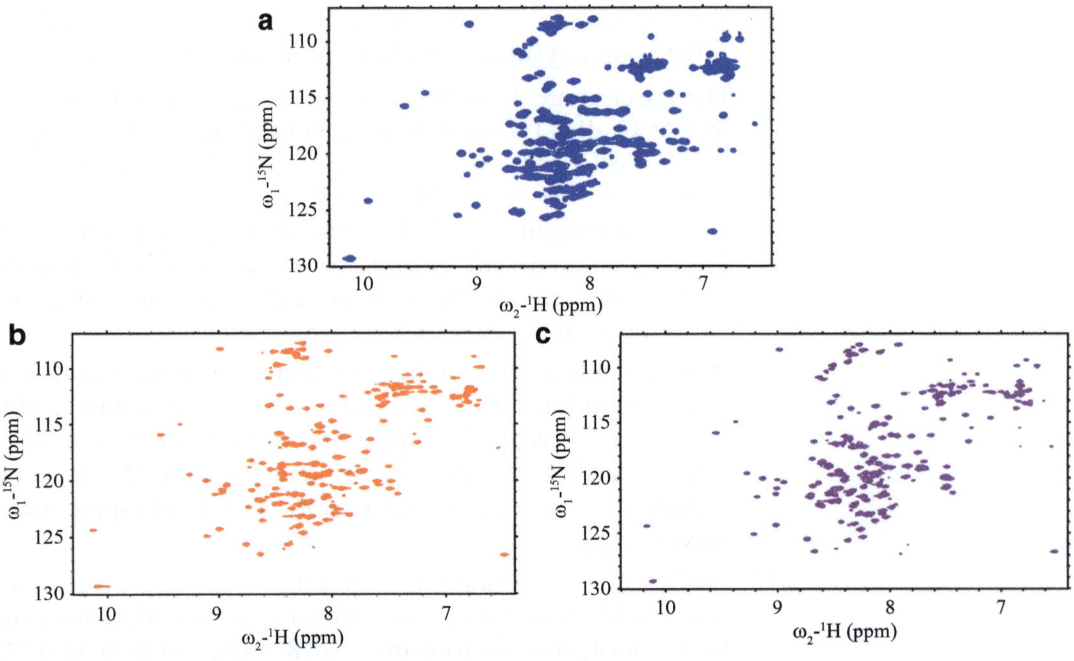

Fig. 1 ^{1}H-^{15}N HSQC spectra of HuPrP variants at pH 5.5. (**a**) HuPrP(90–231, M129, Q212P), (**b**) HuPrP(90–231, M129, V210I), and (**c**) HuPrP(90–231, M129, E219K)

5. Exchange buffer from a protonated to deuterated one to perform ^{13}C heteronuclear experiments needed for side-chain assignments (*see* **Note 5**).
6. For the assignment of ^{1}H and ^{13}C resonances in aliphatic side chains of HuPrP variants, record 2D ^{1}H-^{13}C HSQC (aliphatic and aromatic region), 3D ^{13}C-edited NOESY-HSQC (aliphatic and aromatic region), and (H)CCH-total correlation spectroscopy (TOCSY) experiments (*see* **Note 6**).
7. Process the NMR data by using NMRPipe software [21] or some other suitable software.
8. Analyze recorded spectra with SPARKY [22] and CARA software [23] or other NMR assignment software packages.

3.3 NMR Resonance Assignments

1. Pick peaks in the ^{1}H-^{15}N HSQC spectra of HuPrP variants.
2. The first step is the sequence-specific assignment of the backbone resonances in HuPrP variants. This is accomplished by combining pairs of triple-resonance NMR experiments in which magnetization is transferred across the peptide bond via heteronuclear scalar couplings allowing sequential assignment. One experiment from the pair provides intra-residual correlations between backbone resonances, while the other gives both intra- and inter-residual correlations between the same set of backbone resonances. Start with CA(CO)NH and HNCA experiments. The CA(CO)NH experiment correlates the amide ^{1}H and ^{15}N chemical shifts of residue *i* with the ^{13}Cα chemical shift of previous residue $i - 1$, whereas the HNCA experiment correlates the amide ^{1}H and ^{15}N chemical shifts of residue *i* with the ^{13}Cα chemical shifts of residues *i* and $i - 1$ (Fig. 2a). For each backbone ^{1}H-^{15}N pair in ^{1}H-^{15}N HSQC spectrum, pick and label $^{13}C\alpha_{(i-1)}$ and $^{13}C\alpha_{(i)}$ in HNCA spectrum (*see* **Note 7**).
3. The next pair of 3D NMR experiments to be analyzed are CBCA(CO)NH and HNCACB experiments (*see* **Note 8**). The CBCA(CO)NH experiment correlates the amide ^{1}H and ^{15}N chemical shifts of residue *i* with the ^{13}Cα and ^{13}Cβ chemical shifts of previous residue $i - 1$, while the HNCACB experiment correlates the amide ^{1}H and ^{15}N chemical shifts of residue *i* with the ^{13}Cα and ^{13}Cβ chemical shifts of residues *i* and $i - 1$ (Fig. 2b). For each backbone ^{1}H-^{15}N pair in ^{1}H-^{15}N HSQC spectrum, pick and label $^{13}C\beta_{(i)}$ and $^{13}C\beta_{(i-1)}$ in HNCACB spectrum (*see* **Note 9**).
4. Analyze 3D HNCO spectra. HNCO experiment provides correlations between amide ^{1}H and ^{15}N chemical shifts of residue *i* and ^{13}CO chemical shift of previous residue $i - 1$ (Fig. 2c). For each backbone ^{1}H-^{15}N pair in ^{1}H-^{15}N HSQC spectrum, pick and label $^{13}CO_{(i-1)}$ in HNCO spectrum.

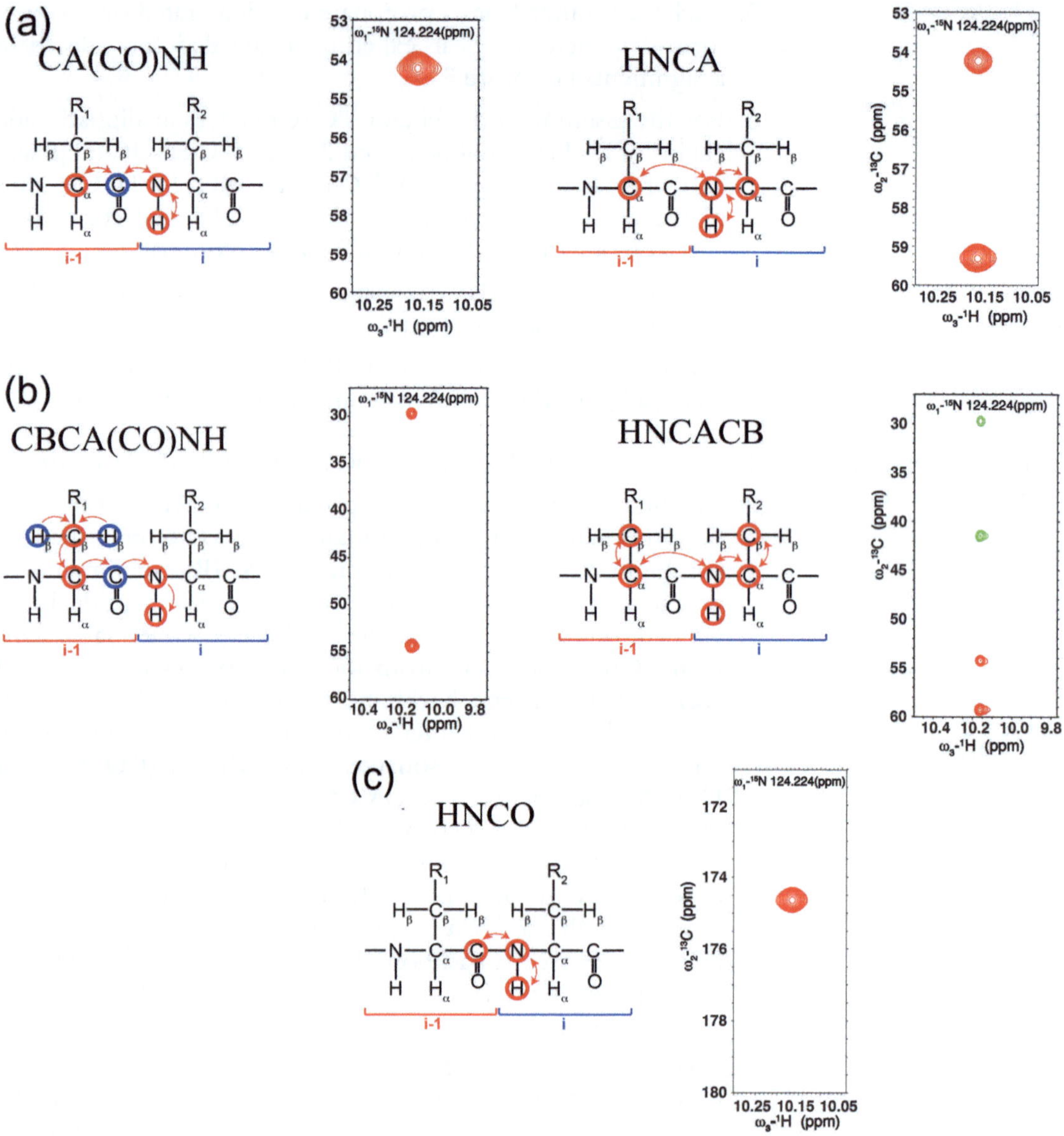

Fig. 2 Schematic view of magnetization transfer in triple-resonance NMR experiments for backbone assignment. (**a**) CA(CO)NH and HNCA, (**b**) CBCA(CO)NH and HNCACB, and (**c**) HNCO

5. Prepare the chemical shift table acquired from 3D CA(CO)NH, HNCA, CBCA(CO)NH, HNCACB, and HNCO experiments for MARS software [24] (*see* **Note 10**). From the data obtained by MARS and subsequent manual validation, assign the backbone resonances ($^{1}H_{N}$, ^{15}N, $^{13}C\alpha$, $^{13}C\beta$, and ^{13}CO) of HuPrP variants (*see* **Note 11**). Additionally verify the assignments by inspection of sequential and medium-range NOEs in the 3D ^{15}N-edited NOESY-HSQC spectra (*see* **Note 12**).

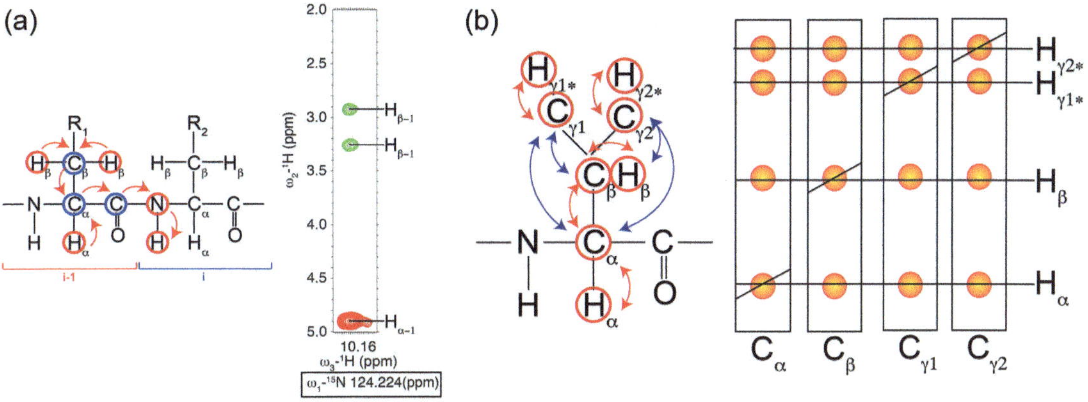

Fig. 3 Schematic view of magnetization transfer in 3D NMR experiments for side-chain assignment. (**a**) HAHB (CO)NH and (**b**) (H)CCH-TOCSY

6. The next step is the assignment of ^{1}H and ^{13}C resonances in aliphatic side chains by using 3D HAHB(CO)NH, (H)CCH-TOCSY, and ^{15}N- and ^{13}C-edited NOESY-HSQC (aliphatic region) experiments. The HAHB(CO)NH experiment correlates the amide ^{1}H and ^{15}N chemical shifts of residue *i* with the ^{1}Hα and ^{1}Hβ chemical shifts of previous residue $i-1$ (Fig. 3a). For each ^{1}H-^{15}N pair in ^{1}H-^{15}N HSQC spectrum, pick and label $^{1}H\alpha_{(i-1)}$ and $^{1}H\beta_{(i-1)}$ in HAHB(CO)NH spectrum. By comparing HAHB(CO)NH and ^{15}N-edited NOESY-HSQC spectra, assign ^{1}Hα and ^{1}Hβ resonances. In that way and from 3D experiments used for backbone assignment, ^{1}Hα/^{13}Cα and ^{1}Hβ/^{13}Cβ pairs are assigned. The rest of the ^{1}H and ^{13}C resonances from the side chains are assigned by analyzing 3D (H)CCH-TOCSY experiment which provides correlations of each CH group with all carbons within the same spin system (Fig. 3b) (*see* **Note 13**).
7. Assign the aromatic ^{1}H and ^{13}C resonances by analyzing the aromatic region of ^{13}C-edited NOESY-HSQC experiment (*see* **Note 14**).
8. Assign the side-chain amide groups of Asn and Gln residues on the basis of NOE signals in 3D ^{15}N- and ^{13}C-edited NOESY-HSQC experiments (*see* **Note 15**).

3.4 Structure Calculation

1. Pick peaks in 3D ^{15}N- and ^{13}C-edited NOESY-HSQC spectra. Create input NOESY peak list suitable for CYANA [25] (*see* **Note 16**).
2. Generate backbone ϕ and ψ dihedral angle restraints from secondary chemical shift data (i.e.,^{1}Hα, ^{13}Cα, ^{13}Cβ, ^{13}CO, and ^{15}N) using program TALOS [26].
3. Use the standard protocol of CYANA for automatic NOE assignment [27] and structure calculations of HuPrP variants

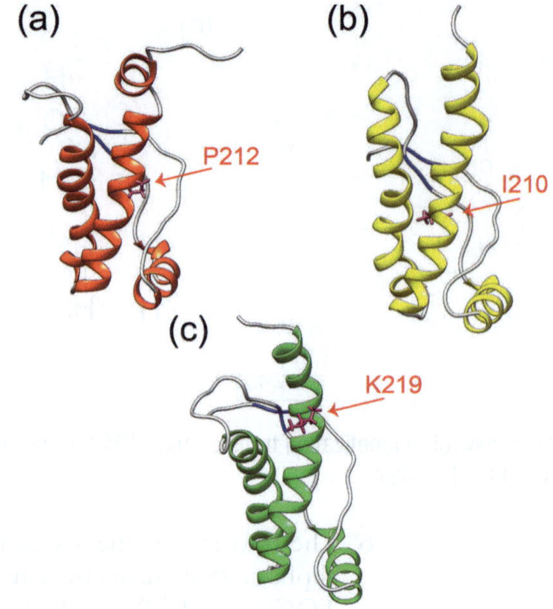

Fig. 4 The lowest energy 3D structures of the folded domains (residues 125–231) of HuPrP variants at pH 5.5. (**a**) HuPrP(Q212P) (pdb id 2KUN), (**b**) HuPrP(V210I) (pdb id 2LEJ), and (**c**) HuPrP(E219K) (pdb id 2LFT)

(*see* **Note 17**). CYANA converts NOE intensities into upper distance restraints according to an inverse sixth power peak volume-to-distance relationship and removes meaningless restraints (*see* **Note 18**).

4. Use YASARA program suite (http://www.yasara.org) [28] for energy minimization in a water shell of the 20 conformers of each protein with the lowest residual target function values.
5. Validate the final ensemble of 3D structures of HuPrP variants shown in Fig. 4 using PROCHECK-NMR [29] and WhatIF [30] (*see* **Note 19**).

4 Notes

1. Different concentrations of the inducer IPTG (0.4, 0.6, 0.8, and 1 mM) and two temperature conditions (30 and 37 °C) were assayed aimed at increasing the PrP expression level in M9 MM.
2. At mildly acidic pH of 5.5, HuPrP variants are stable, there is no aggregation, and fast exchange of amide protons with solvent is reduced. Sodium azide (NaN_3) is added to inhibit bacterial growth.

3. Each cross-peak in the ^{1}H-^{15}N HSQC spectrum corresponds to one backbone or side-chain NH group in the protein. As proline residues do not contain backbone amide protons, they are not visible in ^{1}H-^{15}N HSQC spectra. Inspection of spectral dispersion and number of cross-peaks in the ^{1}H-^{15}N HSQC spectrum provide following information: whether the protein is in the folded state, whether multiple conformers exist, whether some peaks are broadened because of slow conformational exchange processes, or whether parts of the protein sequence are not visible at all.
4. For all experiments, the spectral width for observed proton was 16 ppm with the offset at 4.63 ppm. For ^{1}H-^{15}N HSQC experiment, the spectral width was 27 ppm with the offset at 118 ppm for nitrogen. Data were acquired with 2048 × 256 data points with four scans per fid. The total experimental time with recycle delay of 1.5 s was 1 h. ^{15}N-edited NOESY-HSQC experiment was recorded with mixing time of 150 ms, spectral width of 11.3 ppm, and the offset at 4.6 ppm for indirect detection of proton signals. For nitrogen detection, we used the same parameters as for ^{1}H-^{15}N HSQC experiment. Data were acquired with 2048 × 128 × 32 data points with four scans per fid. The total experimental time with recycle delay of 1.5 s was 28 h. For HNCO experiment, the spectral width was 7.5 ppm with the offset at 176 ppm for carbonyl atom. Data were acquired with 2048 × 48 × 24 data points with four scans per fid. The total experimental time with recycle delay of 1.5 s was 6 h. For HNCA and CA(CO)NH experiments, the spectral width was 25 ppm with the offset at 55 ppm for Cα atoms. For ^{15}N dimension, we used the same parameters as for ^{1}H-^{15}N HSQC spectrum. Data were acquired with 2048 × 64 × 24 data points with four scans per fid. The total experimental time with recycle delay of 1.5 s was 8 h. For CBCA(CO)NH and HNCACB experiments, the spectral width was 80 ppm with the offset at 55 ppm for carbon atoms. For ^{15}N dimension, we used the same parameters as for ^{1}H-^{15}N HSQC spectrum. Data were acquired with 2048 × 64 × 32 data points with eight scans per fid. The total experimental time with recycle delay of 1.5 s was 21 h. For HAHB(CO)NH experiment, the spectral width was 9 ppm with the offset at 4.6 ppm for indirect detection of proton signals. For nitrogen detection, we used the same parameters as for ^{1}H-^{15}N HSQC experiment. Data were acquired with 2048 × 128 × 24 data points with eight scans per fid. The total experimental time with recycle delay of 1.5 s was 22 h.
5. Buffer exchange removes the residual water signal (at ~4.8 ppm) which overlaps with $^{1}H\alpha$ resonances.

6. For all experiments, the spectral width for observed proton was 16 ppm with the offset at 4.63 ppm. For ^{1}H-^{13}C HSQC (aliphatic region) experiment, the spectral width was 80 ppm with the offset at 43 ppm. Data were acquired with 2048 × 256 data points with eight scans per fid. The total experimental time with recycle delay of 1.5 s was 2 h. For ^{1}H-^{13}C HSQC (aromatic region) experiment, the spectral width was 37 ppm with the offset at 123 ppm. Data were acquired with 2048 × 64 data points with eight scans per fid. The total experimental time with recycle delay of 1.5 s was 30 min. ^{13}C-edited NOESY-HSQC (aliphatic region) experiment was recorded with mixing time of 150 ms, spectral width of 11.3 ppm, and the offset at 4.6 ppm for indirect detection of proton signals. For carbon detection, we used the same parameters as for ^{1}H-^{13}C HSQC experiment. Data were acquired with 2048 × 128 × 32 data points with eight scans per fid. The total experimental time with recycle delay of 1.5 s was 60 h. ^{13}C-edited NOESY-HSQC (aromatic region) experiment was recorded with mixing time of 150 ms, spectral width of 11.3 ppm, and the offset at 4.6 ppm for indirect detection of proton signals. For carbon detection, we used the same parameters as for ^{1}H-^{13}C HSQC experiment. Data were acquired with 2048 × 128 × 24 data points with eight scans per fid. The total experimental time with recycle delay of 1.5 s was 45 h. For (H)CCH-TOCSY experiment, the spectral width for both carbon dimensions was 80 ppm with the offset at 43 ppm. Data were acquired with 2048 × 80 × 80 data points with four scans per fid. The total experimental time with recycle delay of 1.5 s was 36 h.
7. Comparison of CA(CO)NH and HNCA experiments allows unambiguous differentiation between the intra- and inter-residual $^{13}C\alpha$ chemical shifts.
8. An unambiguous backbone assignment of HuPrP variants only on the basis of CA(CO)NH and HNCA experiments is difficult due to the degeneracy of $^{13}C\alpha$ frequencies. From that reason, 3D CBCA(CO)NH and HNCACB experiments are also employed. CBCA(CO)NH and HNCACB provide correlations with $^{13}C\beta$ chemical shifts that are spread over a much wider frequency range when compared to the $^{13}C\alpha$ chemical shifts.
9. Combination of CBCA(CO)NH and HNCACB experiments allows one to distinguish between the intra- and inter-residual $^{13}C\beta$ chemical shifts.
10. MARS is a program for an automatic backbone assignment of $^{13}C/^{15}N$-labeled proteins. About 70% of the sequence of HuPrP variants was assigned employing MARS and the rest of it manually.

11. Nearly complete backbone assignments of HuPrP variants were achieved with the exception of backbone amides of Arg164, Tyr169, Asn171, and Phe175 in all three studied HuPrP variants, Gln217 in HuPrP(Q212P), Asp167 in HuPrP(V210I), and Arg220 in HuPrP(E219K). For those residues, the cross-peaks are not observed in ^{1}H-^{15}N HSQC spectra probably due to chemical exchange processes at the intermediate NMR time scales leading to line broadening.
12. NOESY experiments correlate pairs of protons that are in close proximity (within a distance of 5 Å or less) and provide the most important class of restraints used in protein NMR structure calculation.
13. The assignment of aliphatic side chains is completed when aliphatic region of ^{1}H-^{13}C HSQC spectra is fully assigned.
14. The assignment of aromatic side chains is completed when all the peaks in aromatic region of ^{1}H-^{13}C HSQC spectra are assigned.
15. The overall completeness of resonance assignments for all three HuPrP variants was above 95%.
16. The peak list must contain the positions and volumes of the NOESY cross-peaks.
17. Experimentally derived restraints are supplemented with covalent geometry restraints (e.g., bond lengths, bond angles, planar groups, and chirality).
18. The NOE-derived distance restraints are classified according to their intensities into three categories: strong (1.8–2.7 Å), medium (1.8–3.3 Å), and weak (1.8–5 Å). NMR structures of HuPrP(Q212P), HuPrP(V210I), and HuPrP(E219K) proteins were calculated on the basis of over 2200 NOE distance restraints for each protein.
19. NMR structures are usually represented by an ensemble of the 20 lowest-energy conformers. The major criteria for evaluating the quality of NMR structures are (a) the target function which measures discrepancy between the experimental data and the calculated structure giving insights about the accuracy of the structure, (b) the root mean square deviation (r.m.s.d.) relative to the mean coordinates for backbone and heavy atoms providing information about the precision of the determined structure, and (c) the Ramachandran plot which illustrates the distribution of backbone ϕ and ψ angles of a protein giving an overview of allowed and disallowed regions of torsion angle values. PROCHECK-NMR and WhatIF demonstrated that the final family of the 3D structures of HuPrP(Q212P), HuPrP(V210I), and HuPrP(E219K) proteins agree with the distance restraints and offer good geometry and side-chain

packing. The 3D structures of HuPrP(Q212P), HuPrP (V210I), and HuPrP(E219K) are well defined with the average r.m.s.d. values lying in the range from 0.66 to 1.01 Å for backbone atoms and from 1.06 to 1.45 Å for all heavy atoms (residues 125–228).

References

1. Calzolai L, Lysek DA, Guntert P et al (2000) NMR structures of three single-residue variants of the human prion protein. Proc Natl Acad Sci U S A 97(15):8340–8345
2. Christen B, Perez DR, Hornemann S et al (2008) NMR structure of the bank vole prion protein at 20 degrees C contains a structured loop of residues 165-171. J Mol Biol 383 (2):306–312
3. Donne DG, Viles JH, Groth D et al (1997) Structure of the recombinant full-length hamster prion protein PrP(29-231): the N terminus is highly flexible. Proc Natl Acad Sci U S A 94(25):13452–13457
4. Gossert AD, Bonjour S, Lysek DA et al (2005) Prion protein NMR structures of elk and of mouse/elk hybrids. Proc Natl Acad Sci U S A 102(3):646–650
5. James TL, Liu H, Ulyanov NB et al (1997) Solution structure of a 142-residue recombinant prion protein corresponding to the infectious fragment of the scrapie isoform. Proc Natl Acad Sci U S A 94(19):10086–10091
6. Lopez GF, Zahn R, Riek R et al (2000) NMR structure of the bovine prion protein. Proc Natl Acad Sci U S A 97(15):8334–8339
7. Perez DR, Damberger FF, Wuthrich K (2010) Horse prion protein NMR structure and comparisons with related variants of the mouse prion protein. J Mol Biol 400(2):121–128
8. Wen Y, Li J, Yao W et al (2010) Unique structural characteristics of the rabbit prion protein. J Biol Chem 285(41):31682–31693
9. Zahn R, Liu A, Luhrs T et al (2000) NMR solution structure of the human prion protein. Proc Natl Acad Sci U S A 97(1):145–150
10. Kovacs GG, Trabattoni G, Hainfellner JA et al (2002) Mutations of the prion protein gene phenotypic spectrum. J Neurol 249(11):1567–1582
11. Apetri AC, Surewicz K, Surewicz WK (2004) The effect of disease-associated mutations on the folding pathway of human prion protein. J Biol Chem 279(17):18008–18014
12. Liemann S, Glockshuber R (1999) Influence of amino acid substitutions related to inherited human prion diseases on the thermodynamic stability of the cellular prion protein. Biochemistry 38(11):3258–3267
13. Swietnicki W, Petersen RB, Gambetti P et al (1998) Familial mutations and the thermodynamic stability of the recombinant human prion protein. J Biol Chem 273 (47):31048–31052
14. Biljan I, Ilc G, Giachin G et al (2012) Structural Rearrangements at Physiological pH: Nuclear Magnetic Resonance Insights from the V210I Human Prion Protein Mutant. Biochemistry 51:7465–7474
15. Biljan I, Ilc G, Giachin G et al (2011) Toward the Molecular Basis of Inherited Prion Diseases: NMR Structure of the Human Prion Protein with V210I Mutation. J Mol Biol 412:660–673
16. Ilc G, Giachin G, Jaremko M et al (2010) NMR structure of the human prion protein with the pathological Q212P mutation reveals unique structural features. PLoS One 5(7): e11715
17. Biljan I, Giachin G, Ilc G et al (2012) Structural basis for the protective effect of the human prion protein carrying the dominant-negative E219K polymorphism. Biochem J 446:243–251
18. Cavanagh J, Fairbrother WF, Palmer AG et al (2007) Protein NMR spectroscopy: principles and practice, 2nd edn. Elsevier Academic Press, San Diego, CA
19. Kanelis V, Forman-Kay JD, Kay LE (2001) Multidimensional NMR methods for protein structure determination. IUBMB Life 52:291–302
20. Sattler M, Schleucher J, Griesinger C (1999) Heteronuclear multidimensional NMR experiments for the structure determination of proteins in solution employing pulsed field gradients. Prog Nucl Mag Res Sp 34:93–158
21. Delaglio F, Grzesiek S, Vuister GW et al (1995) NMRPipe: a multidimensional spectral processing system based on UNIX pipes. J Biomol NMR 6(3):277–293
22. Goddard TD, Kneller DG (2002) SPARKY 3. University of California, San Francisco, CA

23. Keller R (2004) The computer aided resonance assignment tutorial. CANTINA Verlag, Goldau
24. Jung YS, Zweckstetter M (2004) Mars: robust automatic backbone assignment of proteins. J Biomol NMR 30:11–23
25. Guntert P, Mumenthaler C, Wuthrich K (1997) Torsion angle dynamics for NMR structure calculation with the new program DYANA. J Mol Biol 273(1):283–298
26. Cornilescu G, Delaglio F, Bax A (1999) Protein backbone angle restraints from searching a database for chemical shift and sequence homology. J Biomol NMR 13:298–302
27. Herrmann T, Guntert P, Wuthrich K (2002) Protein NMR structure determination with automated NOE assignment using the new software CANDID and the torsion angle dynamics algorithm DYANA. J Mol Biol 319 (1):209–227
28. Krieger E, Koraimann G, Vriend G (2002) Increasing the precision of comparative models with YASARA NOVA - a self-parameterizing force field. Proteins Struct Funct Genet 47:477–486
29. Laskowski RA, Rullmann JAC, MacArthur MW et al (1996) AQUA and PROCHECK-NMR: Programs for checking the quality of protein structures solved by NMR. J Biomol NMR 8(4):477–486
30. Vriend G (1990) A molecular modelling and drug design program. J Mol Graph 9:52–56

Chapter 5

Immunodetection of PrP^{Sc} Using Western Immunoblotting Techniques

Gerald S. Baron and Gregory J. Raymond

Abstract

Western immunoblotting is a workhorse technique used in the prion field to analyze disease-associated forms of the prion protein, termed PrP^{Sc}. The biochemical stability of PrP^{Sc} aggregates combined with the increased resistance of prion infectivity to inactivation by various treatments that inactivate most other pathogens complicates the use of Western immunoblotting as a means to characterize PrP^{Sc} samples. In this chapter, we describe a method for Western immunoblot analysis of PrP^{Sc} with an emphasis on precautions to address the biochemical and biosafety considerations associated with this procedure.

Key words Amyloid, Prion, Immunoblot, Sodium hydroxide, SDS-PAGE, Enhanced chemifluorescence, AttoPhos, Typhoon, Denaturation

1 Introduction

A hallmark of prion diseases is the conversion of the normal, protease-sensitive prion protein (PrP^{C}) to an aggregated, usually protease-resistant isoform termed PrP^{Sc} (reviewed in [1]). The conversion process involves the polymerization and refolding of the α-helix-rich PrP^{C} to the highly β-sheet-rich PrP^{Sc}. Evidence accumulated over the past several years strongly supports the idea that this conversion process represents replication of the infectious agent, although the molecular composition of that agent remains uncertain. Nevertheless, PrP^{Sc} still remains one of the most reliable molecular markers of prion disease and serves as the basis for most diagnostic tests.

Western immunoblotting is one of the most commonly used techniques to detect PrP^{Sc}. It is fast and simple and provides valuable information about the biochemical properties of PrP^{Sc} (e.g., glycoform profile, apparent molecular mass, profile of protease-resistant fragments) that can address many scientific questions such as aiding discrimination between prion strains. Since PrP^{Sc} is indistinguishable from PrP^{C} by Western immunoblot alone, sample

Victoria A. Lawson (ed.), *Prions: Methods and Protocols*, Methods in Molecular Biology, vol. 1658,
DOI 10.1007/978-1-4939-7244-9_5, © Springer Science+Business Media LLC 2017

processing is required to distinguish between these conformers and enrich the sample for PrP^{Sc} prior to immunoblotting. Protease-sensitive and protease-resistant isoforms of PrP^{Sc} have been reported, termed $sPrP^{Sc}$ and $rPrP^{Sc}$, respectively [2]. Investigators have thus developed a variety of strategies to specifically enrich samples for PrP^{Sc} over PrP^{C} with or without protease treatment steps. For the purpose of this chapter, we will assume that the reader is starting with a sample already enriched for PrP^{Sc} using their method of choice. We refer the reader to the scientific literature for details of some PrP^{Sc} enrichment methods as examples [2–4]. We also assume the reader is familiar with general biosafety considerations and procedures for manipulating prions, such as changing to fresh gloves after handling prion infectivity. The general procedure for Western immunoblotting of PrP^{Sc} is quite similar to standard immunoblotting procedures for other proteins [5, 6]. The main differences are associated with steps necessary to cope with the unusual biochemical properties of PrP^{Sc} aggregates, which are common to many amyloid protein aggregates, and biosafety considerations.

Amyloid aggregates of PrP^{Sc} and other amyloidogenic proteins are highly stable and, in some cases, resist disaggregation even after boiling in standard Laemmli SDS-PAGE sample buffer. Incomplete denaturation can affect immunoblotting results via the appearance of SDS-resistant oligomeric species [7] and, for highly amyloidogenic forms of PrP^{Sc}, a reduced immunoblot signal that underestimates the quantity of PrP ([8, 9] and see below). This is particularly important in the context of GPI-anchorless PrP^{Sc} derived from infection of transgenic mice that express only GPI-anchorless PrP^{C} where PrP^{Sc} deposits form large dense amyloid plaques [10, 11]. GPI-anchorless PrP^{Sc} shows enhanced resistance to denaturation by chaotropes, SDS, and heating compared to wild-type GPI-anchored PrP^{Sc} of the same prion strain [9]. The procedures below describe measures to improve denaturation of PrP^{Sc} and limit the opportunity for problems associated with incomplete denaturation to occur. The measures include the use of a modified sample buffer with increased concentrations of denaturants, thorough solubilization of the sample with extensive vortexing and sonication, extended boiling time, and, in certain applications, a sodium hydroxide (NaOH) denaturation treatment after proteins have been electroblotted to polyvinylidene difluoride (PVDF) membrane.

Western immunoblotting of prion samples requires special biosafety considerations due to the resistance of prion infectivity to thermal and chemical denaturation treatments that inactivate most other infectious agents. For example, there is one albeit unconfirmed report that prion infectivity can persist after boiling in Laemmli sample buffer and polyacrylamide gel electrophoresis (PAGE) [12]. Although the denaturing methods described below are far more aggressive than those used by Brown and co-workers and thus would be expected to achieve greater prion inactivation, we still err on the side of caution and decontaminate all equipment and waste materials from immunoblotting as if low levels of prion infectivity are present. This philosophy stems from the fact that, as noted above, PrP^{Sc}-enriched samples may be generated through various methods. Samples derived via different methods may vary highly in infectivity titer and overall biochemical composition, especially with respect to lipids and non-denaturing detergents. Lipids and non-denaturing detergents can act to neutralize sodium dodecyl sulfate (SDS) molecules and lower the effective SDS concentration, which in turn would diminish the effective denaturing power of the sample buffer. Thus, one must bear these factors in mind when preparing SDS-PAGE samples containing prion infectivity even when using our denaturant-enhanced sample buffer.

2 Materials

All solutions should be prepared using highly purified 18 MΩ distilled, deionized water and filtered through a 0.2 μm filter prior to use. We use a Milli-Q A10 Synthesis water purification system with online UV treatment and filters to minimize pyrogen and total organic carbon content. Solutions are filtered with bottle top and/or syringe filters composed of polyvinylidene fluoride (PVDF) or polyethersulfone (PES). Take care to ensure all tubes, stir bars, spatulas, and beakers are clean and handled with gloves to limit contamination by proteins in the laboratory environment (e.g., human keratin).

2.1 SDS-PAGE Solutions and Equipment

1. 2 M dithiothreitol (DTT): In a clean 15 ml centrifuge tube, dissolve 0.617 g (*see* **Note 1**) in 10 mM sodium acetate (pH 5.0) (*see* **Note 2**) to give a final volume of 2 ml. After filtering the solution, dispense into 50 μl aliquots for storage at −20 °C or below. Thawed aliquots can be refrozen but discard after two freeze-thaw cycles.
2. 1% bromophenol blue: Dissolve 0.4 g bromophenol blue in a final volume of 40 ml of purified water. Filter and store at RT.
3. 2× Urea SDS-PAGE sample buffer stock solution: 125 mM Tris–HCl (pH 6.8), 10% glycerol, 6 mM EDTA, 10% SDS, 0.04% bromophenol blue (optional), and 8 M urea (*see* **Note 3**). Weigh out 24 g of urea pellets and 5 g of SDS into a 150 ml beaker containing a clean stir bar. Add 6.25 ml of 1 M Tris-HCl (pH 6.8), 0.6 ml of 0.5 M EDTA (pH 8.0), and 5 ml of glycerol. Carefully add purified water to a rough final volume of ~40 ml. Stir the solution on a heated stir plate with gentle heating (keep the solution <37 °C) to dissolve the urea (*see* **Note 4**). When more than half the urea has dissolved, add more purified water to a final

volume near but not exceeding 45 ml. Accurately measure the volume once all salts are in solution. If adding bromophenol blue, add 2 ml of 1% bromophenol blue solution at this point (*see* **Note 5**). Finally, adjust the volume of the solution to 47.5 ml by adding an appropriate volume of purified water, stir to ensure thorough mixing, and then filter while warm into a clean 50 ml centrifuge tube. Aliquot the solution into 1.5 ml tubes with 450 or 900 μl per tube and store aliquots at −80 °C (*see* **Note 6**). These are single-use aliquots.

4. 2× Urea SDS-PAGE sample buffer working solution: 125 mM Tris–HCl (pH 6.8), 10% glycerol, 6 mM EDTA, 10% SDS, 0.04% bromophenol blue (optional), 100 mM DTT, and 8 M urea. Prepare this solution fresh (<60 min) before use. Thaw aliquots of 2× Urea SDS-PAGE sample buffer stock solution and 2 M DTT stock solutions at 37 °C. Vortex well, and then add a volume of 2 M DTT to the 2× Urea SDS-PAGE sample buffer stock solution to reach a final concentration of 100 mM DTT. For example, for 500 μl of 2× Urea SDS-PAGE sample buffer working solution, mix 450 μl of 2× Urea SDS-PAGE sample buffer stock solution with 50 μl of 2 M DTT. For 1× Urea SDS-PAGE sample buffer working solution, dilute 2× Urea SDS-PAGE sample buffer working solution with 1 volume of purified water (*see* **Note 7**).
5. 1× NuPAGE MES SDS running buffer: Life Technologies, 20× stock solution. Prepare 800 ml per gel box by adding 40 ml of 20× stock solution to 760 ml of purified water and mixing thoroughly. Dispense 200 ml of 1× NuPAGE MES SDS running buffer into a separate container, add 0.5 ml of NuPAGE antioxidant (Life Technologies), mix thoroughly, and add this solution to the upper electrophoresis chamber only.
6. NuPAGE precast gels and electrophoresis chamber: We typically use 10% NuPAGE Bis-Tris precast gels, 1.0 mm thickness, 15 wells (Life Technologies). Thicker 1.5 mm gels can be used but require longer electroblotting times. XCell *SureLock* Mini-cell (Life Technologies).
7. Bio-Rad Precision Plus All Blue molecular weight standards: Upon receipt mix thoroughly, dispense into 50 μl aliquots, and store at −20 °C. A working aliquot can be stored at 4 °C for several months (*see* **Note 8**). Equilibrate working aliquot to RT and mix well before use. Do not boil.
8. Misonix 3000 sonicator with 3″ diameter cup horn (Misonix, Inc.).
9. 2.0% Environ LpH (Steris): Prion disinfectant solution. Treat prion-contaminated surfaces and liquids for at least 30 min [13] (*see* **Note 9**).

2.2 Immunoblotting Solutions and Equipment

1. Towbin transfer buffer [5]: 25 mM Tris, 192 mM glycine, 20% methanol, pH 8.3. Mix 12.1 g Tris base, 57.65 g glycine, and 800 ml methanol in purified water to a final volume of 4 L. Allow to stir for at least 30–60 min and ensure all salts are dissolved. Do not adjust final pH, which should be approximately pH 8.3.
2. Tris-buffered saline (TBS): 20 mM Tris–HCl (pH 7.5), 150 mM NaCl. Prepare 1× working solutions from a 10× stock solution as needed for each experiment.
3. 0.2 M NaOH: Prepare fresh from NaOH pellets for each experiment, 30 ml per gel. Prepare this solution during the electroblotting step.
4. TBS-Tween (TBS-T): 1× TBS supplemented with 0.1% Tween-20. Prepare fresh for each experiment, 500 ml per blot.
5. Blocking/antibody diluent solution: 5% blotting grade nonfat dry milk blocker (Bio-Rad) in TBS-T. Make 30 ml per blot. Stir this solution for at least 30 min to ensure the blocking agent is completely dissolved.
6. Alkaline phosphatase (AP)-conjugated secondary antibodies: For blots employing enhanced chemifluorescence (ECF) detection, we use alkaline phosphatase conjugates from Jackson ImmunoResearch that have been cross-adsorbed against multiple species other than the host species used to generate the primary antibody in a given experiment. AP conjugates are reconstituted in purified water per the manufacturer's instructions, mixed with one volume of glycerol, and stored for up to 1 year at −20 °C.
7. Working aliquots of primary and secondary antibodies: Aggregates in antibody solutions are a primary source of speckled background on immunoblots. To prepare a working aliquot, pipette 100 μl of antibody into a 0.5 ml microfuge tube, spin at 4 °C in a microfuge at 14,000–15,000 rpm (20,000–22,000 × *g*) for 10 min, then carefully transfer ~90 μl of the supernatant to a fresh tube, and store at 4 °C. Working aliquots are usually stable for at least several months at 4 °C and do not need to be repeatedly transferred to new tubes. However, for every immunoblotting experiment, the working aliquot should be centrifuged at 4 °C in a microfuge at 14,000–15,000 rpm (20,000–22,000 × *g*) for 10 min immediately prior to removing the volume of antibody required for that experiment. If evidence of extensive speckled background appears on a blot, discard the tube and prepare a new working aliquot.
8. Diluted antibody solutions: Antibodies are diluted into blocking/antibody diluent solution and vortexed thoroughly to mix. Antibody concentrations must be optimized for each antibody and detection system. For blots using ECF detection,

typical antibody concentrations are ~100 ng/ml for primary antibody and ~60 ng/ml for AP-conjugated secondary antibody. When using primary antisera of unknown antibody concentration, try using a 1:10,000 dilution as a starting point.

9. AttoPhos ECF substrate: Add the entire contents of one bottle of AttoPhos buffer solution (Promega) to one bottle of AttoPhos substrate (Promega). Mix thoroughly for several minutes and ensure the AttoPhos substrate is completely dissolved. Store at 4 °C. This solution is stable for many months.
10. Biometra Fastblot B34 electroblotting device (*see* **Note 10**).
11. PVDF blotting membranes, precut 7 × 8.4 cm (we use Immobilon-P membranes, Millipore).
12. Blotting filter paper, precut 7 × 8.4 cm (we use Immobilon, Millipore).
13. Clean plastic containers (four), prerinsed with purified water to remove autofluorescent particles.
14. Polypropylene centrifuge tubes (50 ml).
15. Rotating mixer with 50 ml tube rolling rod (we use ATR Biotech, cat # RKVSD and RR-50). Note that this spindle holds the long edge of the tubes parallel to the spindle axis (i.e., tubes are held horizontally).
16. Orbital shaker.
17. Soft rubber brayer.
18. Square petri dishes.
19. Blunt-end filter forceps.
20. Lint-free wipes.

3 Methods

The detection system used below is based on fluorescence detection and is provided as an example that can be adapted to other detection methods (e.g., chemiluminescence). With fluorescence detection, one must pay special attention to sources of contamination of samples and equipment with autofluorescent material (e.g., oils from skin, lint, dust, dirty containers).

3.1 Sample Preparation

Biosafety considerations: Prion-affected samples must be prepared in a biosafety cabinet. Discard all pipette tips and microfuge tubes in a container of 2.0% LpH disinfectant.

1. Prepare a water bath or other heating device adjusted to a temperature just below boiling. If using a water bath, avoid a rolling boil to limit the possibility of water leaking into the sample tubes.

2. Prepare samples in O-ring screw cap tubes to minimize the chance of prion infectivity escaping the tubes and contaminating laboratory equipment. Ensure each cap has a single O-ring and that the cap seats properly on the tube. For prion samples with a PrP^{Sc} pellet [2, 3], resuspend in 2× Urea SDS-PAGE sample buffer with vigorous vortexing until the pellet is dissolved. For prion samples with PrP^{Sc} in suspension (e.g., brain homogenate or bound to beads [4]), add at least one volume of 2× Urea SDS-PAGE sample buffer and vortex vigorously.
3. If samples have been prepared previously and stored frozen, skip **step 4**, remove samples from the freezer, and place directly into the water bath (*see* **Note 11**).
4. Sonicate the sample for 20–30 s on a power level setting appropriate for the size of tube and volume of sample per the manufacturer's recommendations (typically power setting 1, ~30 W output power for samples <0.5 ml). Check the digital display frequently and adjust water flow as necessary to maintain a water level that maximizes sonication output power (*see* **Note 12**).
5. Heat samples at 95–100 °C for 8 min.
6. Vortex sample for at least 15 s (*see* **Note 13**).
7. Briefly spin samples in microfuge for 10–15 s. Inspect the samples to ensure all material has been solubilized.

3.2 SDS-PAGE

1. Fill the electrophoresis chamber with purified water and soak until needed.
2. Remove the gel comb and rinse the wells of the NuPAGE gel three times with 1× MES SDS-PAGE running buffer (*see* **Note 14**).
3. Discard the purified water from the electrophoresis chamber, and assemble the unit with the desired number of gels.
4. Fill the upper electrophoresis chamber with NuPAGE antioxidant-supplemented 1× NuPAGE MES SDS running buffer.
5. Load the gel. Load 10–15 μl of urea SDS-PAGE sample buffer alone in empty lanes. For molecular weight standards, we use 5 μl of Bio-Rad Precision Plus pre-stained All Blue marker (*see* **Note 15**).
6. Fill the lower electrophoresis chamber with 1× NuPAGE MES SDS running buffer.
7. Electrophorese the gel at 200 V for 40 min (*see* **Note 16**). Approximately 5 min before the end of electrophoresis, begin equilibration of PVDF membranes as described below (Subheading 3.3, **step 1**).

8. Carefully separate the gel cassettes with a spatula so that the gel adheres to one plate. Trim off the gel lanes and foot of the gel using the spatula. Invert the plate with the attached gel over a clean plastic box containing ~75 ml of Towbin and use a spatula moistened with 1× MES SDS-PAGE running buffer to gently remove the gel from the plate starting from one corner of the gel (*see* **Note 17**).
9. Shake the gels in Towbin on an orbital shaker for 5 min to equilibrate in transfer buffer. Ensure that the gels float freely and shake at a speed sufficient to achieve good agitation without ripping the gels.
10. Add concentrated Environ LpH to the running buffer in the electrophoresis unit to a final concentration of 2.0%, and mix gently with the spatula to disperse evenly. Spray down all inner surfaces of the unit with a solution of 2.0% LpH. Incubate for 30 min at RT to disinfect the equipment and spatula. Discard the liquid waste in a sink and rinse the unit extensively with water.

3.3 Electroblotting

1. Using clean blunt-end forceps, pre-wet the PVDF membrane by stationary incubation for 30 s in methanol followed by 1 min in purified water. Manually agitate the plastic container until the membrane submerges in the water (*see* **Note 18**).
2. Pick up the membrane by a corner using forceps, briefly drain most of the water, transfer the membrane to a plastic container containing Towbin, and store in Towbin until needed. Do not allow the membrane to dry out at any time.
3. While gels are equilibrating in the transfer buffer, soak blotting filter paper in a clean plastic container containing Towbin. Two stacks of three pieces are required per gel. Using forceps, gently compress each stack to aid the removal of bubbles. Incubate for at least 1 min, ensuring that the blotting paper is saturated.
4. Assemble the blot sandwich (Fig. 1) on the lower electrode of the electroblotting device. Drain excess transfer buffer off PVDF membrane and filter papers for 5 s before placing on sandwich to prevent buffer pooling on the sandwich (*see* **Note 19**).
5. Roll the brayer gently over the assembled sandwich to squeeze out any bubbles between the layers. Use gentle pressure to ensure there is no shifting within the layers of the sandwich.

Fig. 1 Electroblotting sandwich. The gel and PVDF membrane are sandwiched between two groups of three blotting filter papers soaked with Towbin

6. Carefully place the lid on the electroblotting device and press down gently to ensure good contact with the gels. Verify that the lid is level. Place a nonmetallic 0.5–1 kg weight on top of the electroblotting device such as a bottle containing 1 L of water.
7. Electroblot for 22 min at a current of 150 mA per 7 × 8 cm mini-gel.

3.4 Immuno-detection

1. After electroblotting is complete, place the membrane in a plastic container containing Towbin (*see* **Note 20**).
2. Without introducing any sharp bends, curl the membrane using a pair of forceps or a clean gloved hand (Fig. 2). If using a gloved hand, only contact the bottom side of the membrane lacking protein. Transfer the PVDF membrane into a 50 ml polypropylene centrifuge tube with the protein-bound side facing inward. If the optional NaOH treatment procedure will be used, follow **steps 3–5** (*see* **Note 21**). For normal blotting applications, proceed to **step 6**.

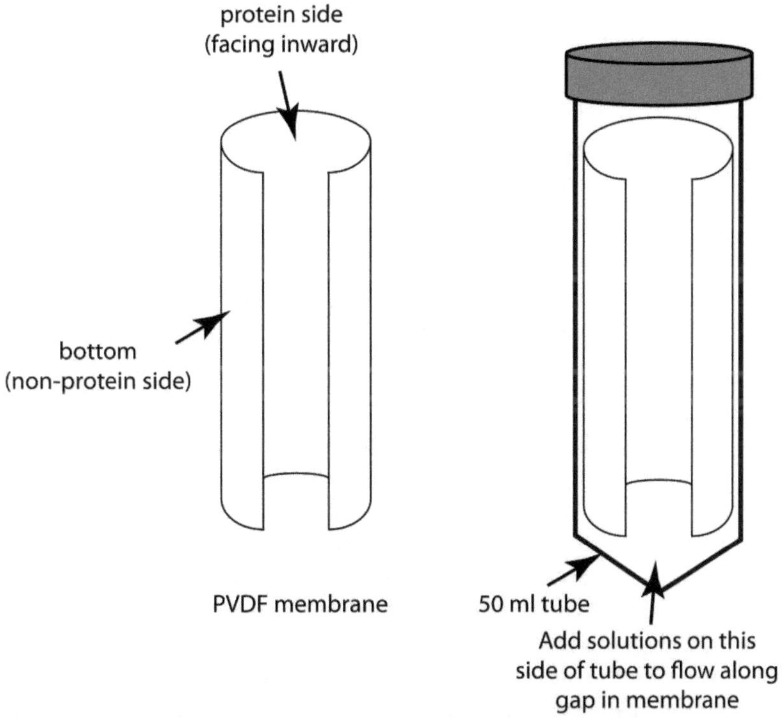

Fig. 2 Incubation configuration for immunodetection of PrP^{Sc} on PVDF membrane. After proteins are electroblotted to PVDF membrane, the membrane is gently curled (*left diagram*) and inserted into a 50 ml polypropylene centrifuge tube (*right diagram*) for incubation in a rotor mixing device during blocking, antibody incubations, and washes. Incubations in this setup are convenient, minimize the volumes of antibody required, and allow up to eight blots to be processed in parallel per rotor mixing device

3. Slowly add 30 ml of 0.2 M NaOH solution. From here on add all solutions slowly along the gap in the membrane (Fig. 2). Incubate on a rotating mixer at ~16–18 rpm for 30 min at RT. Use this rotation speed for all incubations.
4. Slowly pour off the NaOH solution into a 2 L polypropylene waste beaker containing 50 ml of Environ LpH (*see* **Note 22**). Rotate the tube uniformly while discarding the solution (*see* **Note 23**). Use this technique when pouring off all solutions.
5. Slowly add 40 ml of TBS, cap the tube, invert gently 3–4 times, and then discard the solution into the waste beaker. Repeat this step three more times.
6. Add 20 ml of blocking solution and incubate on Rotamix mixer for 60 min at RT.
7. Discard blocking solution into waste beaker. Add 5 ml of diluted primary antibody solution. Incubate on rotating mixer for 60 min to overnight at RT.
8. Discard the primary antibody solution into the waste beaker. Perform one quick wash by slowly adding 40 ml of TBS-T, capping the tube, gently inverting 3–4 times, and pouring the solution in the waste beaker.
9. Wash the blot by adding 40 ml of fresh TBS-T and incubating on Rotamix for 5 min. Discard wash solution. Repeat this wash step two more times.
10. Add 5 ml of diluted secondary antibody solution, and incubate on Rotamix for 60 min at RT.
11. Perform one quick wash as per **step 8** above. Wash as described in **step 9** except perform four washes of 5 min each.
12. After starting the final TBS-T wash, remove the AttoPhos substrate solution from the fridge, and add 1.3 ml to a square petri dish on a flat bench. This allows the substrate to warm to RT before development.
13. Remove the membrane from the tube, drain the liquid off the membrane for ~1 min with one edge touching a paper towel to absorb the liquid, and then carefully place the membrane protein side down into the AttoPhos solution. Lay the membrane down in a manner that avoids the introduction of bubbles, which will interfere with the development process. Observe the blot closely for the appearance of yellow bands. Incubate stationary for 20 s to 12 min as needed.
14. Terminate the AttoPhos incubation by carefully lifting the membrane with forceps and thoroughly draining the liquid off the membrane for 1–2 min by holding the membrane perpendicular to a paper towel and blotting the lower edge on the towel. Place the membrane ***protein side up*** in a clean

square petri dish. Place the dish without a lid in a biosafety cabinet to accelerate initial drying of the membrane. Once the membrane regains a white appearance, usually within 10 min, allow the membrane to dry for 2 h to overnight without a lid while protected from light (e.g., in a cabinet or drawer). The membrane must be completely dry before scanning.

3.5 Fluorescence Scanning of Immunoblot

1. Mark the position of molecular weight markers by placing a spot next to each marker directly on the membrane using a Hi-Liter pen of any bright color. This allows the position of the markers to appear in the scanned image.
2. Wipe all sides of a plastic document sheet protector with crew wipes to remove autofluorescent material (*see* **Note 24**).
3. Place the membrane inside the clean plastic document sheet protector and secure the corners of the membrane to the sheet protector with tape. This prevents direct contact between the prion-affected membrane and the Typhoon instrument. Lay the sheet protector with the membrane facing protein side down onto the Typhoon platen. Use clean crew wipes to cover any glass surface that is not covered by the plastic sheet protector. The crew wipes should overlap the plastic but do not allow the crew wipes to overlap the membrane (*see* **Note 25**).
4. Scan the blot with excitation/emission settings appropriate for AttoPhos. The most commonly used settings that provide the best signal/noise ratio are 457 nm excitation and 555BP20 emission (*see* **Note 26**). An example of a typical result is shown in Fig. 3a. A comparison of blots with and without NaOH treatment is shown in Fig. 3b.

4 Notes

1. Use DTT of the highest purity available. Store powder in a tightly sealed container at 4 °C to prevent oxidation. Allow bottle of DTT powder to warm to room temperature (RT) before opening.
2. DTT solutions are more stable when prepared in acidic sodium acetate buffer.
3. This buffer contains a higher SDS concentration than standard Laemmli sample buffer and is supplemented with urea. This improves the denaturation of highly stable amyloid aggregates, although there is the potential for protein modification via carbamylation. If necessary, the urea can be omitted at the cost of reduced denaturation of PrP^{Sc} aggregates.
4. Avoid excess heating of urea to limit the formation of cyanates.

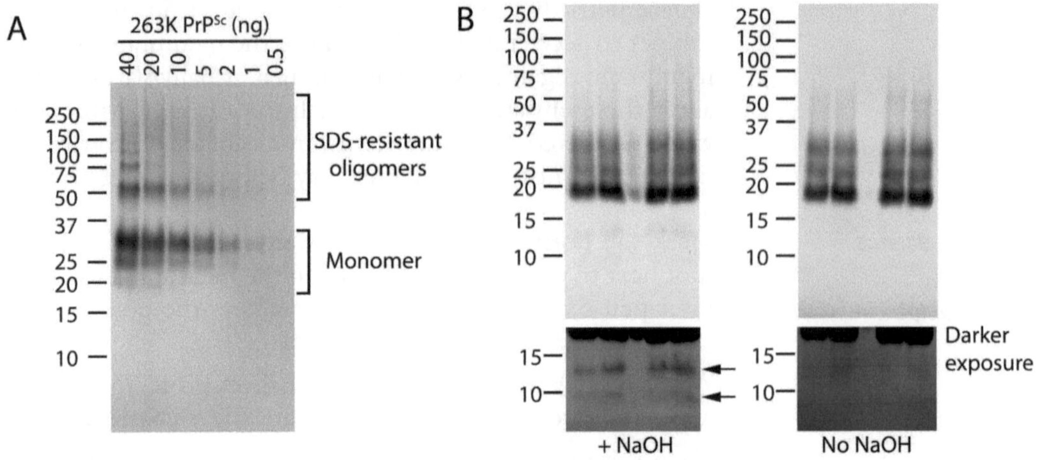

Fig. 3 Immunoblot detection of PrP^{Sc}. (**a**) Dilution series of purified PrP^{Sc} from hamsters infected with the 263K strain. *Brackets* indicate PrP^{Sc} molecules fully solubilized to monomers (monomer) and higher molecular weight oligomers that resisted complete solubilization (SDS-resistant oligomers) even after boiling in urea-containing sample buffer. The three monomer bands represent the three glycoforms of PrP corresponding to diglycosylated, monoglycosylated, and unglycosylated PrP. The ratio of these bands varies between prion strains. PrP^{Sc} was detected using anti-PrP monoclonal antibody 31C6 [15]. Typical detection limit is ~0.2 ng [16]. (**b**) Effect of NaOH treatment post-electroblotting. Samples shown are proteinase K-digested brain homogenates from transgenic mice co-expressing wild-type and GPI-anchorless PrP^{C} infected with synthetic prions that preferentially induce the accumulation of GPI-anchorless PrP^{Sc} amyloid [17]. Gels and immunoblots were performed in parallel from the same samples. Where indicated, PVDF membrane was treated with NaOH after transfer to PVDF and before blocking. Although the differences may appear subtle, we reproducibly observe improved detection of low molecular weight bands (*arrows*) in amyloid-rich PrP^{Sc} samples when membranes have been treated with NaOH after transfer. Lower panels are darker exposures of the corresponding upper gels. PrP^{Sc} was detected using anti-PrP antibody 6D11. The left hand gel was reproduced from [17] with permission from the American Society for Microbiology

5. Adding bromophenol blue at this point allows one to more easily monitor the dissolution of the urea and SDS. Bromophenol blue and other running dyes should not be used for samples run on gels that will be scanned by fluorescence after electrophoresis to visualize fluorophore-tagged proteins or when using direct fluorescence immunodetection. These dyes cause fluorescent background across many wavelengths that interferes with visualization of fluorophore-tagged proteins and reduces sensitivity.
6. Storage at −80 °C is necessary to prevent urea breakdown during long-term storage.
7. Urea precipitation will occur if the 2× urea sample buffer working solution cools to RT, so it must be prepared just prior to use. The solution can be kept at 37 °C for a short period (< 30 min) or briefly reheated at 37 °C if needed. The 1× urea sample buffer is soluble at RT.

8. Note that the dye is non-covalently bound to the protein standards and will dissociate over time during storage at 4 °C.
9. There are several methods used to chemically inactivate prions on surfaces and in liquids, such as exposure to 1–2 M NaOH or 50% household bleach for 1 h. Preferred methods are in part dictated by regional biosafety regulations. We use Environ LpH at a final concentration of 2.0%, which has been shown to be highly effective [13]. Unfortunately, Environ LpH is currently unavailable. The procedures presented in this chapter are intended to serve as an example. The reader is advised to identify the best option available at their facility and if necessary adapt the procedures for use with their chosen disinfectant.
10. Other electroblotting methods can work either in semidry or wet formats. This device uses transfer with high current and has the advantages of achieving rapid efficient transfer and being easy to decontaminate.
11. Placing samples directly into the water bath is especially important for proteinase K (PK)-treated samples. Renaturation of PK can occur after freeze-thaw and cause digestion of denatured PrP^{Sc} fragments. Note that PK-digested samples must be loaded within 60 min after boiling or stored at −20 °C or less. PK-digested samples should be boiled after digestion *before* freezing and must be reboiled after freezing immediately before loading onto the gel.
12. All cup horns have at least one "sweet spot," a location where the ultrasonic energy is higher than most others. Optimal sonication requires that tubes are held in this sweet spot and close to but not touching the probe, which usually requires that tubes be held manually. A common practice of simply placing several tubes in a floater and turning on the sonicator for the desired length of time will not achieve efficient or reproducible sonication of samples.
13. Thorough vortexing is required to ensure highly stable amyloid protein aggregates have been completely solubilized. If samples are known to primarily contain amyloid aggregates, as observed in prion-infected mice expressing GPI-anchorless PrP^{C} or preparations of recombinant PrP amyloid fibrils, then it can be beneficial to sonicate the samples again.
14. It is convenient to use a chemical wash bottle with a gel-loading pipette tip attached for this step. Use a felt pen to mark lane numbers directly on the plastic cassette to aid visualization during loading, especially for samples prepared without loading dye. The gel should be loaded with the electrophoresis unit in a biosafety cabinet.

15. If possible leave a blank lane between the markers and experimental samples. The maximum volume/lane for mini-gels is ~15 μl. Perform gel loading with the electrophoresis unit in a biosafety cabinet covered with absorbent bench pads. Discard all pipette tips and tubes into a disposable waste bottle containing 2.0% LpH. Invert the bottle to ensure LpH contacts all surfaces, and incubate for several hours. Although treatment with LpH alone is very effective at inactivating high levels of infectivity, wherever possible we use additional redundant disinfection treatments beyond chemical inactivation. This includes autoclaving at 270 °F (132 °C) for 90 min and (for solid waste) incineration at ≥1000 °C.
16. The electrophoresis unit can be moved from the biosafety cabinet to a nearby bench top for electrophoresis. The bench should be covered with absorbent bench pads as a precaution against spills.
17. Once a small portion of the gel is separated from the plate, the force of gravity usually causes the remainder of the gel to separate and fall into the box. Two gels can be placed in one box. Minimize direct contact between the gel and gloved hands as background can occur at any point of contact between the gloves and gel. Use a clean spatula to flip or manipulate the gel. Only handle the gel with your hands while assembling the sandwich. When doing so avoid handling the gel in regions where bands are expected. Treat the gel fragments as low-level prion-positive solid waste for disposal.
18. Handle the membrane only on the edges. It is easier to handle the membrane using a pair of forceps. Do not force the membrane to submerge in water using forceps. Any bend or perturbation of the membrane during handling can cause background. This can be a problem when cutting the membranes to size, so it is beneficial to purchase membrane that is precut to the desired size.
19. Handle the gel, PVDF membrane, and filter paper by the edges only. Avoid introducing bubbles when assembling the sandwich. Work quickly to prevent the PVDF membrane from drying out in the rapid airflow of the biosafety cabinet.
20. When removing the lid of the electroblotting device, the sandwich should stick to one of the electrodes. Orient the stack upside-down so that the membrane can be easily removed once the filter papers have been peeled off. Decontaminate the electroblotting device by wiping with a 2.0% LpH solution and incubating for at least 30 min. Once decontaminated, rinse the electroblotting device with distilled water and wipe dry with lint-free tissue. Disinfect forceps periodically throughout the procedure by soaking in 2.0% LpH.

21. NaOH treatment after electroblotting to PVDF can improve immunodetection of amyloid proteins or fragments thereof. Fragments of a yeast prion protein resisted immunoblot detection unless rigorous denaturation treatments were applied before loading on the gel and in conjunction with NaOH treatment after transfer to blotting membranes [14]. Likewise, we have observed instances where NaOH denaturation after transfer can improve the detection of PrP fragments, including those associated with GPI-anchorless and recombinant PrP^C (Fig. 3b). However, NaOH treatment can slightly reduce the signal for glycosylated, GPI-anchored PrP.
22. One must be careful in pouring solutions on/off the sample as these shear forces can contribute to background.
23. Liquid waste from this point on is expected to have at best low levels of infectivity. Therefore, liquid waste from this step on can be disposed down a sink with copious amounts of water after treatment with 2.0% LpH for at least 30 min.
24. Handle the protector with gloves at all times, and always place it on a clean surface such as the Crew wipes. Avoid placing it on any bench because autofluorescent debris can become adsorbed. Autofluorescent debris and oils from skin can be transferred to scanning instruments and cause background in scans. On the Typhoon instrument, autofluorescent material can be removed by multiple rounds of cleaning with Crew wipes and fresh 10% hydrogen peroxide or 70% ethanol. Use extreme caution when cleaning with 70% ethanol as highly autofluorescent material can be extracted from the gasket surrounding the glass platen.
25. We have observed that autofluorescent material can be transferred from the lid of the Typhoon to the glass platen. Crew wipes prevent direct contact between the lid and platen when the sample is pressed as required for scanning a membrane. The Crew wipes should overlap the plastic document protector but not the membrane as the pattern of the Crew wipes is visible when pressed during a scan.
26. AttoPhos can be excited and detected at a variety of wavelengths. Excitation at 457 nm, as on the Typhoon scanner, is ideal as there is minimal excitation of autofluorescent material on membranes and the platen. Use of a 555BP20 emission filter when scanning the blot further limits contributions from autofluorescent signals. In most cases, other filters are acceptable and can increase signal (e.g., 520LP, 520BP40, 580BP30, 560LP).

Acknowledgment

This research was supported by the Intramural Research Program of the National Institute of Allergy and Infectious Diseases, National Institutes of Health. Animal experiments were conducted in an Association for Assessment and Accreditation of Laboratory Animal Care International (AAALAC)-accredited facility in accordance with animal welfare guidelines under an animal study protocol (2009-27) approved by the Animal Care and Use Committee of the Rocky Mountain Laboratories, National Institute of Allergy and Infectious Diseases, National Institutes of Health.

References

1. Caughey B, Baron GS, Chesebro B et al (2009) Getting a grip on prions: oligomers, amyloids, anchors and pathological membrane interactions. Annu Rev Biochem 78:177–204
2. Safar J, Wille H, Itri V et al (1998) Eight prion strains have PrP^{Sc} molecules with different conformations. Nat Med 4:1157–1165
3. Caughey B, Raymond GJ (1991) The scrapie-associated form of PrP is made from a cell surface precursor that is both protease- and phospholipase-sensitive. J Biol Chem 266:18217–18223
4. Korth C, Stierli B, Streit P et al (1997) Prion (PrP^{Sc})-specific epitope defined by a monoclonal antibody. Nature 390:74–77
5. Towbin H, Staehelin T, Gordon J (1979) Electrophoretic transfer of proteins from polyacrylamide gels to nitrocellulose sheets: Procedure and some applications. Proc Natl Acad Sci U S A 76:4350–4354
6. Pluskal MG, Przekop MB, Kavonian MR et al (1986) Immobilon PVDF transfer membrane: a new membrane substrate for western blotting of proteins. BioTechniques 4:272–283
7. Baron GS, Hughson AG, Raymond GJ et al (2011) Effect of glycans and the glycophosphatidylinositol anchor on strain dependent conformations of scrapie prion protein: improved purifications and infrared spectra. Biochemistry 50:4479–4490
8. Mahal SP, Jablonski J, Suponitsky-Kroyter I et al (2012) Propagation of RML prions in mice expressing PrP devoid of GPI anchor leads to formation of a novel, stable prion strain. PLoS Pathog 8:e1002746
9. Bett C, Kurt TD, Lucero M et al (2013) Defining the conformational features of anchorless, poorly neuroinvasive prions. PLoS Pathog 9: e1003280
10. Chesebro B, Trifilo M, Race R et al (2005) Anchorless prion protein results in infectious amyloid disease without clinical scrapie. Science 308:1435–1439
11. Chesebro B, Race B, Meade-White K et al (2010) Fatal transmissible amyloid encephalopathy: a new type of prion disease associated with lack of prion protein membrane anchoring. PLoS Pathog 6:e1000800
12. Brown P, Liberski P, Wolff A et al (1990) Conservation of infectivity in purified fibrillary extracts of scrapie-infected hamster brain after sequential enzymatic digestion or polyacrylamide gel electrophoresis. Proc Natl Acad Sci U S A 87:7240–7244
13. Race RE, Raymond GJ (2004) Inactivation of transmissible spongiform encephalopathy (prion) agents by environ LpH. J Virol 78:2164–2165
14. Schlumpberger M, Wille H, Baldwin MA et al (2000) The prion domain of yeast Ure2p induces autocatalytic formation of amyloid fibers by a recombinant fusion protein. Protein Sci 9:440–451
15. Kim CL, Umetani A, Matsui T et al (2004) Antigenic characterization of an abnormal isoform of prion protein using a new diverse panel of monoclonal antibodies. Virology 320:40–51
16. Baron GS, Magalhaes AC, Prado MA et al (2006) Mouse-adapted scrapie infection of SN56 cells: greater efficiency with microsome-associated versus purified PrP-res. J Virol 80:2106–2117
17. Raymond GJ, Race B, Hollister JR et al (2012) Isolation of novel synthetic prion strains by amplification in transgenic mice coexpressing wild-type and anchorless prion proteins. J Virol 86:11763–11778

Chapter 6

Analysis of miRNA Signatures in Neurodegenerative Prion Disease

Shayne A. Bellingham and Andrew F. Hill

Abstract

Prion diseases or transmissible spongiform encephalopathies are disorders of the central nervous system that affect both humans and animals. The underlying cause of prion diseases is the formation and propagation of the infectious prion protein. Prion diseases are difficult to diagnose and treat due to a prolonged asymptomatic incubation period prior to the onset of clinical symptoms. MicroRNAs (miRNAs) are small noncoding RNA species and have been identified as potential biomarkers that also function to regulate disease-specific pathways and proteins in several neurodegenerative disorders, including prion diseases. Here we describe the quantitative analysis of miRNA isolated from neuronal cells infected with a strain of mouse-adapted human prions. These methods can also be adapted to the discovery of miRNA biomarkers in extracellular vesicles, tissue, and noninvasive biological fluids.

Key words miRNA, Biomarker, Neurodegenerative, Prion disease, Taqman assay, RNA isolation

1 Introduction

Prion diseases are a unique, invariably fatal, and rapidly progressive group of neurodegenerative disorders that are known to affect humans and animals. Human prion diseases include Creutzfeldt-Jakob disease (CJD), fatal familial insomnia, Gerstmann-Sträussler-Scheinker syndrome, and a disease affecting the Fore linguistic group of Papua New Guinea's Eastern Highlands, Kuru. Animal prion diseases include scrapie in sheep, bovine spongiform encephalopathy or "mad cow disease" in cattle, chronic wasting disease in cervids, transmissible mink encephalopathy in mink, and feline spongiform encephalopathy in cats. Human prion disease can arise by either sporadic or familial mutations in the host *PRNP* gene or acquired through exposure to the infectious agent. According to the protein-only model of prion propagation, the fundamental cause of prion diseases involves conformational conversion of the normal host cellular prion protein, PrP^C, into the

Victoria A. Lawson (ed.), *Prions: Methods and Protocols*, Methods in Molecular Biology, vol. 1658,
DOI 10.1007/978-1-4939-7244-9_6, © Springer Science+Business Media LLC 2017

disease-associated infectious isoform, PrP^{Sc}, by a protein-only template-directed mechanism [1].

Prion diseases are characterized by a protracted asymptomatic preclinical period, whereby PrP^{Sc} continually propagates, prior to the rapid onset of dementia, neuronal loss, spongiform change, and ultimately death (reviewed in [2]). Prion diseases are also transmissible with possible routes of infection through dietary exposure, medical procedures, and blood transfusion (reviewed in [2]). Early diagnosis and treatment of prion disease is hampered by the absence of effective noninvasive antemortem diagnostic methods. Due to the unique transmissible infectious nature of prion diseases and the long incubation periods prior to the onset of clinical symptoms, the identification of noninvasive, sensitive, and specific diagnostic markers during the preclinical phase is of major importance. Recent advances for in vitro amplification and detection of prions by protein misfolding cyclic amplification (PMCA) [3] and real-time quaking-induced conversion (RT-QUIC) [4] in blood-based assays have been developed allowing highly sensitive detection of abnormal PrP^{Sc} [5, 6].

The role of noncoding RNAs in functional development, biomarker discovery, and therapeutic intervention strategies for human diseases is of intense interest (reviewed in [7, 8]). MicroRNAs (miRNAs) are a class of noncoding RNA species of ~22 nucleotides in length that functionally repress target mRNA by binding their 3′-untranslated regions [9]. While the functions of many discovered miRNAs so far remain unknown, some miRNAs have been shown to play a role in several biological processes including proliferation, organ development, cell differentiation, apoptosis, and infectious disease [10]. The highest tissue-specific expression of miRNA is found in the brain [11, 12], with estimates that neuronal miRNAs can posttranscriptionally modulate the expression of more than a third of the coding mRNAs [13].

The significance of brain-expressed miRNAs and their conclusive biological functions was gradually discovered using Dicer knockout mouse models that deregulate miRNA processing, leading to defects in neuronal development of the brain, and demonstrated a role for miRNA in neurogenesis [14–16]. Moreover, dysregulated miRNA expression has been detected in brains and neuronal cells for several neurodegenerative-related disorders such as prion diseases, Alzheimer's disease, Parkinson's disease, and tauopathies (reviewed in [17]).

In prion diseases, deregulated miRNA expression has been investigated in both cell culture and animal models. In animal models, studies examined the miRNA profile from brain tissue in terminally prion-infected after the onset of clinical symptoms [18, 19]. Several miRNAs were deregulated in mice intracerebrally inoculated with scrapie strain 22A [18], while a study in primate macaques infected with BSE identified two upregulated miRNAs with miR-342-3p also upregulated in sporadic human CJD brain samples [19]. miRNA has also been shown to be regulated in a biphasic molecular response in the CA1 hippocampal region of mice infected with scrapie, with significant upregulation of several miRNAs throughout preclinical disease that drastically decreased at late stages of infection, in line with the clinical manifestation [20]. Additional noncoding RNA SNORD3A has been identified to be elevated in human CJD E200K patients and prion-infected TgMHu2ME199K mouse models [21].

In cell culture models, infection of mouse neuroblastoma N2A cells with scrapie resulted in the identification of several deregulated miRNAs postulated to be involved in cholesterol homeostasis [19]. We also identified a miRNA signature released from exosomes in mouse hypothalamic neuronal GT1–7 cells infected with the M1000 mouse-adapted human prion strain [22]. This signature comprises significant upregulation of let-7i, miR-21, miR-23b, and downregulated miR-146a levels in infected GT1–7 cells and upregulation of miRNAs let-7b, let-7i, miR-128a, miR-21, miR-222, miR-29b, miR-342-3p, and miR-424 with downregulated miR-146a [22].

In this chapter, we describe detection of miRNA signatures for prion and other neurodegenerative diseases via quantitative PCR (qRT-PCR) using TaqMan microRNA assays. However, other qRT-PCR platforms, including Qiagen miScript system and Exiqon miRCURY LNA system, could easily be adapted. The principal steps involved in analysis of miRNA signature require (1) extraction of high-quality RNA that includes small RNA <200 nt in length; (2) global profiling of the miRNAome by either high-throughput qRT-PCR, microarrays, or small RNA next-generation deep sequencing; and (3) validation of specific miRNA using qRT-PCR.

Here we describe the isolation of miRNA from M1000 infected GT1–7 cells and validation of miRNA signatures via qRT-PCR; the techniques described can be adapted for miRNA analysis from extracellular vesicles, tissue samples, and plasma or serum samples.

2 Materials

Prepare all solutions using 0.2 μM filtered ultrapure Milli-Q RNase-free water and analytical-grade reagents. Prepare and store all reagents at room temperature (unless indicated otherwise). Follow all laboratory regulations when handling, storing, and disposing infectious prion material.

1. Ice.
2. Rubber policeman (Cell Scraper).
3. PBS.

4. miRNeasy Isolation Kit (Qiagen) for Subheading 3.1.
5. QIAzol Lysis Reagent (Qiagen) for Subheading 3.1.
6. mirVana miRNA Isolation Kit (Applied Biosystems) for Subheading 3.2.
7. TRIzol Reagent (Invitrogen) for Subheading 3.2.
8. Chloroform.
9. 100% ethanol, ACS grade.
10. Milli-Q RNase/DNase-free water.
11. Luer Lock syringes.
12. 0.2 μM syringe filters.
13. Heat block set to 95–100 °C.
14. Microcentrifuge.
15. Screw-capped RNase-free Eppendorf tubes.
16. Pipettes.
17. Filter tips.
18. Agilent 2100 Bioanalyzer (Agilent Technologies Inc.).
19. Agilent RNA 6000 Nano Kit (Agilent Technologies Inc.).
20. Agilent Small RNA Kit (Agilent Technologies Inc.).
21. TaqMan MicroRNA Assays for each specific miRNA (Applied Biosystems).
22. microRNA Reverse Transcription Kit (Applied Biosystems).
23. TaqMan Gene Expression Master Mix or Fast Advanced Master Mix (Applied Biosystems).
24. Viia7 Real-Time PCR system (Applied Biosystems) or similar instrument.
25. 0.1 mL fast reaction tubes (Applied Biosystems) or equivalent.
26. 0.1 mL fast reaction 96-well plates (Applied Biosystems) or equivalent.
27. MicroAmp Adhesive Film (Applied Biosystems).
28. Fume hood.

2.1 Optional Materials

1. QIAcube (Qiagen), fully automated sample processing for Subheading 3.1.
2. TaqMan PreAmp Master Mix (Applied Biosystems), in Subheading 3.3.
3. QIAgility (Qiagen), for liquid handling and reaction plate setup in Subheading 3.3.

3 Methods

3.1 Disruption of Cultured Cells and Extraction of Total RNA Including miRNA with miRNeasy Mini Kit

The miRNeasy Mini Kit enables purification of total RNA, which includes RNA from approximately >18 nucleotides from all types of animal tissues and cells, including difficult-to-lyse tissues. The miRNeasy Mini Kit is used for low-throughput RNA purification using spin columns. Co-purification of miRNA and total RNA using the miRNeasy Mini Kit can be automated on the QIAcube (optional).

1. For adherent GT1–7 cells, remove cell culture medium and wash cells in PBS. Remove cells from flask or well surface by spritzing cells with ice-cold PBS. Centrifuge cell suspension at low speed and wash pellet in 1 mL of PBS. Centrifuge again, discard supernatant, and place the cells on ice.
2. Disrupt up to 1×10^7 eukaryotic cultured cells in 700 μL of QIAzol Lysis Reagent. Vortex the sample to lyse cells or vigorously pipette up and down.
3. Incubate the tube containing the homogenate at room temperature (15–25 °C) for 5 min.
4. Add 140 μL chloroform to the tube containing the homogenate in the fume hood and cap the tube securely. Vortex the tube for 15 s.
5. Incubate the tube containing the homogenate at room temperature for 2–3 min in the fume hood.
6. Centrifuge for 15 min at 12,000 × *g* at 4 °C (*see* **Note 1**).
7. Carefully transfer the upper aqueous phase to a new collection tube and record the volume. The volume of the aqueous phase should be approximately 300–350 μL. The lower, red, organic phase contains proteins and hence contains PrP^{Sc} for infected samples. Dispose of using laboratory procedures for decontamination of prion infectious material (*see* **Note 2**).
8. Add 1.5 volumes of 100% ethanol to the volume of "recovered aqueous phase from **step** 7," and mix thoroughly by inverting or pipetting several times. For example, if 350 μL aqueous phase is recovered, add 525 μL of 100% ethanol. Do not centrifuge. Continue without delay.
9. For each sample, assemble a miRNeasy Mini spin column together with a 2 mL collection tube (supplied).
10. Pipette up to 700 μL of the sample, including any precipitate that may have formed, into the assembled miRNeasy Mini spin column, and centrifuge at 8000 × *g* for 15–30 s at room temperature (15–25 °C). Discard the flow-through and reuse collection tube.

11. Repeat **step 10** using the remainder of the sample. Discard the flow-through. Reuse the collection tube in **step 12**.
12. Add 700 μL Buffer RWT to the miRNeasy Mini spin column. Centrifuge for 15–30 s at 8000 × *g* to wash the column. Discard the flow-through.
13. Pipette 500 μL Buffer RPE into the miRNeasy Mini spin column. Centrifuge for 15–30 s at 8000 × *g* to wash the column. Discard the flow-through. Reuse the collection tube for **step 14**.
14. Add another 500 μL Buffer RPE to the miRNeasy Mini spin column. Close the lid gently and centrifuge for 2 min at 8000 × *g* to dry the miRNeasy Mini spin column membrane (*see* **Note 3**).
15. Place the miRNeasy Mini spin column into a new 2 mL collection tube (not supplied) and discard the old collection tube with the flow-through. Centrifuge in a microcentrifuge at full speed for 1 min (*see* **Note 4**).
16. Label 1.5 mL collection tubes and transfer the miRNeasy Mini spin column to 1.5 mL collection tube.
17. Pipette 30–50 μL nuclease-free water directly onto the miRNeasy Mini spin column membrane.
18. Centrifuge for 1 min at 8000 × *g* to elute the RNA (*see* **Note 5**).
19. Eluted RNA should be immediately placed on ice to prevent potential degradation. Keep at −20 °C and −80 °C for short- and long-term storage, respectively.

3.2 Disruption of Cultured Cells and Extraction of Total RNA Including miRNA with mirVana miRNA Isolation Kit

The mirVana™ miRNA Isolation Kit enables the isolation of small RNA-containing total RNA from samples consisting of 10^2–10^7 cultured cells or 0.5–250 mg tissue using a glass fiber filter-based method. The method isolates total RNA ranging in size from kilobases down to ten nucleotides (*see* **Note 6**).

1. When opening a new mirVana™ miRNA Isolation Kit, add 21 mL of ACS grade ethanol 100% to miRNA wash solution 1 and 40 mL of ACS grade ethanol 100% to miRNA wash solution 2/3 (*see* **Note** 7).
2. For adherent GT1–7 cells, remove cell culture medium and wash cells in PBS. Remove cells from flask or well surface by spritzing cells with ice-cold PBS. Centrifuge cell suspension at low speed and wash pellet in 1 mL PBS. Centrifuge again, discard supernatant, and place the cells on ice (*see* **Note 8**).
3. Disrupt up to 1 × 10^7 eukaryotic cultured cells in 300–600 μL of ice-cold lysis/binding buffer. Vortex the sample to lyse cells or vigorously pipette up and down (*see* **Note 9**).

4. Add 1/10 volume of miRNA homogenate additive to lysate and mix well by vortexing or inverting and incubate on ice for 10 min. For example, if the lysate volume is 600 μL, add 60 μL of miRNA homogenate additive.
5. Prepare TRIzol/chloroform reagent as required. For every 1 mL of TRIzol Reagent, add 0.2 mL chloroform and mix by inverting several times (*see* **Note 10**).
6. Add 1 volume of "Lysis" of TRIzol/chloroform reagent to mixture and vortex for 30–60 s to mix. For example, if cells were lysed in 600 μL of ice-cold lysis/binding buffer originally in **step 3**, then add 600 μL TRIzol/chloroform reagent.
7. Centrifuge for 5 min at 10,000 × *g* to separate aqueous and organic phase (*see* **Note 1**).
8. Carefully transfer the upper aqueous phase to a new collection tube and record the volume. The volume of the aqueous phase should be approximately 300–350 μL. The lower, red, organic phase contains proteins and hence contains PrP^{Sc} for infected samples. Dispose of using laboratory procedures for prion infectious material.
9. Turn on heat block set to 95–100 °C.
10. Place Elution Solution in heat block, ~100 μL per sample (*see* **Note 11**).
11. Add 1.25 volumes of 100% ethanol to the "recovered aqueous phase from Subheading 3.2 **step 8**," and invert several times to mix. For example, if 300 μL aqueous phase is recovered, add 375 μL of room temperature 100% ethanol.
12. For each sample, assemble a filter cartridge together with a 2 mL collection tube (supplied).
13. Pipette the sample mixture onto an assembled filter cartridge/collection tube and centrifuge at 10,000 × *g* for ~15–30 s. Discard the flow-through and repeat until all of the sample mixture is passed through the filter cartridge.
14. Apply 700 μL miRNA Wash Solution 1, centrifuge at 10,000 × *g* for ~30 s, and discard flow-through.
15. Apply 500 μL Wash Solution 2/3, centrifuge at 10,000 × *g* for ~30 s, and discard flow-through. Repeat with second 500 μL Wash Solution 2/3, centrifuge at 10,000 × *g* for ~30 s, and discard flow-through.
16. Centrifuge at 10,000 × *g* for 1 min, to remove traces of Wash Solution 2/3.
17. Place filter cartridge into fresh collection tube.
18. Elute RNA with 100 μL of preheated Elution Solution and centrifuge at 10,000 × *g* for 1 min.

19. Eluted RNA should be immediately placed on ice to prevent potential degradation. Keep at −20 °C and −80 °C for short- and long-term storage, respectively.

3.3 MicroRNA Signature Validation

A number of factors govern the choice of miRNAs chosen for qRT-PCR validation. First and most importantly, if cells and/or tissues are of murine origin, the candidate miRNA must have a known human homologue, so that putative neurodegenerative biomarkers identified can be tested in human patient samples. Second, given the goal of producing clinically relevant assays, miRNAs should be highly detectable in both neuronal tissue and noninvasive body fluids such as blood. Finally, prioritization of miRNAs based on their function, such as a known role in neuronal development and disease, may be a sensible approach if the candidate list is large.

3.3.1 MicroRNA Reverse Transcription

The MicroRNA Reverse Transcription Kit provides the necessary components for optimal performance in TaqMan MicroRNA Assays. Components of this kit are used with the RT primer provided with the TaqMan MicroRNA Assay to convert miRNA to cDNA (*see* **Note 12**).

1. Assess the quantity and quality of isolated total RNA including microRNA by running 1 μL of the total RNA sample on the Agilent Bioanalyzer using both the Agilent RNA 6000 kit and Agilent Small RNA assay kit, following the manufacturer's instructions (*see* **Note 13**).
2. Before use, thaw the 5× RT primer and MicroRNA Reverse Transcription Kit components on ice, resuspend the solutions completely by gently vortexing, and then briefly centrifuge 1000 × *g*, ~10 s (*see* **Notes 14** and **15**).
3. In a polypropylene tube, prepare the RT Master Mix on ice by multiplying the volume of components in the table below to the required number of RT reactions (*see* **Note 16**).

Component	RT master mix volume per 15 μL reaction	Volume for ten reactions including excess
10× RT buffer	1.5	18.0
25× dNTP mix (100 mM)	0.15	1.8
RNase inhibitor	0.19	2.28
MultiScribe™ reverse transcriptase	1.0	12.0
Nuclease-free Milli-Q H_2O	4.16	49.92
Total per reaction	7.0	84.0

4. Mix gently. *Do not vortex.* Centrifuge to bring the solution to the bottom of the tube (*see* **Note 15**).
5. Place the RT Master Mix on ice until you prepare the RT reaction tubes or plate.
6. For each 15 μL RT reaction, combine into a 0.2 mL polypropylene RT reaction tube or into a well of a 96-well reaction plate 7 μL RT Master Mix (from **step 3**) with 5 μL total RNA (1–10 ng per reaction as determined from concentration provided by the Bioanalyzer in Subheading 3.3.1, **step 1**) (*see* **Note 17**).
7. Add 3 μL of 5× RT primer from each assay set into the corresponding RT reaction tube or plate well.
8. Seal the tube with optical caps or reaction plate with optical adhesive film and mix thoroughly by inverting the solution. Centrifuge to bring the solution to the bottom of the tube or well.
9. Incubate the tube on ice for 5 min and keep it on ice until you are ready to load the thermal cycler.
10. To perform reverse transcription:
11. Program the thermal cycler conditions.

	Step 1	Step 2	Step 3	Step 4
Temperature (°C)	16	42	85	4
Time	30 min	30 min	5 s	∞

12. Load the reactions into the thermal cycler and start the reverse transcription. When run has finished, remove tubes and centrifuge briefly to spin down contents and eliminate any air bubbles. Store cDNA for short-term (4 °C) or long-term storage (−15 to −25 °C) or proceed immediately to qPCR.

3.3.2 TaqMan Gene Expression MicroRNA Assay

TaqMan® MicroRNA Assays combine a target-specific stem-loop reverse 5× transcription primer that extends the 3′ end of the target to produce a cDNA template that allows specific detection of mature miRNA with a 20× microRNA gene expression assay that can be used in standard real-time PCR.

1. If frozen, thaw cDNA samples by placing them on ice. When thawed, resuspend the samples by vortexing and then centrifuge the tubes briefly.
2. Thaw 20× TaqMan MicroRNA Assays on ice, resuspend samples by vortexing, and then centrifuge the tubes at 1000 × *g*, ~ 10 s (*see* **Notes 14** and **18**).
3. Thaw TaqMan Gene Expression Master Mix or Fast Advanced Master Mix on ice and mix thoroughly by swirling the bottle.

4. Add 210 μL nuclease-free Milli-Q H_2O to each individual cDNA template for a final dilution of 1:15, mix, and centrifuge briefly (*see* **Note 19**).
5. Prepare the PCR reaction mix in a microcentrifuge tube on ice according to the table below before transferring it to the reaction plate for thermal cycling and fluorescence analysis (*see* **Note 20**).

	20 μL reactions		10 μL reactions	
	1 replicate	4 replicates	1 replicates	4 replicates
TaqMan Gene expression (2×) or fast advanced master mix (2×)	10.0	50.0	5.0	25.0
TaqMan MicroRNA assay (20×)	1.0	5.0	0.5	2.5
Diluted cDNA template	9.0	45.0	4.5	22.5
Total volume	20.0	100.0	10.0	50.0

6. Vortex the tube(s) and centrifuge the tube(s) briefly to spin down the contents and eliminate any air bubbles from the solutions.
7. Transfer the appropriate volume of each reaction mixture to each well of the Applied Biosystems 96-well plate, seal plate with optical adhesive film, and spin for 5 min, 3283 × *g* at 4 degrees in centrifuge fitted with plate adapters (*see* **Note 21**).
8. Load and run the plate on real-time quantitative PCR instrument, Viia7 or equivalent instrument either standard mode or fast mode, as indicated in the table below (*see* **Note 22**).

Standard mode thermal cycling conditions				
Step	UDG incubation	Polymerase activation	PCR (40 cycles)	
	Hold	Hold	Denature	Anneal/extend
Time	2 min	10 min	15 s	1 min
Temp	50 °C	95 °C	95 °C	60 °C
Fast mode thermal cycling conditions				
Step	UNG incubation	Polymerase activation	PCR (40 cycles)	
	Hold	Hold	Denature	Anneal/extend
Time	2 min	20 s	1 s	20 s
Temp	50 °C	95 °C	95 °C	60 °C

9. The relative fold change of miRNA levels is determined based on the threshold cycle (Ct) and is normalized by the ΔΔCt method [23]. Normalize your Ct values using the average Ct of the endogenous controls (*see* **Notes 23** and **24**).
10. Use ΔCT (miRNA Ct – averaged endogenous control Ct) or fold change relative to a calibrator or reference sample ($2^{\Delta\Delta CT}$) for relative expression analysis.

4 Notes

1. After centrifugation, the sample separates into three phases: an upper, colorless, aqueous phase containing RNA; a white interphase; and a lower, red, organic phase. Following centrifugation, if the interphase layer is not compact, repeat the centrifugation step.
2. Optional QIAcube for automation of miRNeasy Mini Kit can be continued from this step following the manufacturer's instructions.
3. The long centrifugation dries the spin column membrane, ensuring that no ethanol is carried over during RNA elution. Residual ethanol may interfere with downstream reactions. Following centrifugation, remove the miRNeasy Mini spin column from the collection tube carefully so the column does not contact the flow-through. Otherwise, carryover of ethanol will occur.
4. Perform this step to eliminate any possible carryover of Buffer RPE or if residual flow-through remains on the outside of the miRNeasy Mini spin column after **step 15**.
5. If the expected RNA yield is >30 μg, repeat **step 15** with a second volume of 30–50 μL RNase-free water. Elute into the same collection tube. To obtain a higher total RNA concentration, this second elution step may be performed by using the first eluate (from **step 17**). The yield will be 15–30% less than the yield obtained using a second volume of RNase-free water, but the final concentration will be higher.
6. The small RNA enrichment procedure is not recommended as it is unable to determine quality of total RNA from enriched small RNA alone.
7. An EDTA precipitate may form in Wash Solution 2/3; simply leave crystals in bottle when using wash solution.
8. Cells can be disrupted directly in wells with ice-cold lysis/binding buffer and lysate collected with cell scrapers in clean screw-capped tubes. Vortex cells to obtain homogeneous lysate.

9. Use 300 µL of ice-cold lysis/binding buffer for small number of cells in the 100's range and use 600 µL of ice-cold lysis/binding buffer for large number of cells in the thousands-millions range.
10. The TRIzol Reagent is preferred to the recommended acid-phenol: chloroform in the mirVana Kit due to ease of visualization of the three separate phases.
11. The supplied Elution Solution is 0.1 mM EDTA; if this will interfere with your downstream applications, then elute in 0.2 µM filtered nuclease-free water.
12. MicroRNA RT reactions can be multiplexed with up to 96 individual 5× primers and/or TaqMan MicroRNA Assays with or without preamplification in each reaction without loss of performance [24]. We have successfully multiplexed up to 20 assays per RT reaction following the protocol with or without preamplification.
13. Assessment of RNA quality and integrity is crucial to successful downstream applications for gene expression analysis as microRNA expression cannot be accurately profiled from degraded total RNA samples. Running samples on the Bioanalyzer with the RNA 6000 Kit provides an accurate assessment of both the RNA quantity and quality by determining the RNA integrity number (RIN) of the given sample. A RIN greater than or equal to 7 is recommended for the profiling of microRNAs [25].
14. Centrifugal force greater than 1000 × *g* may result in rubber caps of 5× RT primer and 20× assay tubes being lodged inside tubes.
15. Do not vortex MultiScribe™ Reverse Transcriptase or RT Master Mix containing MultiScribe™ Reverse Transcriptase.
16. Include additional reactions in the calculations to provide 20% excess volume for the loss that occurs during reagent transfers.
17. For each microRNA assay, an individual reverse transcription (RT) reaction must be prepared per sample. Therefore the final 15 µL reaction contains 7 µL RT Master Mix, 5 µL RNA, and 3 µL of 5× RT primer specific for each individual miRNA.
18. Keep all TaqMan reagents protected from light, in the freezer, until you are ready to use them. Excessive exposure to light may affect the fluorescent probes.
19. The final diluted cDNA template volume is sufficient for eight replicates.
20. We recommend quadruplicate replicates for each assay; however, triplicate replicates may be performed if necessary. Calculate the total volume required for each component by the number of replicates for each sample according to the table.

For every four reactions, include volume for a fifth reaction to provide excess volume for the loss that occurs during reagent transfers.

21. Reaction plates can be setup on QIAgility or equivalent liquid handling platform according to manufacturer's instructions.
22. If using TaqMan Gene Expression Master Mix, run the plate in standard mode. Check thermal cycling parameters of your instrument with master mix compatibility as parameters may vary.
23. A universally accepted endogenous control for microRNA has yet to be established. Based on our experience, we find the RNU6B, snoRNA202, and snoRNA135 to show the least variability between mouse hypothalamic GT1–7 samples [22].
24. It is recommended that a set of several endogenous small nuclear RNA (snRNA) and/or small nucleolar RNA (snoRNA) endogenous control genes be selected and screened based on the species, tissues, or cell lines used in your study. Alternatively, or in addition to, use specific miRNAs that demonstrate the least variability across experimental conditions. Based upon data generated by a wide variety of tissues and cell lines that show the least variability, Applied Biosystems recommends the following candidate control genes: for human samples, use RNU48, RNU44, U47, and RNU6B; and for mouse samples, use snoRNA202 and snoRNA234 [26].

References

1. Prusiner SB (1982) Novel proteinaceous infectious particles cause scrapie. Science 216:136–144
2. Aguzzi A, Heikenwalder M (2006) Pathogenesis of prion diseases: current status and future outlook. Nat Rev Microbiol 4:765–775
3. Saborio GP, Permanne B, Soto C (2001) Sensitive detection of pathological prion protein by cyclic amplification of protein misfolding. Nature 411:810–813
4. Atarashi R, Satoh K, Sano K et al (2011) Ultrasensitive human prion detection in cerebrospinal fluid by real-time quaking-induced conversion. Nat Med 17:175–178
5. Lacroux C, Comoy E, Moudjou M et al (2014) Preclinical detection of variant CJD and BSE prions in blood. PLoS Pathog 10:e1004202
6. Edgeworth JA, Farmer M, Sicilia A et al (2011) Detection of prion infection in variant Creutzfeldt-Jakob disease: a blood-based assay. Lancet 377:487–493
7. Esteller M (2011) Non-coding RNAs in human disease. Nat Rev Genet 12:861–874
8. Dimmeler S, Nicotera P (2013) MicroRNAs in age-related diseases. EMBO Mol Med 5:180–190
9. Bartel DP (2004) MicroRNAs: genomics, biogenesis, mechanism, and function. Cell 116:281–297
10. Croce CM, Calin GA (2005) miRNAs, cancer, and stem cell division. Cell 122:6–7
11. Cao X, Yeo G, Muotri AR et al (2006) Noncoding RNAs in the mammalian central nervous system. Annu Rev Neurosci 29:77–103
12. Sempere LF, Freemantle S, Pitha-Rowe I et al (2004) Expression profiling of mammalian microRNAs uncovers a subset of brain-expressed microRNAs with possible roles in murine and human neuronal differentiation. Genome Biol 5:R13
13. Kosik KS (2006) The neuronal microRNA system. Nat Rev Neurosci 7:911–920
14. Schaefer A, O'carroll D, Tan CL et al (2007) Cerebellar neurodegeneration in the absence of microRNAs. J Exp Med 204:1553–1558

15. Kawase-Koga Y, Otaegi G, Sun T (2009) Different timings of Dicer deletion affect neurogenesis and gliogenesis in the developing mouse central nervous system. Dev Dyn 238:2800–2812
16. Huang T, Liu Y, Huang M et al (2010) Wnt1-cre-mediated conditional loss of Dicer results in malformation of the midbrain and cerebellum and failure of neural crest and dopaminergic differentiation in mice. J Mol Cell Biol 2:152–163
17. Bellingham SA, Guo BB, Coleman BM et al (2012) Exosomes: vehicles for the transfer of toxic proteins associated with neurodegenerative diseases? Front Physiol 3:124
18. Saba R, Goodman CD, Huzarewich RL et al (2008) A miRNA signature of prion induced neurodegeneration. PLoS One 3:e3652
19. Montag J, Hitt R, Opitz L et al (2009) Upregulation of miRNA hsa-miR-342-3p in experimental and idiopathic prion disease. Mol Neurodegener 4:36
20. Majer A, Medina SJ, Niu Y et al (2012) Early mechanisms of pathobiology are revealed by transcriptional temporal dynamics in hippocampal CA1 neurons of prion infected mice. PLoS Pathog 8:e1003002
21. Cohen E, Avrahami D, Frid K et al (2013) Snord 3A: a molecular marker and modulator of prion disease progression. PLoS One 8: e54433
22. Bellingham SA, Coleman BM, Hill AF (2012) Small RNA deep sequencing reveals a distinct miRNA signature released in exosomes from prion-infected neuronal cells. Nucleic Acids Res 40:10937–10949
23. Pfaffl MW (2001) A new mathematical model for relative quantification in real-time RT-PCR. Nucleic Acids Res 29:e45
24. Applied biosystems (2013) Procedure for multiplexing the RT step with or without preamplification while using TaqMan MicroRNA Assays In: Applied Biosystems User Bulliten 4465407. http://www.appliedbiosystems.com
25. Ibberson D, Benes V, Muckenthaler MU et al (2009) RNA degradation compromises the reliability of microRNA expression profiling. BMC Biotechnol 9:102
26. Wong L, Lee K, Russell I et al. (2007) Endogenous controls for real-time quantitation of miRNA using TaqMan® MicroRNA assays. In: Applied biosystems application note. http://www.appliedbiosystems.com

Part II

Methods for Cellular Analysis of Prion Disease

Chapter 7

Cell Biology Approaches to Studying Prion Diseases

Suzette A. Priola

Abstract

During the course of prion infection, the normally soluble and protease-sensitive mammalian prion protein (PrP^C) is refolded into an insoluble, partially protease-resistant, and infectious form called PrP^{Sc}. The conformational conversion of PrP^C to PrP^{Sc} is a critical event during prion infection and is essential for the production of prion infectivity. This chapter briefly summarizes the ways in which cell biological approaches have enhanced our understanding of how PrP contributes to different aspects of prion pathogenesis.

Key words Prion, Prion cell culture, Scrapie cell assay, Transmissible spongiform encephalopathies, Scrapie

1 Introduction

PrP^{Sc} is the primary, and possibly sole, component of the infectious prion responsible for the transmissible spongiform encephalopathies (TSEs) or prion diseases. Prion diseases are a group of fatal and transmissible neurodegenerative diseases that naturally occur in mammals and include Creutzfeldt-Jakob disease in humans, scrapie in sheep, and bovine spongiform encephalopathy (BSE). In the laboratory, mouse and hamster models of rodent-adapted sheep scrapie are primarily used to study prion infection. Disease can take from months to years to develop in an infected host, making it particularly difficult to study the mechanisms underlying different aspects of prion pathogenesis using in vivo models. Multiple cell lines have now been identified that can support prion infection [1], and new models of prion infection including primary neuronal cultures [2], organotypic slice cultures [3], and differentiable neurospheres [4] have been developed. When combined with techniques to specifically label both PrP^C and PrP^{Sc} and the ability to express wild-type and mutant forms of PrP^C in cells, these in vitro systems have greatly contributed to our understanding of the mechanisms underlying multiple aspects of prion pathogenesis.

Victoria A. Lawson (ed.), *Prions: Methods and Protocols*, Methods in Molecular Biology, vol. 1658,
DOI 10.1007/978-1-4939-7244-9_7, © Springer Science+Business Media LLC 2017

Moreover, they have proven invaluable in the development of in vitro-based titration assays and diagnostics as well as for the identification of potential prion therapeutic compounds.

2 Prion Protein Cell Biology

The expression of PrP^{C} in almost all established and primary cell lines coupled with the discovery that an established mouse neuroblastoma cell line of neuronal origin was susceptible to infection with brain homogenate-derived mouse prions (Fig. 1a) [5, 6] enabled the study of the cell biology of both PrP^{C} and PrP^{Sc}. These studies have shown that synthesis of the two forms of PrP is significantly different. Wild-type PrP^{C} is synthesized via the secretory pathway, passing through the endoplasmic reticulum and Golgi where it is glycosylated [7] and a glycophosphatidylinositol (GPI) membrane anchor added [8]. While the majority of intact PrP^{C} is trafficked to the cell membrane where it is anchored via its GPI anchor [9], some PrP^{C} can be either secreted [10] or located in the cytosol [11]. A portion of PrP^{C} is also proteolytically processed at the N-terminus [12]. The half-life of PrP^{C} is between 4 and 6 h [13], and it is either endocytosed into endolysosomes for degradation [14] or trafficked back to the cell surface via recycling endosomes [15, 16].

In contrast to PrP^{C}, studies using prion-infected mouse neuroblastoma cells have shown that PrP^{Sc} is made at or near the cell surface from its GPI-anchored precursor PrP^{C} [13, 17], with more recent work suggesting that the plasma membrane, early endosomes, and the endocytic recycling pathway are all sites of PrP^{Sc} formation [18, 19]. Following the conformational conversion of GPI-anchored PrP^{C} to PrP^{Sc}, newly synthesized PrP^{Sc} is endocytosed after which it may be either recycled back to the cell membrane or stored in an endolysosomal compartment [18, 19] where it is likely exposed to cellular proteases and N-terminally truncated [17, 20]. Unlike PrP^{C}, the greater protease resistance of PrP^{Sc} extends its cellular life-span, and it has a half-life of greater than 24 h [13, 21].

In humans, genetic forms of prion disease are associated with insertions, deletions, and single amino acid mutations within PrP^{C}. However, it is unclear how these mutations lead to the generation of PrP^{Sc} and disease. Cell lines expressing mutant forms of PrP^{C} have been used to show that these mutations can change how PrP^{C} is processed as well as its biophysical properties. For example, insertional mutations in PrP^{C} result in PrP molecules with properties intermediate between those of PrP^{C} and PrP^{Sc}, including a propensity to aggregate and become partially protease-resistant [22, 23]. Some single amino acid mutations associated with heritable human prion diseases can also lead to aberrantly processed PrP^{C}

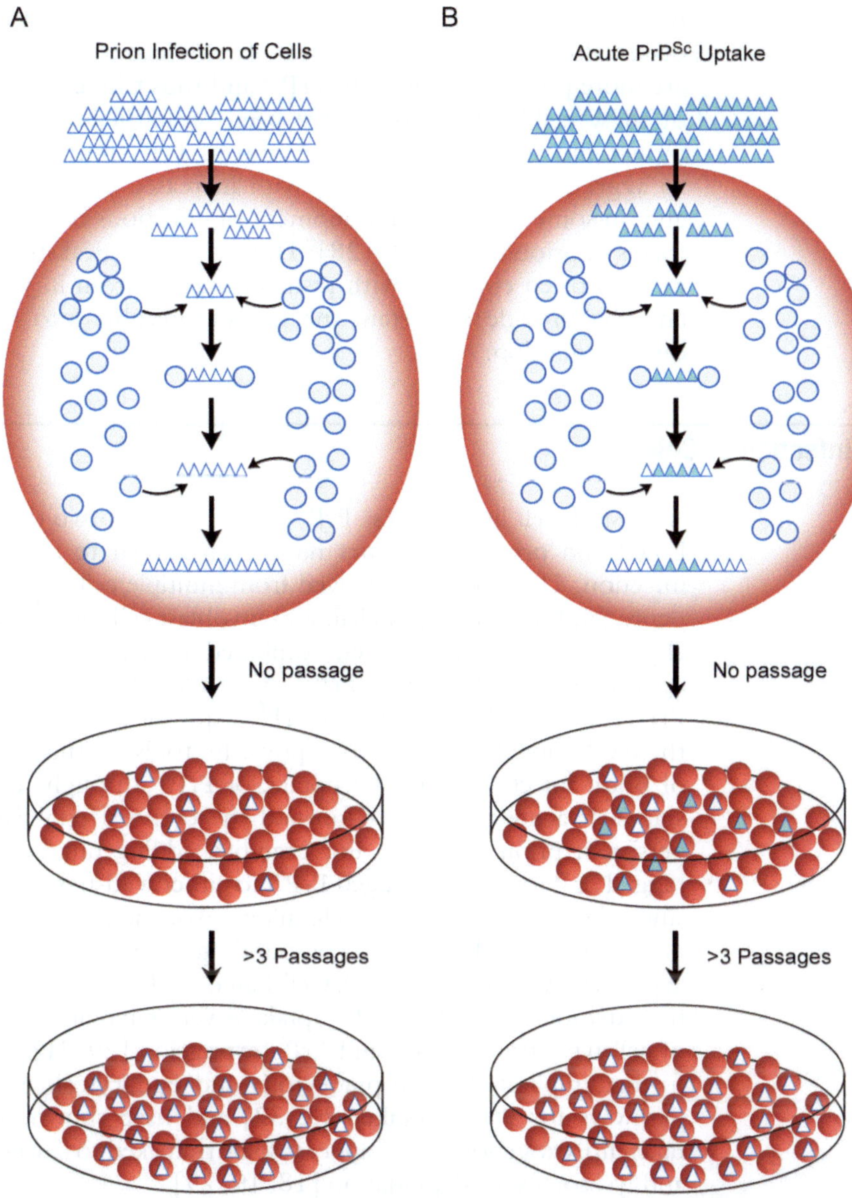

Fig. 1 Cell models of prion infection. The process of prion infection in a single cell is shown at the top of each panel, while the middle and bottom part of each panel represents a monolayer of cells in a tissue culture plate or dish either prior to passage (no passage) or after multiple passages (>3 passages). (**a**) Prion infection of susceptible cells. Following exposure of a cell monolayer to a prion-infected brain homogenate containing PrP^{Sc} (open triangles), PrP^{Sc} is taken up by the cell where it binds PrP^{C} (*open circles*) and converts it to PrP^{Sc}. Prior to cell passage, newly made PrP^{Sc} cannot be distinguished from the exogenous PrP^{Sc} initially taken up by the cell (*red circles*). Cells where PrP^{Sc} is accumulating (represented by *open triangles* within *red circles*) become infected, and that infection becomes persistent if, following at least 3 or more passages, PrP^{Sc} is still easily detectable in either the whole well or individual subclones from that well. Cell systems such as this can be used to study the cell biology of PrP and to identify potential anti-prion therapeutics. (**b**) PrP^{Sc} uptake during acute prion infection. This model of prion infection is identical to that shown in (**a**) except that the PrP^{Sc} added

molecules [24, 25]. Studies such as these suggest that mutation-driven alterations in the properties of PrP^C lead to molecules which are more prone to refold into PrP^{Sc} and thus trigger disease. Some of these cell-based studies have been recapitulated in transgenic mouse models where overexpression of PrP^C mutants also leads to aggregated PrP^C molecules with increased protease resistance and, in some cases, the development of infectious or noninfectious forms of prion disease [26–28]. Thus, cell biological studies of PrP in vitro often correlate with results obtained from in vivo studies and can provide important mechanistic insights into how prion diseases develop.

3 Prion Infection of Cells

Almost all cell lines express PrP^C and more than two dozen established and primary cell lines have been shown to support prion infection. These lines are derived from multiple cell types including neuronal, fibroblast, epithelial, and myoblast cells as well as mesenchymal stem cells [1]. When combined with the development of epitope-tagged PrP^C and PrP^{Sc} molecules, fluorescently labeled PrP^{Sc}, and cell lines in which PrP^C expression has been ablated, these cell models have provided powerful tools to study the process of prion infection from the uptake of PrP^{Sc} through to the early stages of acute prion infection and the eventual establishment of persistent prion infection. For example, adding uniquely epitope-tagged or fluorescently tagged PrP^{Sc} to either wild-type PrP^C expressing cells or PrP^C knockout cells allows exogenously added PrP^{Sc} to be followed without any confounding effects from the PrP^C expressed by the cell (Fig. 1b). Cell-based studies using this approach have demonstrated that PrP^{Sc} uptake is very rapid and independent of cell type, prion strain, and PrP^C expression [29–31]. They have further implicated macropinocytosis and endocytosis as potential pathways for PrP^{Sc} uptake [29, 32] and have shown that, once taken up, exogenous PrP^{Sc} appears to be trafficked to late endosomes and lysosomes for degradation [18, 19, 29].

Similarly, expression of uniquely epitope-tagged PrP^C in cells followed by exposure of those cells to wild-type PrP^{Sc} has for the first time enabled the study of the acute stages of prion infection and de novo PrP^{Sc} formation (Fig. 2a) [33]. These studies have shown

Fig. 1 (continued) to the cells is tagged, either with a fluorescent tag or with a unique antibody epitope (*blue triangles*). Prior to passage, exogenous PrP^{Sc} taken up by the cells (*blue triangles* in *red circles*) can thus be distinguished from any newly made PrP^{Sc} in the cell (*open triangles* in *red circles*). After multiple passages, the tagged PrP^{Sc} is diluted out and can no longer be detected in the infected culture (*bottom of panel*). Cell culture systems such as this can be used to study how PrP^{Sc} is taken up and trafficked by the cell early during acute prion infection

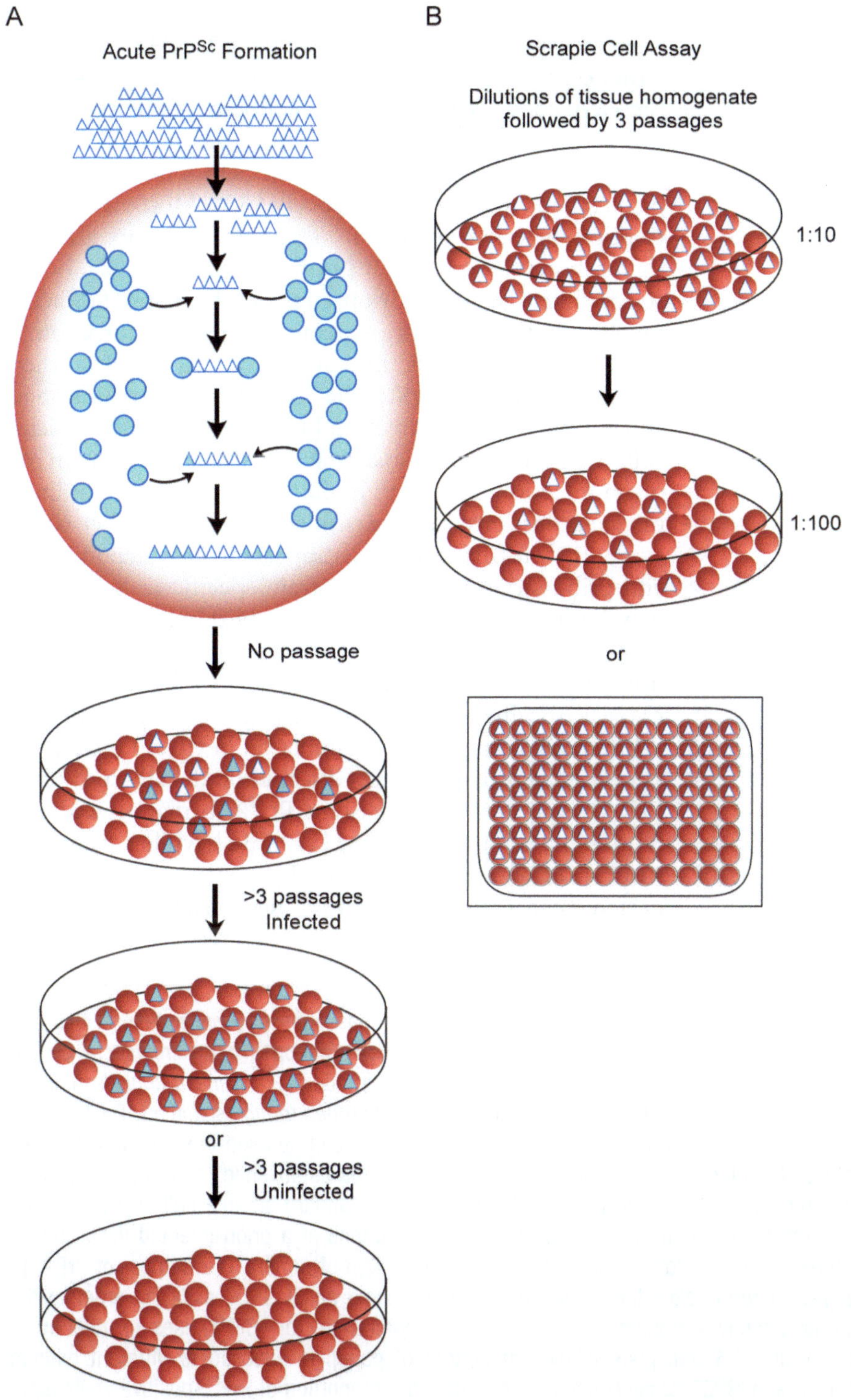

Fig. 2 Cell-based systems for studying acute prion infection and determining prion titer. (**a**) Acute PrP^{Sc} formation during prion infection. The process of prion infection in a single cell is shown at the top of the panel, while the lower part of the panels represents a monolayer of cells in a tissue culture plate or dish either prior to

that de novo or acute PrP^{Sc} formation occurs almost immediately after exposure of cells to prions [19, 33, 34]. Acute PrP^{Sc} is made either on the plasma membrane or in the early endosome and can be recycled back to the plasma membrane to initiate further PrP^{Sc} formation [19, 34, 35]. Surprisingly, de novo PrP^{Sc} formation is independent of cell type and prion strain and can be induced by prion strains that do not infect cells [33]. Moreover, in cells resistant to prion infection, acute PrP^{Sc} formation does not always lead to the establishment of persistent prion infection (Fig. 2a, bottom panel), suggesting that other prion strain or host-specific factors may be involved in the switch from acute to persistent infection [33].

One of the major limitations of cell-based studies of prion infection is that, despite the identification of multiple cell lines susceptible to prion infection [1], it remains very difficult to infect cells with prions, and most established cell lines remain refractory to prion infection. A second limitation is that only a limited subset of prion strains have been identified which can infect cells [1]. Since acute PrP^{Sc} formation in cells appears to be independent of cell type and prion strain [33], prion strain-specific or host-specific factors may be important primarily in establishing persistent prion infection in vitro. Lack of these factors may help to explain the limited number of cell lines susceptible to prion infection. There are at least two cell-specific requirements that may be important for the establishment of persistent prion infection in vitro. One study using PrP knockout cells expressing GPI-anchored PrP^{C} and/or anchorless PrP^{C} has shown that persistent PrP^{Sc} formation and cell infection require membrane-anchored PrP^{C} [36], while a second study using susceptible and resistant mouse neuroblastoma cell clones has suggested that genes involved in homeostasis of the extracellular

Fig. 2 (continued) passage (no passage) or after multiple passages (>3 passages). In this model of prion infection, PrP^{C} is engineered to contain a unique antibody epitope (*blue circles*). In combination with denaturation and/or protease digestion, this allows newly made PrP^{Sc} (*blue triangles*) to be distinguished from both the PrP^{Sc} present in the prion-infected homogenate (*open triangles*) and any PrP^{Sc} already present in the cell (*open triangles* in *red circles*). This cell culture model can be used to study de novo PrP^{Sc} formation and trafficking, to determine how persistent prion infection is established, and to identify compounds that inhibit prion infection. (**b**) Scrapie cell panel assay. In the standard scrapie cell assay (SSCA), cells highly susceptible to prion infection are exposed to different dilutions of a prion-infected tissue homogenate and passaged three times to allow amplification of infectivity (top of panel). The number of PrP^{Sc}-positive cells (*open triangles* in *red circles*, top of panel) is counted using an ELISPOT assay in order to determine the amount of prion infectivity present. The scrapie cell endpoint assay (or SCEPA) uses a similar approach except that each row of a 96-well plate (shown at bottom of panel) is exposed to different dilutions of tissue homogenate. The ELISPOT assay is used to determine the proportion of PrP^{Sc}-positive wells (*open triangle* in *red circle*) to negative wells (*red circles*) for each dilution. These values are then used to calculate the endpoint titer of prion infectivity in the homogenate. SCEPA is therefore analogous to an in vivo endpoint bioassay. The scrapie cell assays can be used to titer prions from multiple different species, to study prion strains, and for prion inactivation and diagnostic studies

matrix may be involved [37]. Specific identification of the genes involved in the susceptibility of a cell to prion infection may lead not only the development of anti-prion drugs but may also enable the development of reliable cell culture models of human prion infection, the lack of which has hampered investigations into the mechanisms underlying the different forms of human prion disease.

4 Prion Species Barriers, Strains, and Structure

Species barriers in prion diseases are defined by extremely long incubation times when prions from one species infect another species. In vivo, studies on prion species barriers take years to complete and often require the generation of transgenic mice expressing either different species of PrP^C or chimeric PrP^C molecules composed of PrP^C sequences from different species. By contrast, cell-based studies are much less time-consuming and expensive. They involve expressing chimeric PrP^C molecules tagged with unique antibody epitopes in prion-infected cells (as in Fig. 2a) and analyzing whether or not those molecules are converted into PrP^{Sc}. Studies such as these, although limited in scope to those prion strains which can infect cells, have identified the regions of PrP^C involved in the mouse-hamster prion species barrier down to the single amino acid level [38, 39] and have correlated well with in vivo transgenic mouse data [40, 41]. The recent discovery that rabbit kidney epithelial cells, which don't express PrP^C, can be infected with mouse, sheep, vole, and elk prions if they express the corresponding PrP^C sequence, provides very strong evidence that PrP^C sequence is a primary factor in controlling prion species barriers [42–44]. Moreover, cell-based studies have shown that, mechanistically, species barriers are mainly driven by heterologous PrP^C molecules from one species interfering with or even completely blocking the formation of PrP^{Sc} from the other species [45], a process also known as dominant-negative inhibition [46]. This significantly slows PrP^{Sc} formation and disease progression and likely accounts for the prolonged disease incubation times observed in vivo following first passage of prions from one species into another species.

The same approach used to study the mouse-hamster prion species barrier in cells has also been used to study the regions of PrP^C involved in the formation of PrP^{Sc}. In this instance deletion, insertion, or point mutations are introduced into epitope-tagged PrP^C molecules which are then expressed in prion-infected cells (as in Fig. 2a) and assayed for their ability to form epitope-tagged PrP^{Sc} and/or inhibit the formation of non-epitope-tagged PrP^{Sc}. These studies have identified the structural regions of PrP^C involved in prion formation [47, 48], the sites where PrP^C and PrP^{Sc} bind [48–50], and the influence of glycosylation [51–53] and the GPI anchor [36] on PrP^{Sc} formation.

Different prion strains, which are defined by their stable phenotypes when inoculated into a defined host, have been characterized and isolated from multiple animal species. Strain phenotypes are believed to be determined by conformational differences between PrP^{Sc} molecules. Although insufficient PrP^{Sc} is generated in cells infected with different prion strains for use in structural studies, prion-infected cells have been used to study how prion strain properties are maintained and evolve. For example, when mouse prion-infected cells were cured of infection by the original strain and then re-infected with a second strain of mouse prion, the phenotype of the second strain was faithfully maintained [54]. These studies clearly demonstrated that prion strain determinants were stable and host independent, suggesting that they were encoded by the prion itself. Another set of studies took advantage of the fact that some cells are susceptible to a limited number of mouse scrapie strains while other cells are not. In this instance, an assay called the scrapie cell panel assay (SCPA), which differentiates mouse prion strains based upon their cell tropism and susceptibility to different drugs [55, 56], was used to look at how prion strains mutate and evolve. The results suggested that each prion strain was composed of many variants or quasi-species of PrP^{Sc} which likely differed conformationally and could be selected for using different selective pressures [57]. Thus, cell-based studies were used to demonstrate that, similar to other infectious organisms, prions can mutate and evolve and they further provided a potential mechanism for that process.

5 Practical Assays

Prion-infected cells have been widely utilized as a screen to identify potential anti-prion drugs [58]. In most of these studies, cells persistently infected with prions are exposed to potential inhibitors of PrP^{Sc} formation for various times and the level of PrP^{Sc} assayed. Multiple potential anti-prion compounds have been identified this way, many of which also showed efficacy when tested against prion-infected animals (e.g., [59, 60]). Assaying for inhibitors of de novo PrP^{Sc} formation and/or the early stages of prion infection can also be done by treating prion susceptible cells at the time of infection with candidate compounds and monitoring the production of PrP^{Sc} and establishment of prion infection. Although none of the many compounds identified have proved efficacious in the treatment of clinical cases of human prion disease [61], cell culture assays for potential prion therapeutics remain one of the primary tools for prion drug discovery.

For many years, the only way to accurately titer prion infectivity was via in vivo bioassay, a costly and time-consuming process. The ability to infect cells with prions led to the development of the first in vitro prion titer system, the standard scrapie cell assay (SSCA) [62]. In this assay, a mouse neuroblastoma cell clone highly susceptible to mouse prion infection is exposed to different dilutions of mouse prion-infected brain homogenate, passaged three times, and the number of prion-infected cells is analyzed using an ELISPOT assay (Fig. 2b). The assay is as sensitive as a bioassay and can be used to determine the amount of infectivity in an infected brain homogenate. It has even greater sensitivity when used for endpoint titration, an assay known as the scrapie cell endpoint assay (SCEA) [62]. In this assay, the titer is determined not by counting individual cells but by using the ELISPOT assay to determine the proportion of PrP^{Sc}-positive tissue culture wells associated with a given dilution of brain homogenate (Fig. 2b, bottom panel). Although the SSCA is limited to a handful of mouse strains, the discovery that rabbit kidney epithelial cells were susceptible to multiple species of prion allowed the development of titration assays for sheep, elk, mouse, and vole prions [42–44, 63]. Organotypic slice cultures have also been developed as a means to titer multiple species of prions [3]. These cell culture-based prion titration systems represent a significant advance in prion research and have allowed for the quantification of prion infectivity using an approach that is significantly faster and far less expensive than in vivo bioassay.

Although the primary focus of the development of diagnostic tests for prions has been on cell-free systems which utilize sonication or agitation to amplify low levels of prions (e.g., [64, 65], prion-infected cell cultures have also been utilized as potential diagnostic tests for prion disease. A recent study compared cell culture-based prion detection systems to in vivo bioassay and the in vitro prion amplification technique protein misfolding cyclic amplification (PMCA). The study showed that assays using prion-susceptible organotypic slice cultures or rabbit kidney epithelial cells were in some aspects as sensitive as the in vivo bioassay and were able to detect prion infectivity in the blood of preclinical sheep infected with scrapie [66]. The SSCA has also been used to assay the removal of prions from blood using prion reduction filters [67] and other processes [68], demonstrating that prion cell culture assays can be useful in studying the robustness of prion removal techniques. Thus, even with the limited number of available prion susceptible cell lines and prion strains capable of infecting those cells, studies like these clearly show that cell-based assays can provide a means of diagnosing preclinical prion infection and assaying prion inactivation or removal protocols.

Acknowledgments

This work was supported by the Intramural Research Program of the National Institutes of Health, National Institute of Allergy and Infectious Disease. The author would like to thank Anita Mora for the graphical assistance.

References

1. Grassmann A, Wolf H, Hofmann J, Graham J, Vorberg I (2013) Cellular aspects of prion replication in vitro. Virus 5:374–405
2. Cronier S, Laude H, Peyrin JM (2004) Prions can infect primary cultured neurons and astrocytes and promote neuronal cell death. Proc Natl Acad Sci U S A 101:12271–12276
3. Falsig J, Julius C, Margalith I, Schwarz P, Heppner FL, Aguzzi A (2008) A versatile prion replication assay in organotypic brain slices. Nat Neurosci 11:109–117
4. Giri RK, Young R, Pitstick R, Dearmond SJ, Prusiner SB, Carlson GA (2006) Prion infection of mouse neurospheres. Proc Natl Acad Sci U S A 103:3875–3880
5. Butler DA, Scott MR, Bockman JM et al (1988) Scrapie-infected murine neuroblastoma cells produce protease-resistant prion proteins. J Virol 62:1558–1564
6. Race RE, Caughey B, Graham K, Ernst D, Chesebro B (1988) Analyses of frequency of infection, specific infectivity, and prion protein biosynthesis in scrapie-infected neuroblastoma cell clones. J Virol 62:2845–2849
7. Haraguchi T, Fisher S, Olofsson S et al (1989) Asparagine-linked glycosylation of the scrapie and cellular prion proteins. Arch Biochem Biophys 274:1–13
8. Hope J, Morton LJ, Farquhar CF, Multhaup G, Beyreuther K, Kimberlin RH (1986) The major polypeptide of scrapie-associated fibrils (SAF) has the same size, charge distribution and N-terminal protein sequence as predicted for the normal brain protein (PrP). EMBO J 5:2591–2597
9. Stahl N, Borchelt DR, Hsiao K, Prusiner SB (1987) Scrapie prion protein contains a phosphatidylinositol glycolipid. Cell 51:229–240
10. Borchelt DR, Rogers M, Stahl N, Telling G, Prusiner SB (1993) Release of the cellular prion protein from cultured cells after loss of its glycoinositol phospholipid anchor. Glycobiology 3:319–329
11. Rane NS, Yonkovich JL, Hegde RS (2004) Protection from cytosolic prion protein toxicity by modulation of protein translocation. EMBO J 23:4550–4559
12. Vincent B, Paitel E, Saftig P et al (2001) The disintegrins ADAM10 and TACE contribute to the constitutive and phorbol ester-regulated normal cleavage of the cellular prion protein. J Biol Chem 276:37743–37746
13. Caughey BG, Raymond J (1991) The scrapie-associated form of PrP is made from a cell surface precursor that is both protease- and phospholipase-sensitive. J Biol Chem 266:18217–18223
14. Sunyach C, Jen A, Deng J et al (2003) The mechanism of internalization of glycosylphosphatidylinositol-anchored prion protein. EMBO J 22:3591–3601
15. Shyng SL, Huber MT, Harris DA (1993) A prion protein cycles between the cell surface and an endocytic compartment in cultured neuroblastoma cells. J Biol Chem 268:15922–15928
16. Magalhaes AC, Silva JA, Lee KS et al (2002) Endocytic intermediates involved with the intracellular trafficking of a fluorescent cellular prion protein. J Biol Chem. 277:33311–33318
17. Caughey B, Raymond GJ, Ernst D, Race RE (1991) N-terminal truncation of the scrapie-associated form of PrP by lysosomal protease (s): implications regarding the site of conversion of PrP to the protease-resistant state. J Virol 65:6597–6603
18. Marijanovic Z, Caputo A, Campana V, Zurzolo C (2009) Identification of an intracellular site of prion conversion. PLoS Pathog 5:e1000426
19. Yamasaki T, Baron GS, Suzuki A, Hasebe R, Horiuchi M (2014) Characterization of intracellular dynamics of inoculated PrP-res and newly generated PrP(Sc) during early stage prion infection in Neuro2a cells. Virology 450-451:324–335
20. Dron M, Moudjou M, Chapuis J et al (2010) Endogenous proteolytic cleavage of disease-associated prion protein to produce C2 fragments is strongly cell- and tissue-dependent. J Biol Chem 285:10252–10264
21. Borchelt DR, Scott M, Taraboulos A, Stahl N, Prusiner SB (1990) Scrapie and cellular prion proteins differ in their kinetics of synthesis and topology in cultured cells. J Cell Biol 110:743–752
22. Lehmann S, Harris DA (1996) Two mutant prion proteins expressed in cultured cells acquire biochemical properties reminiscent of the scrapie isoform. Proc Natl Acad Sci U S A 93:5610–5614
23. Priola SA, Chesebro B (1998) Abnormal properties of prion protein with insertional mutations in different cell types. J Biol Chem 273:11980–11985
24. Lehmann S, Harris DA (1996) Mutant and infectious prion proteins display common biochemical properties in cultured cells. J Biol Chem 271:1633–1637
25. Petersen RB, Parchi P, Richardson SL, Urig CB, Gambetti P (1996) Effect of the D178N mutation and the codon 129 polymorphism on the metabolism of the prion protein. J Biol Chem 271:12661–12668

26. Hsiao KK, Scott M, Foster D, Groth DF, Dearmond SJ, Prusiner SB (1990) Spontaneous neurodegeneration in transgenic mice with mutant prion protein. Science 250:1587–1590
27. Chiesa R, Piccardo P, Quaglio E et al (2003) Molecular distinction between pathogenic and infectious properties of the prion protein. J Virol 77:7611–7622
28. Jackson WS, Borkowski AW, Faas H et al (2009) Spontaneous generation of prion infectivity in fatal familial insomnia knockin mice. Neuron 63:438–450
29. Magalhaes AC, Baron GS, Lee KS et al (2005) Uptake and neuritic transport of scrapie prion protein coincident with infection of neuronal cells. J Neurosci 25:5207–5216
30. Paquet S, Daude N, Courageot MP, Chapuis J, Laude H, Vilette D (2007) PrPc does not mediate internalization of PrPSc but is required at an early stage for de novo prion infection of Rov cells. J Virol 81:10786–10791
31. Greil CS, Vorberg IM, Ward AE, Meade-White KD, Harris DA, Priola SA (2008) Acute cellular uptake of abnormal prion protein is cell type and scrapie-strain independent. Virology 379:284–293
32. Wadia JS, Schaller M, Williamson RA, Dowdy SF (2008) Pathologic prion protein infects cells by lipid-raft dependent macropinocytosis. PLoS One 3:e3314
33. Vorberg I, Raines A, Priola SA (2004) Acute formation of protease-resistant prion protein does not always lead to persistent scrapie infection in vitro. J Biol Chem 279:29218–29225
34. Goold R, Rabbanian S, Sutton L et al (2011) Rapid cell-surface prion protein conversion revealed using a novel cell system. Nat Commun 2:281
35. Goold R, McKinnon C, Rabbanian S, Collinge J, Schiavo G, Tabrizi SJ (2013) Alternative fates of newly formed PrPSc upon prion conversion on the plasma membrane. J Cell Sci 126:3552–3562
36. McNally KL, Ward AE, Priola SA (2009) Cells expressing anchorless prion protein are resistant to scrapie infection. J Virol 83:4469–4475
37. Marbiah MM, Harvey A, West BT et al (2014) Identification of a gene regulatory network associated with prion replication. EMBO J 33:1527–1547
38. Scott MR, Kohler R, Foster D, Prusiner SB (1992) Chimeric prion protein expression in cultured cells and transgenic mice. Protein Sci 1:986–997
39. Priola SA, Chesebro B (1995) A single hamster PrP amino acid blocks conversion to protease-resistant PrP in scrapie-infected mouse neuroblastoma cells. J Virol 69:7754–7758
40. Scott M, Foster D, Mirenda C et al (1989) Transgenic mice expressing hamster prion protein produce species-specific scrapie infectivity and amyloid plaques. Cell 59:847–857
41. Scott M, Groth D, Foster D et al (1993) Propagation of prions with artificial properties in transgenic mice expressing chimeric PrP genes. Cell 73:979–988
42. Vilette D, Andreoletti O, Archer F et al (2001) Ex vivo propagation of infectious sheep scrapie agent in heterologous epithelial cells expressing ovine prion protein. Proc Natl Acad Sci U S A 98:4055–4059
43. Courageot MP, Daude N, Nonno R et al (2008) A cell line infectible by prion strains from different species. J Gen Virol 89:341–347
44. Bian J, Napier D, Khaychuck V, Angers R, Graham C, Telling G (2010) Cell-based quantification of chronic wasting disease prions. J Virol 84:8322–8326
45. Priola SA, Caughey B, Race RE, Chesebro B (1994) Heterologous PrP molecules interfere with accumulation of protease-resistant PrP in scrapie-infected murine neuroblastoma cells. J Virol 68:4873–4878
46. Zulianello L, Kaneko K, Scott M et al (2000) Dominant-negative inhibition of prion formation diminished by deletion mutagenesis of the prion protein. J Virol 74:4351–4360
47. Vorberg I, Chan K, Priola SA (2001) Deletion of beta-strand and alpha-helix secondary structure in normal prion protein inhibits formation of its protease-resistant isoform. J Virol 75:10024–10032
48. Taguchi Y, Mistica AM, Kitamoto T, Schatzl HM (2013) Critical significance of the region between helix 1 and 2 for efficient dominant-negative inhibition by conversion-incompetent prion protein. PLoS Pathog 9:e1003466
49. Williamson RA, Peretz D, Pinilla C et al (1998) Mapping the prion protein using recombinant antibodies. J Virol 72:9413–9418
50. Abalos GC, Cruite JT, Bellon A et al (2008) Identifying key components of the PrPC-PrPSc replicative interface. J Biol Chem 283:34021–34028
51. Korth C, Kaneko K, Prusiner SB (2000) Expression of unglycosylated mutated prion protein facilitates PrP(Sc) formation in neuroblastoma cells infected with different prion strains. J Gen Virol 81:2555–2563
52. Neuendorf E, Weber A, Saalmueller A et al (2004) Glycosylation deficiency at either one of the two glycan attachment sites of cellular prion protein preserves susceptibility to bovine spongiform encephalopathy and scrapie infections. J Biol Chem 279:53306–53316

53. Salamat MK, Dron M, Chapuis J, Langevin C, Laude H (2011) Prion propagation in cells expressing PrP glycosylation mutants. J Virol 85:3077–3085
54. Birkett CR, Hennion RM, Bembridge DA et al (2001) Scrapie strains maintain biological phenotypes on propagation in a cell line in culture. EMBO J 20:3351–3358
55. Mahal SP, Baker CA, Demczyk CA, Smith EW, Julius C, Weissmann C (2007) Prion strain discrimination in cell culture: the cell panel assay. Proc Natl Acad Sci U S A 104:20908–20913
56. Browning S, Baker CA, Smith E et al (2011) Abrogation of complex glycosylation by swainsonine results in strain- and cell-specific inhibition of prion replication. J Biol Chem 286:40962–40973
57. Li J, Browning S, Mahal SP, Oelschlegel AM, Weissmann C (2010) Darwinian evolution of prions in cell culture. Science 327:869–872
58. Trevitt CR, Collinge J (2006) A systematic review of prion therapeutics in experimental models. Brain 129:2241–2265
59. Caughey WS, Raymond LD, Horiuchi M, Caughey B (1998) Inhibition of protease-resistant prion protein formation by porphyrins and phthalocyanines. Proc Natl Acad Sci U S A 95:12117–12122
60. Priola SA, Raines A, Caughey WS (2000) Porphyrin and phthalocyanine antiscrapie compounds. Science 287:1503–1506
61. Forloni G, Artuso V, Roiter I, Morbin M, Tagliavini F (2013) Therapy in prion diseases. Curr Top Med Chem 13:2465–2476
62. Klohn PC, Stoltze L, Flechsig E, Enari M, Weissmann C (2003) A quantitative, highly sensitive cell-based infectivity assay for mouse scrapie prions. Proc Natl Acad Sci U S A 100:11666–11671
63. Arellano-Anaya ZE, Savistchenko J, Mathey J et al (2011) A simple, versatile and sensitive cell-based assay for prions from various species. PLoS One 6:e20563
64. Tattum MH, Jones S, Pal S, Collinge J, Jackson GS (2010) Discrimination between prion-infected and normal blood samples by protein misfolding cyclic amplification. Transfusion 50:996–1002
65. Orru CD, Bongianni M, Tonoli G et al (2014) A test for Creutzfeldt-Jakob disease using nasal brushings. N Engl J Med 371:519–529
66. Halliez S, Jaumain E, Huor A et al (2014) White blood cell-based detection of asymptomatic scrapie infection by ex vivo assays. PLoS One 9:e104287
67. Sowemimo-Coker SO, Demczyk CA, Andrade F, Baker CA (2010) Evaluation of removal of prion infectivity from red blood cells with prion reduction filters using a new rapid and highly sensitive cell culture-based infectivity assay. Transfusion 50:980–988
68. You B, Aubin JT, Le-Hir G, Arzel A, Laude H, Flan B (2010) In vitro infectivity assay for prion titration for application to the evaluation of the prion removal capacity of biological products manufacturing processes. J Virol Methods 164:1–6

Chapter 8

Expression of Heterologous PrP and Prion Propagation in RK13 Cells

Zaira E. Arellano-Anaya, Alvina Huor, Pascal Leblanc, Olivier Andréoletti, and Didier Vilette

Abstract

Cultured cells are valuable models to study prion infections at the cellular level. Unfortunately, the vast majority of cell lines are resistant to the propagation of prion agents. The rabbit epithelial RK13 cell line is among the few cell lines permissive to prion infection. When genetically engineered to express heterologous PrP proteins, RK13 cells become permissive to several strains of prions from various animal species. Here, we describe the generation of stable RK13 cell clones expressing a heterologous PrP protein in an inducible manner, the establishment and maintenance of chronically infected cultures, and the selection of cell clones suitable for cell-based titration of prions.

Key words Cell models, Prion, Prion titration, PrP protein, Inducible expression

1 Introduction

Cellular models of prion propagation have greatly improved our understanding of the molecular and cellular events controlling the conversion of the normal PrP protein into the abnormally folded PrP^{sc}, and the cell-to-cell spreading of prions [1, 2] and primary neurons [3] may be crucial to identify how PrP^{sc} toxic species damage neurons. In addition, robust and sensitive cell models have been demonstrated to be potent alternative to animals to titrate prions [4–6]. For unknown reasons, most of PrP-expressing cells are resistant to prion infection, and thus, permissiveness of cultured cells must be determined empirically. The most popular cellular model is the mouse neuroblastoma N2a cell line that proved to be permissive to a variety of different mouse-adapted prion strains [2]. However, one limitation of this model is that N2a cells express endogenous mouse PrP and are only permissive to murine prion strains.

Victoria A. Lawson (ed.), *Prions: Methods and Protocols*, Methods in Molecular Biology, vol. 1658,
DOI 10.1007/978-1-4939-7244-9_8,

Epithelial RK13 cells do not express detectable endogenous rabbit PrP. Importantly, when genetically engineered to express PrP from various species, including ovine [7, 8], mouse [9, 10], bank vole [9], and cervids [4], they become permissive to at least a subset of prion strains from the corresponding species. Even if we and others failed to propagate human and bovine prion strains in RK13 cells expressing, respectively, human and bovine PrP, RK13 became a very good model (1) to propagate prions from different animal species in a single cellular background, (2) for reverse genetic approaches to assess the importance of PrP amino-acid substitution on prion multiplication, and (3) to isolate cell clones for sensitive and robust prion titration. In addition, the use of inducible expression vectors to drive PrP expression in RK13 cells allows studying PrP biological properties [11, 12].

In the present chapter, we will describe the cloning of the PrP coding sequence in suitable expression vectors and their functional assessment in RK13 cells, the transfection of RK13 cells with selection and characterization of stably transfected individual cell clones, the use of RK13 cells as cell-based assay to titrate prion preparations, and the establishment and maintenance of chronically infected cultures.

2 Materials

2.1 Reagents for Cell Culture

1. OptiMEM-Glutamax culture medium (Invitrogen).
2. PBS without calcium and magnesium.
3. Trypsin 0.25% EDTA.
4. Fetal calf serum (FCS) from Eurobio.
5. Penicillin and streptomycin solution.
6. Freezing medium: 10% DMSO in FCS.
7. 12-well cell culture plates.
8. 100 mm^2 cell culture plates.
9. 25 cm^2 cell culture plates.
10. 6-well cell culture plates.

2.2 Reagents for Molecular Biology and Biochemistry

1. Plasmid Midi Kit (Qiagen).
2. Lipofectamine 2000 reagent (Invitrogen).
3. Doxycycline hyclate (Sigma Aldrich), 5 mg/mL in ethanol, stored at −20 °C.
4. Puromycin dihydrochloride, 10 mg/mL in H_2O, filter sterilized, stored at −20 °C.
5. Lysis buffer: 50 mM Tris–HCl, pH 7.4, 0.5% Triton X-100, 0.5% sodium deoxycholate. Stored at 4 °C.
6. BCA Protein Assay kit (Pierce Biotechnology).

7. High-speed homogenizer (TeSeE Precess 48 system) with grinding tubes (Bio-Rad).
8. Proteinase K, recombinant, 20 mg/mL.
9. Pefabloc SC (Sigma Aldrich), 1 M in H_2O, stored at −20 °C.
10. Monoclonal antibody against PrP: Sha31 (Bertin Pharma).

3 Methods

3.1 Construction of PrP-Expressing Vectors

We routinely use the tetracycline Tet-On expressing system from Clontech laboratories. It includes the pTRE plasmid in which the PrP open reading frame (ORF) will be subcloned (*see* **Note 1**) and the pTet-On plasmid, which encodes for the rtTA regulator protein required for tetracycline-induced expression from pTRE. Typically, we also use a third plasmid conferring puromycin resistance to transfected cells (puro^R plasmid, commercially available from various suppliers including InvivoGen, Addgene, Clontech):

1. Subclone the PrP protein ORF into an appropriate cloning site of pTRE plasmid by standard molecular techniques. Identify the correct recombinant plasmid by appropriate restriction analysis of plasmid minipreps.
2. Amplify and purify the pTRE-PrP plasmid (*see* **Note 2**).
3. If the PrP ORF has been PCR-amplified before subcloning, check by standard DNA sequencing that no mutation has been introduced by the DNA polymerase (*see* **Note 3**).
4. To ensure that the pTRE-PrP plasmid is functional in RK13 cells, proceed to a transient transfection. Plate 0.25×10^6 RK13 cells per individual wells of a 12-well plate (*see* **Note 4**).
5. When the cells are confluent (1 or 2 days later), proceed to transfection with the lipofectamine reagent according to the manufacturer's recommendations. We typically use 1 μg of each pTRE-PrP and pTet-On plasmids per well. One day later, add 1 μg/mL of doxycycline (dox) into the culture medium to induce the expression of PrP in transfected cells.
6. Two days later, solubilize the transfected cells and the untransfected controls with lysis buffer (*see* **Note 5**) and proceed to the detection of PrP by standard immunoblotting to ensure functionality of the pTRE-PrP plasmid.

3.2 Generation, Selection, and Characterization of Stably Transfected Cell Clones

1. Plate RK13 cells into 12-well plates. When confluent, transfect cells from at least 2 wells with 1 μg of each pTRE-PrP and pTet-On plasmids and 0.1 μg of puro^R plasmid and cells from 2 other wells in which the puro^R plasmid is omitted. Six hours later, rinse the cells once with medium without FCS and antibiotics and then incubate overnight the cells in regular medium (*see* **Note 4**).

2. Trypsinize and transfer the cells from each well into two 100 mm plates in regular medium containing 2 μg/mL of puromycin to select stably transfected cells (*see* **Note 6**). As massive cell death will occur, change the medium regularly, not omitting to add puromycin.
3. Cell clones will typically be visible from 10 to 14 days after seeding (*see* **Note 7**).
4. Pick up and transfer individual cell clones to individual wells of 12-well plates and grow them in puromycin-containing medium (*see* **Note 8**). Grow the cells until confluence.
5. Trypsinize all cell clones and transfer each clone into three wells of three separate 12-well plates in puromycin-containing medium. Two sets of plates (plate A and plate B) will be analyzed for dox-inducible PrP expression, while the third one (plate C) will be kept as the master plates. Add 1 μg/mL of dox in all wells of plate A, and leave B and C plates untreated.
6. When the cell clones are confluent, solubilize individual cell clones of plates A and B, quantify cellular proteins with BCA, and analyze each cell clone for dox-inducible expression of PrP by immunoblotting (*see* **Note 5**).
7. Based on the results (*see* **Note 9** and Fig. 1), amplify the appropriate cell clones from the master plates C in puromycin-containing medium. We typically seed cells from a single well into a 25 cm^2 flask, and we grow them until confluence.

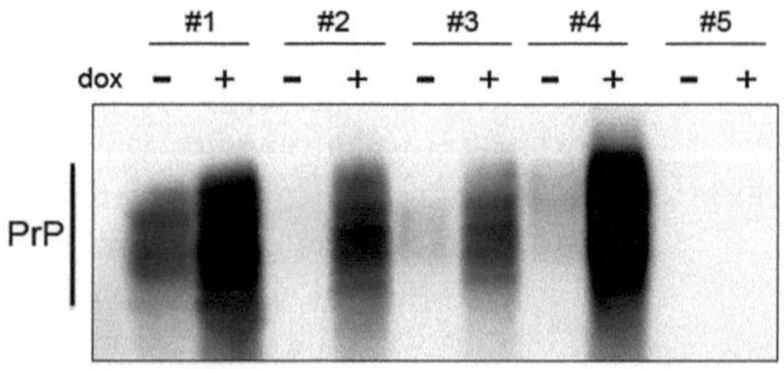

Fig. 1 Immunoblotting analysis of PrP in stably transfected RK13 cell clones. Individual stably transfected cell clones (#1 to #5) were selected in puromycin-containing medium. Each cell clone was then grown in the presence (+) or the absence (-) of doxycycline to induce PrP expression. Identical amounts of cell lysates (10 μg of proteins) were analyzed by immunoblotting for dox-inducible expression of PrP (mAb: sha31). Clone #5 does not show detectable PrP, while PrP is expressed at various levels in clones #1, #2, #3, and #4. PrP is detected in untreated (dox) clone #1, indicating that dox-dependent expression of PrP is less tightly regulated in this particular clone

8. Split each clone from the single flask into two new flasks at ¼ dilution (i.e., discard half of the cells) in puromycin-containing medium.
9. One flask is used to freeze the cells (*see* **Note 10**). Split the other flask and seed six wells of a 6-well plate for testing permissiveness of each clone to prion infection. At this stage, adding puromycin is no longer required. Add dox into 3 of the 6 wells to induce PrP expression.

3.3 Identification of the most Sensitive Clones by Cell-Based Assay of Prion Infection

1. For each 6-well plate (i.e., for each clone to be tested), prepare the infectious inocula when the cells are confluent. Add culture medium with dox into 3 × 15 mL tubes (3 mL per tube). To each tube add 30 μL, 3 μL, or 0.3 μL of 10% wt/vol infectious brain homogenate (*see* **Note 11**). The inoculum is thus diluted 100-, 1000-, and 10,000-fold, respectively. Aspirate the medium of the three wells in which cells were grown with dox and add the three diluted inocula to the cells. Repeat the same procedure for the three wells of non-dox-treated cells, being aware not to add dox in the diluted inocula. Then, proceed to the inoculation of the other cell clones.
2. One week later, aspirate all the infectious inocula and refresh the cells with new culture medium containing or not dox, as appropriate.
3. One week later, change the culture medium as before (*see* **Note 12**).
4. One week later (i.e., 3 weeks postinoculation of the cells), wash the cells from each well with 3 mL of cold PBS and solubilize with 1 mL of cold lysis buffer for 10 min. Collect the post-nuclear supernatants and quantify cellular proteins as indicated in **Note 5**. Typical yields are 1–1.5 mg of cellular proteins per well. Dilute all the lysates to 1 mg/mL with lysis buffer.
5. To isolate PrP^{res} (the PK-resistant form of abnormal PrP, as a marker of successful prion infection), add 250 μL of each lysate in a microtube and add 1 μg of proteinase K (PK). Incubate 2 h at 37 °C. Stop the reaction by adding Pefabloc (2 mM final). Orientate the tubes in the rotor (to identify the pellet locations) and centrifuge the samples for 30 min at 20,000 × g at room temperature. Carefully aspirate the supernatant (*see* **Note 13**). Add 20 μL of Laemmli buffer to the barely visible pellets. Do not try to resuspend as the pellet may stick to the tip. Heat at 100 °C for 10 min, vortex, briefly spin down, and freeze at −20 °C until immunoblotting analysis.
6. Immunoblot the samples to detect PrP^{res}. From the six samples per cell clone, three are from cells inoculated with increasing amounts of inocula (10^{-4}, 10^{-3}, 10^{-2}) in the presence of dox, and they should show increasing amount of PrP^{res}. The three

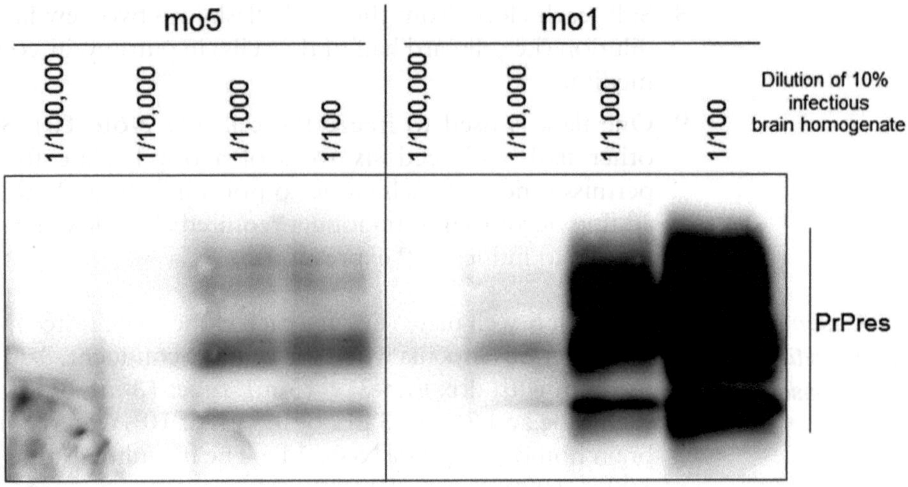

Fig. 2 Comparison of cell clone sensitivity to prion infection. Two different cell clones (mo1 and mo5) of RK13 expressing the mouse PrP were infected with the indicated dilutions of a 10% infectious brain homogenate (22L strain). Three weeks postinoculation, PrP^{res} was isolated and analyzed by immunoblotting (Mab: sha31). The mo1 clone is more sensitive than mo5 as the 10,000-fold diluted inoculum led to detectable PrP^{res}

others are from dox-untreated cells inoculated with the same dilutions of inocula. They should contain much lower levels of PrP^{res}, if any (*see* **Note 14**).

7. From the results, identify the most sensitive cell clones (i.e., those that accumulate the higher levels of PrP^{res} upon inoculation with the highest dilution of inoculum) (*see* **Note 15** and Fig. 2). Thaw the corresponding vials from Subheading 3.2, **step 9** and amplify the cultures to prepare a large batch of cells for liquid nitrogen freezing.

3.4 Establishment of Chronically Infected Cultures

1. Split the cells and seed an aliquot into several wells of a 12-well plate. Add dox in the culture medium and grow the cells until confluent.
2. Remove the medium from 1 well and add 1 mL of dox-containing medium supplemented with 10 μL of 10% wt/vol prion-infected brain homogenate. Remove the medium from another well and add 1 mL of dox-containing medium supplemented with 10 μL of 10% uninfected brain homogenate (mock-infected). Incubate for 3–5 days at 37 °C.
3. Trypsinize the infected and mock-infected cells, transfer the cells from the whole well into a 25 cm^2 flask, and grow them in dox-containing medium for 1 week. This is the first passage postinfection (pi).
4. Trypsinize each culture and transfer ¼ of the cells into two news flasks, and grow them into dox-containing medium. This will be the second passage pi.

5. One week later, one of the two flasks will be passaged as before. The cells from the other flask will be solubilized with 1.5 mL of lysis buffer and the postnuclear supernatant (*see* **Note 5**) stored at −20 °C for further analysis.
6. Subcultivate the prion-infected and mock-infected cultures for three to four passages and prepare a cellular lysate at each passage.
7. Check for successful infection by analyzing PrP^{res} in cellular lysates from prion-infected and mock-infected cultures at passages 2, 3, and 4 pi by immunoblotting (*see* Subheading 3.3, **step 5**). If stably infected, cultures should show increasing or stable amounts of PrP^{res} over the three analyzed passages. No PrP^{res} must be detected in the mock-infected culture lysates.
8. Further subcultivate the prion-infected and mock-infected cell clones for freezing (*see* **Note 16**).

4 Notes

1. Depending on the prion strain or on the natural prion isolate under study, select the appropriate animal species from which PrP will be cloned.
2. We usually use QIAGEN kits for preparing plasmid DNA used for cell transfections.
3. This step is important since we occasionally have found point mutations in constructs obtained with DNA amplified with high-fidelity DNA polymerases.
4. Parental RK13 cells and derived cell clones are cultured in Opti-MEM supplemented with 10% FCS, penicillin, streptomycin at 37 °C, and 5% CO_2. Cultures are trypsinized and subcultivated once a week at a ¼ dilution.
5. Wash the cells with cold PBS, add 500 μL of lysis buffer for 10 min, collect the lysate into a microtube, and centrifuge at room temperature at 425 × *g* for 2 min. Collect the postnuclear supernatant and quantify the proteins by bicinchoninic acid (BCA). One should typically get about 200 μg of cellular proteins. Immunoblotting analysis of 5–10 μg of proteins allows the detection of dox-inducible PrP in transfected cells.
6. Death of cells that have not integrated the $puro^R$ plasmid is visible as soon as 2 days following addition of puromycin.
7. The number of cell clones depends on transfection efficiency that may somewhat vary from well to well. However, a few hundreds of cell clones are typically expected in each 100 mm plate. Cell clones of densely packed cells are heterogeneous in size, from about a few ten or so to a few hundred cells. No cell

clones must be present in untransfected RK13 controls or in RK13 transfected without the puro^R plasmid, even if a few isolated viable cells may still be detected at this stage.

8. Cell clones are visualized under the microscope, and their location can be conveniently indicated with a pencil dot on the outside of the plate. Carefully rinse the plate with PBS and remove PBS as much as possible. A few μL of trypsin are then added onto each cell clone. After a few min, check under the microscope that cells of each cell clone round up. With a micropipettor, add 10 μL of medium on the cell clone, immediately aspirate the cell suspension, transfer to a well of a 12-well plate, and proceed similarly with the other cell clones. Alternatively, cloning cylinder may be used. We typically pick up 20–30 different cell clones.
9. A minority of clones will show no detectable PrP upon dox induction and will be discarded. In most of the dox-induced cell clones, variable levels PrP should be detected. From these, discard those in which PrP expression is not tightly regulated by dox, i.e., those in which PrP is also detected in the absence of dox stimulation (typical results are shown in the Fig. 1). From the remaining, select six to eight clones expressing different amounts of PrP that grow well and that look healthy under the microscope.
10. Trypsinize the cells, dilute the cells with regular medium (*see* **Note 4**), centrifuge at 120 × *g* for 7 min, and resuspend the pelleted cells into 2 mL of freezing medium. Transfer into two cryovials (1 mL *per* vial) and place at 4 °C for 1 h or so. Transfer the vials into a storage polystyrene box filled with paper and put the box at −80 °C for at least 24 h. Transfer the vials to liquid nitrogen for long-term storage.
11. Use brain tissue from an animal at the terminal stage of prion disease and weigh it in a sterile microtube. Cut it in small pieces with a sterile razor blade. Remove the solution from the high-speed homogenizer grinding tubes and rinse once with sterile PBS. Transfer brain pieces into the grinding tubes (about 200 mg of tissue per tube), add 1 mL of sterile PBS per tube, and homogenize with the high-speed homogenizer (two cycles of 45 s, wait 10 min for cooling of the samples and proceed to two more cycles of 45 s). Collect and pool the homogenates in a single tube, adjust to 10% wt/vol with PBS, aliquot, and store at −80 °C.
12. RK13, as many other epithelial cell lines, can be kept for weeks as confluent monolayers of polarized cells, provided the cell culture medium is renewed every week. However, during a 7-day period of time, some RK13 cells are extruded from the monolayer and will remain floating in the medium until the next medium renewal. Very occasionally, more floating cells

can be observed. In this case we change the medium more frequently, and we never observed that this affects the outcome of the cell-based assay.

13. The small pellet containing aggregated, PK-resistant proteins may not be visible and is not firmly attached to the plastic. Aspirate most of the supernatant, leaving about 50 μL. Centrifuge again for 5 min with the same orientation. Carefully aspirate the remaining supernatant by placing the tip opposite to the pellet. Including 0.025% bromophenol blue in the digestion reaction may help the visualization of the small, light blue pellet.
14. There are two possible explanations for the detection of low levels of PrP^{res} in samples from non-dox-treated, inoculated cells. It could be residual PrP^{res} from the inoculum used for cell inoculation. Alternatively, or additionally, if expression of PrP is not repressed enough in the absence of dox (or if the batch of FCS contains traces of tetracycline), a few cells from a particular cell clone may become infected leading to detectable PrP^{res}. An easy way to settle the two possibilities is to quantify residual, inoculum-derived PrP^{res} by including the parental (i.e., untransfected) RK13 cells in each cell-based assay.
15. It is not the clones expressing the higher levels of PrP that will necessarily accumulate the higher PrP^{res}. We have routinely identified low PrP expressor cell clones producing high amount of PrP^{res} upon infection.
16. Prion replication in RK13-derived cell clones is usually very stable. For instance, we have propagated ovine and murine prions in RK13 cultures expressing the ovine and mouse PrP, respectively, for more than 6 months with no loss of infectivity and PrP^{res}. However, stability of prion propagation over the passages must be checked for any newly generated and uncharacterized cell clone.

Acknowledgments

This work was partly supported by a grant from ANR-08-MIEN-010 (France) and Z.E. Arellano-Anaya was a recipient of a fellowship from Mexican CONACYT.

References

1. Krauss S, Vorberg I (2013) Prions ex vivo: what cell culture models tell us about infectious proteins. Int J Cell Biol 2013:704546
2. Vilette D (2008) Cell models of prion infection. Vet Res 39:10
3. Cronier S, Laude H, Peyrin JM (2004) Prions can infect primary cultured neurons and astrocytes and promote neuronal cell death. Proc Natl Acad Sci U S A 101:12271–12276

4. Bian J, Napier D, Khaychuck V, Angers R, Graham C et al (2010) Cell-based quantification of chronic wasting disease prions. J Virol 84:8322–8326
5. Arellano-Anaya ZE, Savistchenko J, Mathey J, Huor A, Lacroux C et al (2011) A simple, versatile and sensitive cell-based assay for prions from various species. PLoS One 6: e20563
6. Klohn PC, Stoltze L, Flechsig E, Enari M, Weissmann C (2003) A quantitative, highly sensitive cell-based infectivity assay for mouse scrapie prions. Proc Natl Acad Sci U S A 100:11666–11671
7. Neale MH, Mountjoy SJ, Edwards JC, Vilette D, Laude H et al (2010) Infection of cell lines with experimental and natural ovine scrapie agents. J Virol 84:2444–2452
8. Vilette D, Andreoletti O, Archer F, Madelaine MF, Vilotte JL et al (2001) Ex vivo propagation of infectious sheep scrapie agent in heterologous epithelial cells expressing ovine prion protein. Proc Natl Acad Sci U S A 98:4055–4059
9. Courageot MP, Daude N, Nonno R, Paquet S, Di Bari MA et al (2008) A cell line infectible by prion strains from different species. J Gen Virol 89:341–347
10. Lawson VA, Vella LJ, Stewart JD, Sharples RA, Klemm H et al (2008) Mouse-adapted sporadic human Creutzfeldt-Jakob disease prions propagate in cell culture. Int J Biochem Cell Biol 40:2793–2801
11. Paitel E, Alves da Costa C, Vilette D, Grassi J, Checler F (2002) Overexpression of PrPc triggers caspase 3 activation: potentiation by proteasome inhibitors and blockade by anti-PrP antibodies. J Neurochem 83:1208–1214
12. Rachidi W, Vilette D, Guiraud P, Arlotto M, Riondel J et al (2003) Expression of prion protein increases cellular copper binding and antioxidant enzyme activities but not copper delivery. J Biol Chem 278:9064–9072

Chapter 9

Generation of Infectious Prions and Detection with the Prion-Infected Cell Assay

Laura J. Vella, Bradley Coleman, and Andrew F. Hill

Abstract

Cell lines propagating prions are an efficient and useful means for studying the cellular and molecular mechanisms implicated in prion disease. Utilization of cell-based models has led to the finding that PrP^C and PrP^{Sc} are released from cells in association with extracellular vesicles known as exosomes. Exosomes have been shown to act as vehicles for infectivity, transferring infectivity between cell lines and providing a mechanism for prion spread between tissues. Here, we describe the methods for generating a prion-propagating cell line with prion-infected brain homogenate, cell lysate, conditioned media, and exosomes and also detection of protease-resistant PrP with the prion-infected cell assay.

Key words Prion, PrP, RK13, GT1-7, Cell culture, Exosomes, Extracellular vesicles, Infection, Prion-infected cell assay, PICA

1 Introduction

According to the protein-only hypothesis, an abnormal isoform of the host-encoded prion protein (PrP^C), referred to as PrP^{Sc}, is the sole or major component of the infectious agent (the "prion") [1]. The mechanism of neurodegeneration, transmission, and even the normal biological function of PrP^C is not completely clear; however, considerable progress has been made in the last 20 years due in part to the establishment of cell culture models of prion propagation. The main advantages of in vitro models have been the ability to study PrP^C and PrP^{Sc} propagation at a cellular level and provide an alternative to lengthy bioassays in mice, enabling efficient drug screening and prion strain characterization.

The first prion-infected cell culture was reported in 1970 [2], and many groups since then have successfully propagated a variety of prion strains in cell culture [3–7], the most common being Rocky Mountain Laboratory (RML) strain in N2a cells. Other commonly used cell lines include N2a, GT1-7, PC12, Mov, RK13, 3T3, MG20, and CAD [5, 6, 8–16]. It is not known why

Victoria A. Lawson (ed.), *Prions: Methods and Protocols*, Methods in Molecular Biology, vol. 1658,
DOI 10.1007/978-1-4939-7244-9_9, © Springer Science+Business Media LLC 2017

these cells are susceptible to prion infection while many others are not [17–19], with many cell lines able to take up PrP^{Sc} but few maintaining the infection or capable of de novo prion propagation [20]. It was originally thought PrP expression played a large role in susceptibility; however, many high PrP-expressing cell lines are unable to support prion propagation.

Utilization of cell culture models led to the finding that RK13 cells overexpressing PrP from the appropriate species can propagate prions from that species [12]. Several murine strains (Fukuoka-1, RML, MU-02, RML, Chandler), a bank vole strain, sheep scrapie, and CWD strain have successfully propagated in genetically engineered RK13 cells [21–25].

Despite several cell culture models of prion propagation established for non-human strains, an in vitro system for studying human prion strains has been difficult to develop. By introducing human PrP into the generally permissive RK13 cells, we attempted to create a human prion-susceptible cell line (huRK13); however, huRK13 cells proved invariably resistant to infection by sporadic Creutzfeldt-Jakob disease (CJD) prions with the T2MM PrP^{Sc} pattern, the most common human prion subtype [26]. Transmission of the same sporadic CJD prions to wild-type mice generated a strain of mouse-adapted human prions, which efficiently propagated in RK13 cells expressing mouse PrP, demonstrating infection with human strains is more successful once adapted into mice.

As infection rates and infectivity titers in cell culture are usually low, cloning of infected cells is sometimes necessary to increase the percentage of infected cells within a cell population [10, 27]. The growth medium and culture conditions are important for not only establishing infection but also maintenance. With this in mind, if inexperienced in propagating prions in cell culture, it may be best to utilize cell lines previously recognized for supporting prion propagation.

Prion-propagating cell lines can transfer infectivity to recipient cells in the absence of cell-to-cell contact. Schatzl et al. [5] showed the transfer of infectivity from GT1-7 cell media to N2a cells, while the SN56 cell line initiated infection in recipient SN56 cells [28]. Both PrP^{C} and PrP^{Sc} have been identified in association with exosomes with exosomal PrP^{Sc} infectious in both animal and cell bioassays [29–34]. Exosomes are small membrane vesicles formed by invagination of the membrane of multi-vesicular bodies (MVBs) [35] and released into the extracellular environment upon fusion of the MVB with the plasma membrane. Exosomes originating from either non-neuronal or neuronal cells can transmit infection between cell lines validating the notion that exosomes may have a role in trafficking of infectious prions to different tissue types in vivo [30]. This chapter outlines a method for generating a prion-infected cell line with brain homogenate, cell lysate and exosomes. To quantitate and screen for prion infectivity, we developed a high-throughput, cell-based prion infectivity assay referred to as the "prion-infected cell assay (PICA)" which is outlined in this chapter. This assay is an adaptation of the standard scrapie cell assay (SSCA) [13, 36]. PICA allows the determination of differences in prion infectivity between various samples by performing titrations of the samples and then inoculating susceptible cells. It is relatively fast (compared to animal bioassay), high throughput, and low cost.

2 Materials

2.1 Reagents

1. Fetal calf serum (FCS) (we use Australian origin #10099-141, Life Technologies).
2. OCCM—Opti-MEM (Life Technologies) containing FCS (10% w/v), penicillin/streptomycin 1% (100 μg/mL), glutamine 1% (1 mM).
3. Phosphate-buffered saline (PBS)(Life Technologies).
4. Puromycin.
5. Lysis buffer: 0.5% sodium deoxycholate, 0.5% Triton X-100, 150 mM NaCl, 50 mM Tris–HCl, pH 7.5.
6. Methanol.
7. Denaturation buffer (3 M guanidine thiocyanate, 10 mM Tris pH 7.4).
8. Superblock (Thermo Scientific).
9. Proteinase K (PK); 10 mg/mL in water. Make 100 μL aliquots and store at −20 °C.
10. PK digestion buffer—proteinase K (0.6 μg/μL) in lysis buffer. Add 0.6 μL PK stock solution (10 mg/mL) to 10 mL of lysis buffer.
11. Phenylmethylsulfonyl fluoride (PMSF). Make up 50 mL stock solution (100 mM) in methanol. PMSF is sensitive to light so wrap tube in aluminum foil and store at −20 °C.
12. PMSF PICA solution—500 μL PMSF Stock solution in 25 mL PBS.
13. TBS-T: 150 mM NaCl, 20 mM Tris, pH 7.4, 0.1% Tween-20.
14. Anti-mouse IgG HRP (1:5000 in TBS-T) (GE Healthcare).
15. QuantaBlu fluorogenic peroxidase substrate (Pierce—Thermo Scientific). Nine parts of QuantaBlu substrate solution were mixed with one part of QuantaBlu stable peroxide solution and allowed to equilibrate to RT prior to use (a total of 11 mL is required per 96-well plate).
16. BCA Protein Quantitation Kit (Pierce).
17. Prion protein antibodies: ICSM18 (#0130-01810 D-Gen Ltd., UK), SAF32 (#189720 Cayman, USA).

2.2 Equipment

1. Biological safety cabinet; class II.
2. Tissue culture plasticware (Nalgene Nunc); six-well plates, T75cm2, and T300cm2 vented flasks.
3. CO_2 humidified incubator.
4. Benchtop refrigerated centrifuge.
5. Ultracentrifuge with fixed-angle and swinging-bucket rotors (Beckman).
6. Polyallomer tubes or polycarbonate bottles, appropriate for the ultracentrifuge rotor.
7. 0.22 μm filters (Millipore, Millex-GP).
8. Rotating heat block (e.g., thermomixer, Eppendorf, Germany).
9. Liquid nitrogen.
10. Water bath.
11. Syringes and needles.
12. 96-well tissue culture plate (Nunc).
13. 96-well MultiScreen HTS™ IP Elispot plates (Millipore).
14. 96-well BCA microplate (Greiner Bio-One).
15. MultiScreen® HTS vacuum manifold (Merck Millipore).
16. Varioskan multimode plate reader (for fluorescence excitation 325 nm and emission 420 nm).
17. Vacuum pump, double vacuum traps, hoses, and attachments.
18. Twelve-channel pipette.
19. Eight-channel pipette.
20. Eight-channel vacuum aspirator pipette.
21. Plate shaker.

3 Methods

3.1 Maintaining Cultured Cell Lines

For the purpose of this methods chapter, we will be describing maintenance and infection of the mouse hypothalamic cell line GT1-7.

GT1 cells are susceptible to a range of prion strains including RML [5], 22L, 139A [37], Fukuoka and mouse-adapted sporadic CJD (SY) [38], New Guinea kuru, Fukuoka-1 [30, 39], and sheep scrapie [40], making them advantageous over other cell lines.

Prion-propagating GT1-7 cells typically will not lose infectivity if carefully maintained, even after a year or more of continuous passaging, and de novo infection can generally be detected at an early passage post-infection (p.i). GT1-7 cells divide slowly (doubling time of ~48–72 h) and require more maintenance and care than other hardier cell lines such as RK13. A protocol for passaging GT1-7 cells is listed below.

Ensure fresh media and PBS are prewarmed to 37 °C.

1. Remove spent media from the cells. Wash adhered cells twice with PBS. Use the pressure generated by a pipettor and 10 mL pipette to force cells from the flask surface with 10 mL of fresh PBS (spritzing). Collect the cells in a sterile tube. Repeat this process with fresh PBS. Repeat until all cells have detached and been collected (*see* **Note 1**).
2. Spin the cells at 600 × *g* for 5 min.
3. Remove PBS and gently resuspend cells in media replating into a new flask (*see* **Note 2**).
4. Passage the cells every 3–4 days (*see* **Note 3**).

3.2 Infection of the GT1-7 Cell Line with Prions

The most common approach for establishing a prion-propagating cell line is to incubate infectious brain homogenate with recipient cells for 1–3 days, then wash extensively, and passage the cells at least three times before testing for de novo prion propagation (*see* **Note 4**). Other methods include incubation of recipient cells with infected cell lysate, cell media, or exosomes which are outlined below. To successfully maintain prion propagation long term, *see* **Notes 5** and **6**.

3.2.1 Infection of Cell Lines with Brain Homogenate

1. Plate cells (1×10^5 per well) into six-well plate (*see* **Note 7**) the day prior to infection. Cells should be ~50% confluent on the day of infection.
2. On the day of infection, wash the cells with PBS and incubate with 600 μL of 1% brain homogenate, either prion infected or uninfected (*see* **Notes 8** and **9**). Gently rock the plate back and forth to ensure the surface of the well is covered and place the plate in an incubator (37 °C 5% CO_2) for 5 h.
3. Following incubation, add 2.0 mL of fresh Opti-MEM to the cells, and incubate for a further 72 h.
4. Remove brain homogenate, wash cells twice with PBS, and split cells into a new six-well plate. Passage an additional three to five times before testing for the expression of protease-resistant PrP (*see* **Note 10**).

3.2.2 Infection of Cell Lines with Cell Lysate or Media

One way to generate infected cell lines without resorting to brain homogenate is to utilize cell lysate or conditioned media from established prion-propagating cell lines (*see* **Note 11**).

1. Grow non-infected and prion-infected cells to 80% confluency in a T75cm2 flask.
2. Wash cells twice with PBS and then remove cells from flask surface with ice-cold PBS.
3. Pellet cells and remove PBS.

4. To prepare the cell lysate, expose the cell pellet to repeated freeze thawing in liquid nitrogen. Place the cell pellet into liquid nitrogen for 30 s, then into a 37 °C water bath until thawed. Repeat four to five times.
5. Pass the lysate through a 28-gauge needle to homogenize cells.
6. Spin lysate 2000 × *g* for 3 min to pellet large insoluble or any remaining intact cells.
7. Take the supernatant, dilute in 5 mL of fresh Opti-MEM, then incubate 2.5 mL with GT1-7 cells (prepared the day prior) in one well of a six-well plate, and incubate for 72 h in a 37 °C 5% CO_2 incubator.
8. Remove inoculum, wash cells twice with PBS, and split cells into a new six-well plate. Passage cells an additional three to five times before testing for the expression of protease-resistant PrP (*see* **Note 10**).

3.2.3 Infection of Cells with Conditioned Media

1. Plate out cultures of non-infected and prion-infected cells and incubate for 4 days.
2. Remove cell media and spin at 2000 × *g* for 10 min to remove detached cells and cellular debris.
3. Take the supernatant and incubate it for 72 h in a 37 °C 5% CO_2 incubator with GT1-7 cells prepared the day prior.
4. Passage cells an additional five times before testing for the expression of protease-resistant PrP (*see* **Note 10**).

3.2.4 Infection of Cells with Exosomal PrP^{Sc}

Exosome-associated PrP^{Sc} is transmissible between homologous cell types [29] in addition to heterologous cells [30]. A method for transferring exosomal prion infectivity is outlined below.

1. Isolate exosomes (from 2×10^7 cells) under sterile conditions in Opti-MEM.
2. Incubate the Opti-MEM containing exosomes with cells (50% confluent) plated the day prior. Exosomes isolated from 1×10^7 cells should be ample to establish highly efficient conversion of PrP^C on 8×10^{4} cells plated in one well of a six-well tissue culture plate.
3. Incubate for 72 h in a 37 °C 5% CO_2 incubator.
4. Wash and passage the cells as normal.
5. Following five to ten passages, test recipient cells for protease-resistant PrP (*see* **Note 10**).

3.3 Bioassay of Prion-Infected Cell Lysates

1. Resuspend cells (3×10^6) in 1 mL of PBS and subject to five consecutive cycles of freeze-thawing and sonication. Inoculate 30 μL into the left parietal region of anesthetized mice.

2. Monitor mice for typical signs of murine prion disease such as hunched posture, hind limb paresis, ataxia, and reduced motor activity and euthanize mice in the terminal stage of disease.
3. Examine the brain of every animal for the presence of PrP^{Sc} by immunoblotting or immunohistochemistry (*see* **Note 8**).

3.4 Detecting Protease-Resistant PrP Using the Prion-Infected Cell Assay

We developed PICA, which is a high-throughput in vitro assay based on the standard scrapie cell assay (SSCA) [27] (Fig. 1). PICA uses a homologous population of RK13 cells stably expressing PrP transgenes that are subcloned then screened to identify a clonal population highly susceptible to prion infection. This can be achieved via the following.

3.4.1 Generation of Cell Clones Highly Susceptible to Mouse-Adapted Sporadic Human Creutzfeldt-Jakob Disease

1. Stably transfect RK13 cells expressing wild-type mouse PrP (MoRK13) and then subclone by limited dilution (*see* **Note 12**). Duplicate each clone into two sets and inoculate one set with 0.1% (w/v) M1000 or MU-02 and culture for five passages.
2. Using PICA (*see* Subheading 3.4.3), screen the cells for their sensitivity to infection with M1000 prions as detected by de novo PrP^{Sc} and determine the cell line with the greatest sensitivity. In the example shown (Fig. 2), 21 clones were isolated and 3 clones (D1, B4, and A2) identified as producing the highest levels of PICA PrP^{Sc} immunoreactivity. The A2 clone

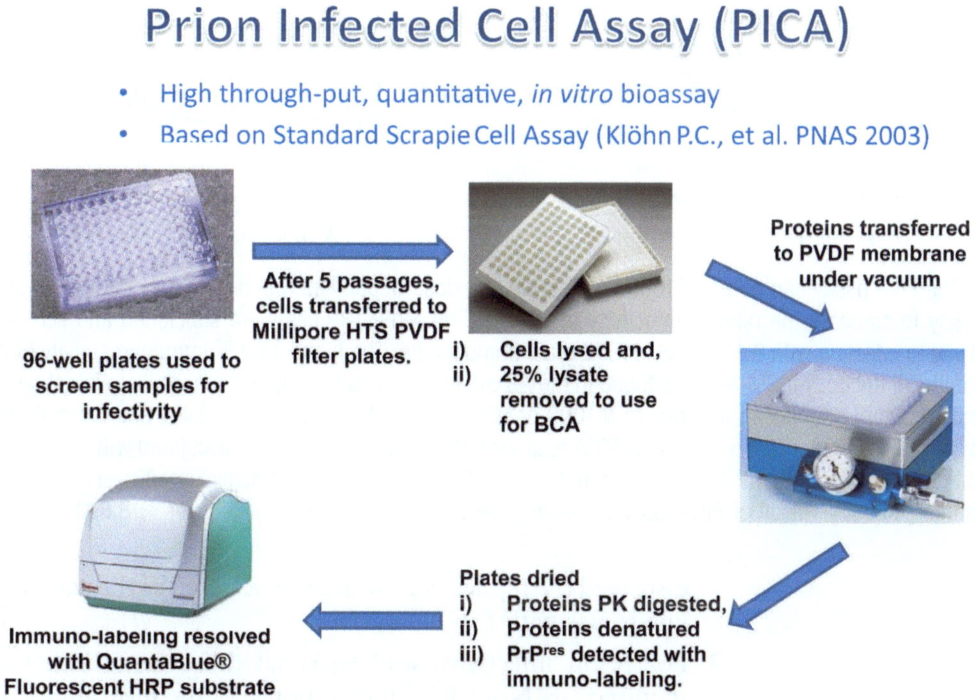

Fig. 1 The high-throughput quantitative in vitro prion-infected cell assay (PICA) method

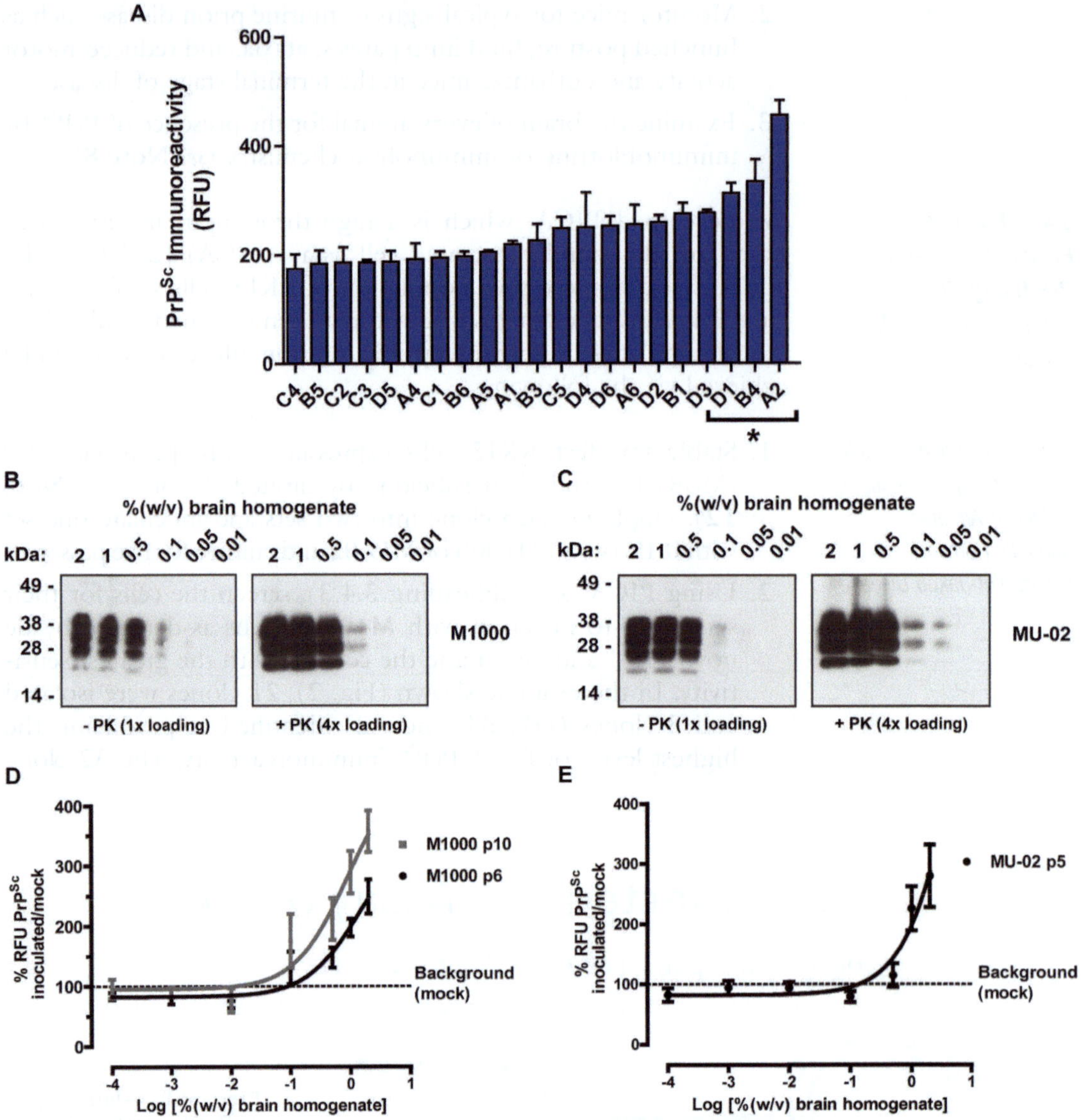

Fig. 2 The PICA assay can detect M1000 and MU-02 in a dose-dependent manner demonstrating the ability of this assay to compare the relative infectivity of samples. (**a**) MoRK13 cells were subcloned and screened for sensitivity to infection with 0.1% (w/v) M1000 brain homogenate. The highest PrP^{Sc} immunoreactivity (detected using anti-PrP antibody ICSM-18) was found in the A2 clone indicating the highest sensitivity to infection. (**b** and **c**) Western blot of decreasing dilutions of M1000 or MU-02 $\pm$ PK to compare sensitivity and correlate relative levels of PrP and rPrP^{Sc} in inoculum to A2-PICA response. (**d** and **e**) A2 cells were inoculated with 2% to 0.001% dilutions of M1000 or MU-02 brain homogenate and PICA assay performed at indicated passage number to detect infected cells. Result of PICA indicates clear dose-response activity. $n = 3$. Error = S.E.M.

displayed the highest signal and was selected as the standard cell line to use with PICA.

3. For prion infectivity studies, serial dilutions (titrations) are required (*see* **Note 13**). The response of the most highly sensitive cells (hs-RK13) should be tested by exposing them to

dilutions of prion-infected brain homogenate, such as M1000 or MU-02 ranging from 2 to 0.001% (w/v).

4. For comparative purposes, the titrated samples should be examined by western blot for total PrP and PrP^{Sc} at passage 6 p.i. as shown in Fig. 2 when normalized to background signal from A2 cells not inoculated (100%), de novo PrP^{Sc} is readily detected above background in cells exposed to 2, 1, and 0.5% dilutions of M1000 infected brain homogenate, but not any lower dilutions. The decrease in signal corresponding with decreasing dilutions allowed a sigmoidal dose-response curve to be fit to this dilution series.
5. It is recommended to perform another PICA passage 10 p.i. to determine if further passaging of cells improves the sensitivity of the assay (as shown in Fig. 2) (*see* **Note 14**).

3.4.2 Preparation of Assay Cells

1. Plate hs-RK13 cells in a 96-well tissue culture plate (10,000 cells/well) and incubate for 48 h, 37 °C 5% (v/v) CO_2.
2. Resuspend samples for in vitro bioassay using PICA in OCCM at desired concentration and use to inoculate the hs-RK13 containing wells in triplicate (30 μL per well) (*see* **Note 15**).
3. Incubate the hs-RK13 cells for 16 h, 37 °C 5% (v/v) CO_2, followed by the addition of 120 μL OCCM, and incubate for a further 48 h.
4. Remove inoculum by aspirating with eight-channel vacuum aspirator pipette, add 100 μL warm, sterile PBS to each well using the eight-channel pipette, swirl gently, remove PBS by aspiration, and repeat. Wash twice as above but with OCCM.
5. After final aspiration, add of 150 μL OCCM containing 0.5 μg/mL puromycin to each well and incubate until cells are confluent.
6. When confluent, remove media by aspiration, wash once with 200 μL warm PBS, and add 120 μL warm OCCM.
7. Passage cells, using the 12-channel manual pipette to gently detach the cells, by pipetting up and down five times. Discard 80 μL of cell suspension and add 110 μL fresh, warm OCCM containing 0.5 μg/mL puromycin. This is a 1:3 split.
8. Incubate cells at 37 °C 5% (v/v) CO_2 until confluent.
9. Repeat **steps 6** and 7 to passage cells at 1:5 split.
10. Repeat **step 9** until passage 6 is reached.

3.4.3 PICA Assay

1. At passage 6 p.i., remove OCCM by aspiration and wash cells with 200 μL ice-cold PBS.

2. Remove PBS and add 120 μL/well cell immunoblot assay lysis buffer.
3. Incubate at room temperature for 10 min on a rocker (Ratek).
4. During incubation activate the 96-well MultiScreen HTS™ IP Elispot plates by adding 30 μL methanol to each well. Allow to incubate for 2 min and remove by vacuum using the 96-well plate vacuum manifold.
5. Wash wells twice with 200 μL PBS and, after final vacuum, add 30 μL PBS to keep membrane hydrated.
6. Transfer the lysates to the methanol-activated 96-well MultiScreen HTS™ IP Elispot plates (Millipore) containing the 30 μL PBS.
7. Transfer 30 μL from each well to a 96-well microplate (Greiner Bio-One) for performing a BCA protein quantitation assay.
8. Transfer the remaining lysate in the MultiScreen filter plates onto the filter membrane by vacuum and then dry the membranes at 37 °C for 1 h.
9. Add 60 μL of PK digestion buffer to each well and incubate plates with rocking at 37 °C for 1.5 h.
10. Remove the PK containing lysis buffer by vacuum aspiration from the top using the vacuum pipette, wash each well with 160 μL of PBS, and remove by vacuum aspiration.
11. Stop proteolysis with the addition of 150 μL/well PICA-PMSF buffer (PBS containing 2 mM PMSF) and incubate for 15 min with rocking. Remove PBS/PMSF by vacuum and wash wells twice with 160 μL PBS.
12. To denature the protein add 150 μL/well of denaturation buffer (3 M guanidine thiocyanate) (Amresco), 10 mM Tris pH 7.4) for 10 min with rocking.
13. Remove denaturation buffer by vacuum aspiration and wash wells four times with 160 μL/well PBS using vacuum aspiration to remove the solutions.
14. Add 150 μL/well superblock to block membranes and incubate for 1 h on rocker.
15. Remove superblock by vacuum.
16. Add 60 μL/well primary anti-PrP antibody ICSM-18 (1:5000 in TBS-T). Incubate plates for 1 h on a rocker at room temperature.
17. Remove primary antibody by aspiration and wash wells with TBS-T: Add 150 μL TBS-T to each well using the eight-channel digital pipette, swirl for 1 min, and aspirate off wash solution between washes. Repeat three more times.
18. Add 60 μL of secondary antibody anti-mouse IgG HRP (1:5000 in TBS-T) to each well and incubate plates for 1 h on a rocker at RT.

19. Remove secondary antibody by aspiration and wash wells four times with TBS-T, aspirating between washes as described in **step 27**.
20. Use QuantaBlu fluorogenic peroxidase substrate (Pierce – Thermo Scientific) to develop the HRP signal according to manufacturer's instructions: Nine parts of QuantaBlu substrate solution were mixed with one part of QuantaBlu stable peroxide solution and allowed to equilibrate to RT prior to use (a total of 11 mL is required per 96-well plate). Add 100 μL/well of the QuantaBlu solution, cover plates with aluminum foil, and incubate for 30 min at 37 °C. 100 μL/well of the QuantaBlu stop solution is then added using the eight-channel digital pipette, and QuantaBlu fluorescence was measured on a Varioskan multimode plate reader (excitation 325 nm and emission 420 nm).

3.4.4 PICA Assay Data Analysis

1. Download results from BCA protein assay and QuantaBlue fluorescence assay to Microsoft Excel.
2. Divide the QuantaBlue result for each well by the BCA assay result to normalize the reading to the total protein levels.
3. Divide the results of each inoculated well by the negative control wells and multiply by 100. This provides a percentage-fold change above the negative control. A positively infected well should be at least 150% (with negative control being 100%).
4. For serial dilutions and quantitation, use GraphPad or similar program to plot the results of each well (in triplicate) against the log concentration of the respective inoculum. Using GraphPad, fit a standard dose-response curve and use nonlinear regression analysis to calculate log PICA-ID50.

4 Notes

1. Repeated pipetting of GT1-7 cells will result in extensive cell death as will the use of cell scrapers. Trypsinization is optional; however, it is not recommended when performing experiments because PrP is cleaved by trypsin.
2. GT1-7 cells are split at a 1:3 ratio. If the cells are plated at too low confluency, they will not grow and should then be replated into a smaller flask. GT1-7 cells release a lot of debris, so it may be necessary to replace the media 1–2 days after plating.
3. Ideally GT1-7 cells should be no more than 80% confluent at the time of passaging. Cells should not be left without passaging for longer than 5 days.
4. It is crucial to perform mouse bioassays to confirm that your cell line is producing infectious prions. Use infected and control lysate derived from cells passaged at least three times to ensure the original inoculum is not present.

5. Cell lines can vary in their susceptibility to infection and can lose PrP^{Sc} expression easily. We and others [4] have noted the importance of media and fetal calf serum on the establishment of infection and also the longevity of PrP^{Sc} expression. Bosque et al. grew [4] scrapie-susceptible N2a cells in high-glucose DMEM, low-glucose DMEM, or MEM (Gibco, Invitrogen). Although the cells were morphologically identical and PrP^{Sc} was detected in all exposed cultures, N2a cells grown in low-glucose media grew more slowly and had lower amounts of PrP^{Sc} than cells grown in high-glucose DMEM. The study concluded that the concentration of glucose was critical for sensitivity to inoculated prions. In addition, we have tested GT1-7 and mouse PrP expressing RK13 (moRK13) cells grown in DMEM versus Opti-MEM and found that cells grown in DMEM and subsequently inoculated with M1000 prions exhibit less susceptibility to the initial infection and are unable to propagate PrP^{Sc} for as many passages as cells grown and infected in Opti-MEM media.
6. In our experience the quality of the fetal calf serum (FCS) used in the media is very important. Several brands of FCS should be tested, and a few batches of each brand tested. It is best to use the same batch number once a suitable FCS brand is chosen to avoid testing the suitability of every bottle.
7. To avoid using large volumes of brain homogenate, infections are done in a six-well plate.
8. Brain homogenates are prepared from BALB/c mice in the terminal phase of prion infection with the M1000 (Fukuoka) strain or matched un-infected mice. It is important to maintain sterility following removal of the brain from the skull to avoid contamination of cell cultures. Brain, 10% homogenates, are prepared in sterile PBS by passing through 18-, 22-, and 28-gauge needles in succession (~20 times through each needle). Homogenates are snap frozen in liquid nitrogen and stored at −80 °C, and repeated freeze-thawing is avoided. After thawing, the appropriate dilution of brain homogenate is made in Opti-MEM.
9. Varying dilutions of brain homogenate can be used. However we would recommend no more than 1% for GT1-7 cells due to associated toxicity.
10. This step ensures removal of the original inoculum. Methods for detecting PrP^{Sc} include western immuno-blotting and immuno-fluorescence. Following detection of PrP^{Sc}, some cell lines may need to be subcloned to isolate high-PrP^{Sc}-expressing cells; however, subcloning is not necessary for GT1-7 cells.
11. In our hands this method is very efficient and results in less toxicity than brain homogenate infections.

12. For details on vector construction and the mouse-adapted sporadic human prion strains, *see* ref. 26.
13. By performing a titration, the response to a range of sample concentrations can be assayed which increases the accuracy and sensitivity when comparing different samples. If only one sample concentration is assayed, the concentration may be high enough that it saturates the assay and results in an inaccurate comparison. More importantly, the lowest dilution to cause a response can be determined which provides a more accurate measure of how infectious a sample is – if a very low concentration causes an assay response, then this sample has a high level of infectivity.
14. In our hands the PrP^{Sc} signal of the three highest concentrations all increased by approximately 100%, and infectivity was also detected in the cells inoculated with 0.01% (w/v) M1000, resulting in a shift of the dose-response curve leftward (Fig. 2). This indicated that further passaging increases the sensitivity of the PICA assay. PICA is compatible for use with other strains such as MU-02.
15. If the intention is to quantitate infectivity, make triplicate serial dilutions of inoculum in OCCM. For brain homogenates, this can range from 2% (w/v) to 0.0001% (w/v). Note concentrations higher than 2% (w/v) may elicit toxicity to cells. For cell lysates or exosomes, use a BCA protein assay to determine concentration of stock solution. Then prepare dilutions ranging from 2 μg/mL to 0.001 μg/mL and inoculate cells with this. Volumes and dilution range will depend on amount of stock solution available.

References

1. Prusiner SB (1982) Novel proteinaceous infectious particles cause scrapie. Science 216:136–144
2. Clarke MC, Haig DA (1970) Evidence for the multiplication of scrapie agent in cell culture. Nature 225:100–101
3. Race RE, Fadness LH, Chesebro B (1987) Characterization of scrapie infection in mouse neuroblastoma cells. J Gen Virol 68(Pt 5):1391–1399
4. Bosque PJ, Prusiner SB (2000) Cultured cell sublines highly susceptible to prion infection. J Virol 74:4377–4386
5. Schatzl HM, Laszlo L, Holtzman DM et al (1997) A hypothalamic neuronal cell line persistently infected with scrapie prions exhibits apoptosis. J Virol 71:8821–8831
6. Vorberg I, Raines A, Story B et al (2004) Susceptibility of common fibroblast cell lines to transmissible spongiform encephalopathy agents. J Infect Dis 189:431–439
7. Grassmann A, Wolf H, Hofmann J et al (2013) Cellular aspects of prion replication in vitro. Virus 5:374–405
8. Priola SA, Caughey B, Raymond GJ et al (1994) Prion protein and the scrapie agent: in vitro studies in infected neuroblastoma cells. Infect Agents Dis 3:54–58
9. Supattapone S, Wille H, Uyechi L et al (2001) Branched polyamines cure prion-infected neuroblastoma cells. J Virol 75:3453–3461
10. Butler DA, Scott MR, Bockman JM et al (1988) Scrapie-infected murine neuroblastoma cells produce protease-resistant prion proteins. J Virol 62:1558–1564
11. Rubenstein R, Carp RI, Callahan SM (1984) In vitro replication of scrapie agent in a neuronal model: infection of PC12 cells. J Gen Virol 65 (Pt 12):2191–2198
12. Vilette D, Andreoletti O, Archer F et al (2001) Ex vivo propagation of infectious sheep scrapie agent in heterologous epithelial cells expressing

ovine prion protein. Proc Natl Acad Sci U S A 98:4055–4059
13. Mahal SP, Baker CA, Demczyk CA et al (2007) Prion strain discrimination in cell culture: the cell panel assay. Proc Natl Acad Sci U S A 104:20908–20913
14. Mediano DR, Sanz-Rubio D, Ranera B et al (2015) The potential of Mesenchymal stem cell in prion research. Zoonoses Public Health 62 (3):165–178
15. Mahal SP, Browning S, Li J et al (2010) Transfer of a prion strain to different hosts leads to emergence of strain variants. Proc Natl Acad Sci U S A 107:22653–22658
16. Iwamaru Y, Takenouchi T, Ogihara K et al (2007) Microglial cell line established from prion protein-overexpressing mice is susceptible to various murine prion strains. J Virol 81:1524–1527
17. Polymenidou M, Trusheim H, Stallmach L et al (2008) Canine MDCK cell lines are refractory to infection with human and mouse prions. Vaccine 26:2601–2614
18. Clarke MC, Millson GC (1976) Infection of a cell line of mouse L fibroblasts with scrapie agent. Nature 261:144–145
19. Gibson PE, Bell TM, Field EJ (1972) Failure of the scrapie agent to replicate in L5178Y mouse leukaemic cells. Res Vet Sci 13:95–96
20. Krejciova Z, De Sousa P, Manson J et al (2014) Human tonsil-derived follicular dendritic-like cells are refractory to human prion infection in vitro and traffic disease-associated prion protein to lysosomes. Am J Pathol 184:64–70
21. Kim HJ, Tark DS, Lee YH et al (2012) Establishment of a cell line persistently infected with chronic wasting disease prions. J Vet Med Sci 74:1377–1380
22. Courageot MP, Daude N, Nonno R et al (2008) A cell line infectible by prion strains from different species. J Gen Virol 89:341–347
23. Neale MH, Mountjoy SJ, Edwards JC et al (2010) Infection of cell lines with experimental and natural ovine scrapie agents. J Virol 84:2444–2452
24. Sabuncu E, Petit S, Le Dur A et al (2003) PrP polymorphisms tightly control sheep prion replication in cultured cells. J Virol 77:2696–2700
25. Arellano-Anaya ZE, Savistchenko J, Mathey J et al (2011) A simple, versatile and sensitive cell-based assay for prions from various species. PLoS One 6:e20563
26. Lawson VA, Vella LJ, Stewart JD et al (2008) Mouse-adapted sporadic human Creutzfeldt-Jakob disease prions propagate in cell culture. Int J Biochem Cell Biol 40:2793–2801
27. Klohn PC, Stoltze L, Flechsig E et al (2003) A quantitative, highly sensitive cell-based infectivity assay for mouse scrapie prions. Proc Natl Acad Sci U S A 100:11666–11671
28. Baron GS, Magalhaes AC, Prado MA et al (2006) Mouse-adapted Scrapie infection of SN56 cells: greater efficiency with Microsome-associated versus purified PrP-res. J Virol 80:2106–2117
29. Fevrier B, Vilette D, Archer F et al (2004) Cells release prions in association with exosomes. Proc Natl Acad Sci U S A 101:9683–9688
30. Vella LJ, Sharples RA, Lawson VA et al (2007) Packaging of prions into exosomes is associated with a novel pathway of PrP processing. J Pathol 211:582–590
31. Alais S, Simoes S, Baas D et al (2008) Mouse neuroblastoma cells release prion infectivity associated with exosomal vesicles. Biol Cell 100:603–615
32. Arellano-Anaya ZE, Huor A, Leblanc P et al (2015) Prion strains are differentially released through the exosomal pathway. Cell Mol Life Sci 72(6):1185–1196
33. Saa P, Yakovleva O, De Castro J et al (2014) First demonstration of transmissible spongiform encephalopathy-associated prion protein (PrPTSE) in extracellular vesicles from plasma of mice infected with mouse-adapted variant Creutzfeldt-Jakob disease by in vitro amplification. J Biol Chem 289:29247–29260
34. Coleman BM, Hanssen E, Lawson VA et al (2012) Prion-infected cells regulate the release of exosomes with distinct ultrastructural features. FASEB J 26:4160–4173
35. Stoorvogel W, Kleijmeer MJ, Geuze HJ et al (2002) The biogenesis and functions of exosomes. Traffic 3:321–330
36. Brown CA, Schmidt C, Poulter M et al (2014) In vitro screen of prion disease susceptibility genes using the scrapie cell assay. Hum Mol Genet 23:5102–5108
37. Nishida N, Harris DA, Vilette D et al (2000) Successful transmission of three mouse-adapted scrapie strains to murine neuroblastoma cell lines overexpressing wild-type mouse prion protein. J Virol 74:320–325
38. Arjona A, Simarro L, Islinger F et al (2004) Two Creutzfeldt-Jakob disease agents reproduce prion protein-independent identities in cell cultures. Proc Natl Acad Sci U S A 101:8768–8773
39. Miyazawa K, Emmerling K, Manuelidis L (2011) Replication and spread of CJD, kuru and scrapie agents in vivo and in cell culture. Virulence 2:188–199
40. Miyazawa K, Okada H, Iwamaru Y et al (2014) Susceptibility of GT1-7 cells to mouse-passaged field scrapie isolates with a long incubation period. Prion 8:313–306

Chapter 10

Analysis of Cellular Prion Protein Endoproteolytic Processing

Victoria Lewis

Abstract

Like numerous proteins of various structural and functional classes, the glycosylphosphatidylinositol (GPI)-anchored cellular prion protein (PrP^C) has been recognized to undergo endoproteolytic processing for decades, a phenomenon observed in various cultured cell lines, as well as human and several animal tissue extracts. Despite this, the physiological significance of PrP^C proteolytic cleavage has not yet been entirely elucidated. Experimental evidence suggests independent normal biological functions of the full-length and truncated PrP^C species, as well as probable links of endoproteolysis to prion disease transmission susceptibility, pathogenesis, and toxicity. The accurate characterization of constitutive PrP^C processing, through the method outlined in this chapter, is therefore an important tool in order to investigate the biological relevance of the alternative cleavage events.

Key words Prion protein, PrP^C, Endoproteolysis, Cleavage, Processing, Protease, Deglycosylation

1 Introduction

The ubiquitously expressed cellular prion protein, PrP^C, is a cell surface secretory protein, which following synthesis and en route to the cell surface is subject to a number of posttranslational modifications. These modifications include co-translational cleavage of the amino- and carboxy-terminal signal peptides (the latter resulting in the attachment of a GPI anchor), asparagine (N)-linked glycosylation at up to two different sites, and formation of a single disulfide bond [1]. PrP^C is also among the vast group of membrane-bound proteins that undergo posttranslational constitutive and regulated proteolytic processing [2], in theory enabling a functionally diverse protein without the requirement for separate biosynthetic pathways [3, 4]. Importantly, there are frequently more truncated PrP^C species than full-length PrP^C observed in cells and tissues [5–7], highlighting the likely biological requirement for the occurrence of these processes.

Victoria A. Lawson (ed.), *Prions: Methods and Protocols*, Methods in Molecular Biology, vol. 1658, DOI 10.1007/978-1-4939-7244-9_10,

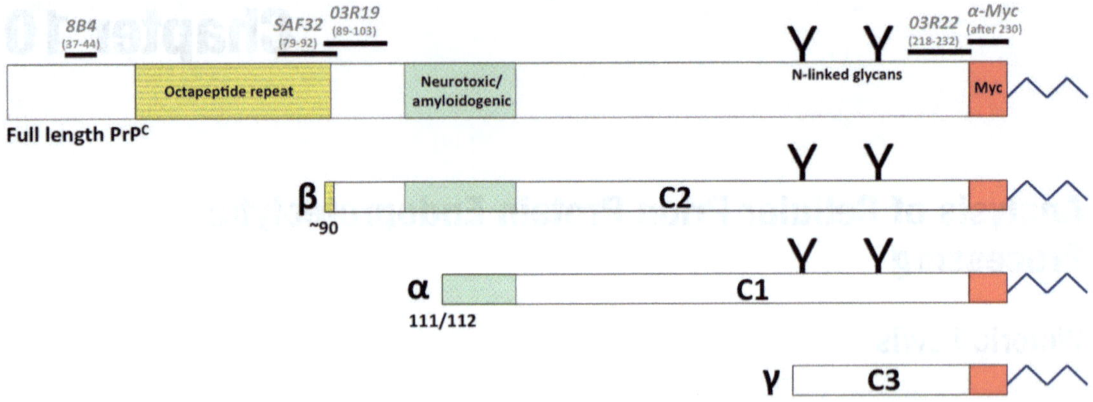

Fig. 1 Schematic representation of the PrP^C α-, β- and, putative γ-cleavage sites and the C-terminal fragments produced from each cleavage event, with epitope alignment of the antibodies used to detect full-length MycPrP^C and its cleavage fragments in this study

PrP^C can be endoproteolytically processed at two internal sites, by the well-described α- or β-cleavage events (Fig. 1). Additionally we have recently uncovered and characterized the existence of a third cleavage event, which following on from the existing nomenclature we have coined PrP^C "γ-cleavage" (Fig. 1) [8]. The predominant α-cleavage event occurs within the potentially neurotoxic and amyloidogenic region of PrP^C after residues 111/112 (human PrP numbering) [9] and produces the C1/N1 fragment pair, with the C1 fragment readily detectable in tissue extracts and cell lysates [5]. The precise mechanism behind PrP^C α-cleavage is not yet resolved, with several conflicting reports presenting evidence for and against the involvement of the *a disintegrin and metalloprotease* (ADAMs) family of proteases [10–13]. Adding complexity, a recent study also provides evidence that α-cleavage is multifaceted, suggesting several close neighboring cleavage sites are targeted by different proteases [14]. Traditionally, β-cleavage has been linked with processing of misfolded prion protein conformers (PrP^{Sc}) and prion disease [9, 15], although it is known to also occur to PrP^C in uninfected cells and tissues [5, 6, 9, 16, 17]. This generally less common β-processing event, mediated by reactive oxygen species (ROS) [18, 19], is a "ragged" cleavage at the end of the metal-binding octapeptide repeat region, around residue 90, producing the C2/N2 fragment pair [6, 16]. Finally, our recent discovery of a small C-terminal cleavage fragment, C3 (Fig. 2), verified the existence of the PrP^C γ-cleavage event [8], which occurs in the highly conserved [20] far C-terminus of PrP^C.

The physiological significance of PrP^C proteolysis has not yet been entirely elucidated, although there is emerging, and sometimes conflicting, evidence of separate biological functions for full-length PrP^C [21–23] and the various fragments derived from endogenous PrP^C cleavage [24–29]. Despite uncertainties regarding the precise biological function of PrP^C and its fragments, it is known that PrP^C

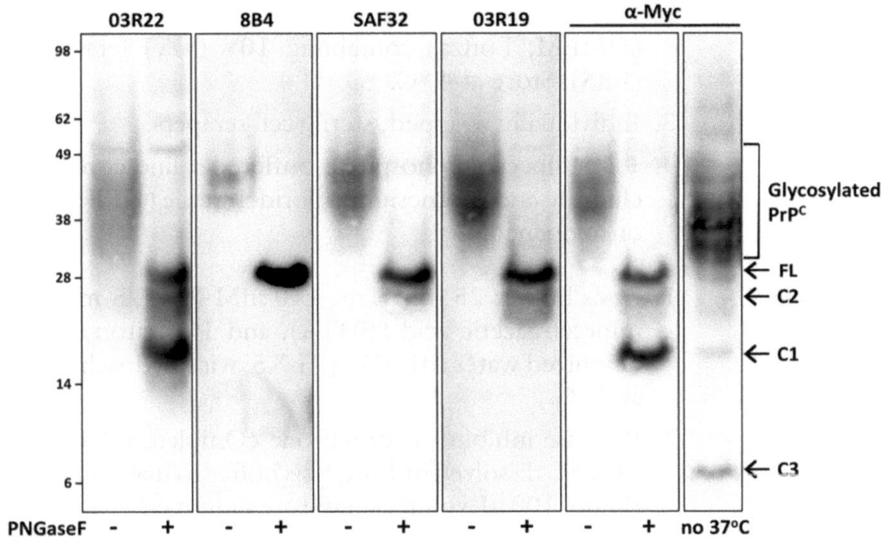

Fig. 2 Representative PAGE and western blot epitope mapping of PrP^C. Lysates (50 μg total protein per lane) generated from MycRK cells were incubated in the presence (+) or absence (−) of PNGaseF overnight at 37 °C, or lysate was mixed with sample buffer immediately after preparation (i.e., not PNGaseF digested or incubated at 37 °C - far right lane). The different PrP^C species, including full-length PrP^C (FL) and C-terminal fragments (C1, C2, C3), are indicated

expression is an absolute requirement for successful prion disease transmission and pathogenesis [30–32], where it aberrantly folds into PrP^{Sc}. The initiation of PrP^C misfolding is not yet understood; however, convention proposes a continued interaction and conversion of PrP^C to PrP^{Sc} with eventual production of sufficient neurotoxic species, ultimately causing cell death. Furthermore, it is likely that PrP^C cleavage/truncation has an influence in prion disease transmission susceptibility, pathogenesis, and the efficiency of conversion/misfolding of PrP^C [5, 33–37]. Therefore the investigation and accurate characterization of PrP^C endoproteolytic processing will provide insight into the physiological relevance of the alternative cleavage events to normal PrP^C function and prion diseases.

The following chapter describes the methods for examination of endoproteolytic cleavage in cultured "MycRK" cells; however, it is applicable to the analysis of other cell lines and tissue protein extracts.

2 Materials

2.1 Cell Culture

1. MycRK cells: rabbit kidney epithelial (RK13) cells, stably overexpressing murine PrP^C that contains a C-terminal (after residue 230) human c-myc epitope tag [38].

2. Cell culture medium: Dulbecco's modified Eagle's medium (DMEM, Lonza) containing 10% (v/v) fetal bovine serum (FBS). Store at 4 °C.
3. Individually wrapped, sterile cell scrapers.
4. 1× Dulbecco's phosphate buffered saline containing calcium chloride and magnesium chloride, sterile (DPBS). Store at 4 °C once opened.

2.2 Cell Lysis and Protein Determination

1. Lysis buffer: 25 mM Tris, 150 mM NaCl, 5 mM ethylenediaminetetraacetic acid (EDTA), and 1% Triton X-100 (v/v) in deionized water (dH_2O), pH 7.5, with hydrochloric acid. Store at 4 °C.
2. Protease inhibitor cocktail: one cOmplete ULTRA mini tablet (Roche) dissolved in 1 mL lysis buffer to make a 10× stock. Sub-aliquot 100 μL volumes and store aliquots at −20 °C (*see* **Note 1**).
3. Protein determination: BCA Protein Assay Kit (Pierce), reagents A and B, and albumin standards are provided. Store reagents A and B at room temperature. From the 2 mg/mL albumin standard dilute with dH_2O to make 1 mg/mL, 0.5 mg/mL, 0.4 mg/mL, 0.2 mg/mL and 0.1 mg/mL, and store at 4 °C.
4. Clear, flat-bottom, 96-well microplate.

2.3 Deglycosylation of PrP^C

1. Ten percent solution (w/v) of sodium dodecyl sulfate (SDS) in dH_2O. Store at room temperature (*see* **Notes 2** and **3**).
2. 10× denaturation buffer: 5% (v/v) SDS (diluted from the 10% SDS aqueous solution), 5% (v/v) β-mercaptoethanol (BME), in dH_2O. Store at room temperature (*see* **Note 3**).
3. 10× detergent buffer: 10% (v/v) Nonidet P-40 (also called Igepal CA-630) in dH_2O. Store at room temperature.
4. 10× incubation buffer: 500 mM sodium phosphate monobasic (NaH_2PO_4) in dH_2O. Store at room temperature.
5. N-Glycosidase F (PNGaseF). Store at −20 °C.

2.4 PAGE

1. 4× sample buffer: 12% 2-mercaptoethanol (BME) in 4 × NuPAGE sample buffer (Novex, Life Technologies). Store at 4 °C once BME and sample buffer are mixed.
2. NuPAGE mini gels (Novex, Life Technologies): 4–12% Bis-Tris, 1.0 mm or 1.5 mm thick, 10 or 15 wells depending on number/volume of samples (*see* **Note 4**). Store gels at room temperature.
3. SeeBlue®Plus2 prestained protein standard (Life Technologies), store at 4 °C.

4. 20× MES running buffer (Novex, Life Technologies): to make up 1× transfer buffer, dilute 50 mL of 20× stock with 1950 mL water, mix well, and store at room temperature or 4 °C (preferable) (*see* **Note 5**).

2.5 Western Blotting

1. 10× Transfer buffer (no methanol): 250 mM Tris, 2 M glycine. Stored at room temperature. To make up 1× transfer buffer, dilute 100 mL of 10× stock with 700 mL water and add 200 mL methanol (to give (20% v/v)), mix well, and store at 4 °C (*see* **Note 6**).
2. Immobilon-P polyvinylidene fluoride (PVDF), 0.45 μM pore membrane (Millipore).
3. Whatman™ blotting/filter paper.
4. 10× PBS: 1.4 M NaCl, 27 mM KCl, 18 mM KH_2PO_4, 100 mM Na_2HPO_4 in dH_2O. To make 1× PBS, dilute 100 mL of 10× stock with 900 mL water. Store both solutions at room temperature.
5. 1× PBS.T: 0.5% (v/v) Tween 20 in 1× PBS. Store at room temperature.
6. Block buffer: 5% (w/v) skim milk powder in 1× PBS.T, prepared immediately prior to use.
7. Antibodies (Table 1, Fig. 1), stored at 4 °C.
   (a) 8B4 – PrP residues 37–44, mouse monoclonal (Alicion).
   (b) SAF32 – PrP residues 79–92, mouse monoclonal (Cayman Chemical).
   (c) 03R19 – PrP residues 89–103, rabbit polyclonal, a generous gift from Associate Professor Victoria Lawson [39].
   (d) 03R22 – PrP residues 218–232, rabbit polyclonal, a generous gift from Associate Professor Victoria Lawson [39].
   (e) Anti-Myc tag, clone 4A6 – Myc tag inserted after PrP residue 230, mouse monoclonal (Millipore).

Table 1
Optimized antibody dilutions

Primary antibody	Dilution	Secondary antibody (HRP conjugated)	Dilution
8B4	1:5000	Anti-mouse	1:7000
SAF32	1:3000	Anti-mouse	1:7000
03R19	1:5000	Anti-rabbit	1:8000
03R22	1:5000	Anti-rabbit	1:8000
Anti-Myc tag, clone 4A6	1:5000	Anti-mouse	1:7000

(f) Sheep anti-mouse and donkey anti-rabbit, horseradish peroxidase (HRP) conjugated secondary antibodies (ECL™ range, GE Healthcare).

8. Amersham™ ECL™ Prime Western Blotting Detection System (GE Healthcare) stored at 4 °C.

3 Methods

PrP^C can be visualized by PAGE and western blotting as an anti-PrP antibody immunoreactive "smear" between the approximate molecular weights of 27–50 kDa (Fig. 2). Depending on which antibody is utilized, this wide PrP^C band comprises full-length and truncated PrP^C species, which can also be un-, mono-, or di-glycosylated. Sometimes lower molecular weight PrP^C species are also detected, including the recently characterized C3 fragment, when probing with an antibody raised to the PrP far C-terminus or to an epitope tag that has been engineered into in the far C-terminus of PrP (Fig. 2). To simplify visualization of the dominant C-terminal cleavage fragments, C1 and C2, the PrP^C proteins can be enzymatically stripped of their sugars using the bacterial N-glycosidase F enzyme (also known as PNGaseF). The C1 and C2 fragments can then be readily differentiated using epitope mapping with antibodies to specific regions of PrP (Table 1, Fig. 1). Depending on the immunoreactivity profile (Fig. 2), assumptions can be made about the PrP^C region where cleavage has occurred and the relative preference for each cleavage event.

3.1 Cell Lysis and Protein Determination

1. Culture MycRK cells in standard tissue culture treated flasks and maintain in cell culture medium, in a humidified incubator at 5% CO_2, until confluent.
2. Rinse cells twice with approximately 5–10 mL ice-cold sterile 1× DPBS to remove media proteins.
3. Scrape cells into a further 5–10 mL ice-cold PBS and transfer into a 15 mL Falcon tube. Centrifuge tubes at 500 × *g*, for 3 min in a bench top centrifuge, and remove the supernatant.
4. Resuspend the cell pellet in 500 μL PBS and transfer to an appropriately labeled 1.5 mL microfuge tube. Centrifuge tubes at 500 × *g* for 3 min in a microcentrifuge. Again remove the supernatant (*see* **Note 7**).
5. Make up fresh cell lysis buffer by adding 900 μL lysis buffer stock to 100 μL of 10× protease inhibitor cocktail solution (final concentration of 1× protease inhibitors).
6. To lyse cells, add an appropriate volume (usually 100–200 μL depending on the size of the pellet (*see* **Note 8**)) of cell lysis buffer to cell pellet and triturate through end of pipette tip to break up pellet.

7. Incubate crude lysate at 4 °C for 30 min, triturate again, and then centrifuge at 1000 × *g* for 5 min at room temperature or 4 °C (preferable) to pellet cellular debris. Remove the supernatant (lysate) to a new microfuge tube and discard pellet.
8. Perform a BCA assay on the lysate to determine protein concentration. Briefly, pipette in duplicate, 10 μL of each BSA standard, 10 μL of dH_2O diluent, and 10 μL of diluted (usually 1:5 or 1:10 dilution in dH_2O) lysate into wells of a 96-well microplate. Mix BCA reagents A:B at a ratio of 50:1 and add 200 μL per well. Cover the plate and incubate at 37 °C for 20–30 min and then read absorbance at 560 nm. After correction for background (dH_2O) readings, calculate the protein concentration of the lysate based on a standard curve generated by the absorbance values of the BSA standards, multiplied by the dilution factor.

3.2 De-Glycosylation of PrP^C (See Note 9)

1. Preheat a heat block (100 °C) and water bath (37 °C) (*see* **Note 10**).
2. Pipette 10 μL of lysate (diluted with lysis buffer to final total protein content of 30–50 μg in the 10 μL) into thin-walled 0.5 mL PCR tube (*see* **Note 11**).
3. Add 1 μL of 10× denaturation buffer and heat samples for 5 min in the 100 °C heat block.
4. Spin samples (max speed on microcentrifuge) briefly to bring down any condensation and cool lysate to room temperature approximately 5–10 min.
5. To denatured samples add 1.5 μL 10× detergent solution, 1.5 μL 10× incubation buffer, and 1 μL (1 unit) of PNGaseF (*see* **Note 12**).
6. Incubate samples overnight (approximately 14–16 h) in the 37 °C water bath.
7. Spin samples (max speed in microcentrifuge) briefly to bring down any condensation.

3.3 PAGE

The following instructions assume the use of the Novex, Life Technologies NuPAGE gels and XCell SureLock™ Mini-Cell Electrophoresis system, although adaptation to other SDS-PAGE gel systems is possible.

1. Add 5 μL of 4× sample buffer to lysate and place samples at 4 °C until ready for PAGE (*see* **Note 13**).
2. Prepare the NuPAGE mini gel by carefully removing the comb and white seal tape at the bottom of the gel. Rinse the wells thoroughly with dH_2O to remove any residual acrylamide.
3. Assemble the gel electrophoresis system using two gels or one gel and a buffer dam per SureLock™ tank: place the gel

cassette/dam on the electrode core with the shorter cassette edge facing inward toward the core and then place the core and gels into the SureLock™ Mini-Cell tank and lock the gel tension wedge to secure the gels such that two chambers (inner and outer) are formed.

4. Using 1× MES running buffer (chilled or at room temperature), fill the inner chamber above the level of the wells in the gel and the outer chamber such that at a minimum the wires from the electrode core are submerged.
5. Heat samples briefly (approximately 5–10 min) at 100 °C, and pulse spin the samples in a microfuge to bring down any condensation.
6. Load 20 μL (i.e., all that is in the tube) of prepared samples carefully into wells, avoiding overflow from well. Allow one well for loading 5 μL of SeeBlue®Plus2 (or other suitable) molecular weight standard, usually at one end of the gel for sample orientation purposes.
7. Attach the lid of the electrophoresis tank and connect to the power supply. Run gel at 200 V for approximately 45 min to 1 h or until the dye front has reached the bottom of the gel (*see* **Note 14**).

3.4 Western Blotting

The following instructions assume the use of the Bio-Rad Criterion™ Blotter, which can accommodate up to two full mini gels per transfer cassette, although adaptation to other western blotting systems is possible.

1. Just prior to the end of PAGE, set up for western blotting: presoak PVDF membrane (cut to be just larger than one or two mini gels, depending on requirements) in 100% methanol, cut six pieces of blotting/filter paper, just larger than the size of the PVDF membrane, and in a tray large enough for one open transfer cassette, add enough 1× transfer buffer to presoak two sponges.
2. Disconnect the gel electrophoresis unit and disassemble the core/tension wedge/gel cassette.
3. Using the spatula tool provided, carefully open the gel cassette and excise the wells and approximately 5 mm from the bottom of the gel to release the gel from the cassette.
4. Set up transfer sandwich in the tray, on top of the black side of the open transfer cassette as follows, taking care not to allow any bubbles to form between each consecutive layer by using the provided roller tools: sponge, three pieces of filter paper, gel, and PVDF membrane, three pieces of filter paper, sponge. Carefully close the transfer cassette (*see* **Note 15**).

5. Place the transfer cassette into the Criterion™ tank, such that the black side of the cassette is alongside the black wall of the electrode module. This ensures that the gel and membrane are correctly positioned so that the direction of electricity flow transfers proteins resolved in the gel electrophoretically onto the membrane.
6. Add a magnetic stirrer bar and ice pack to the buffer tank, and then fill the tank with cold 1× transfer buffer. Place tank on a magnetic stirrer plate, and turn on to activate the bar, so as to ensure continued circulation of the cold transfer buffer around the tank. Place the lid on the buffer tank, connect to power supply, and transfer at 100 V for 30 min.
7. Immediately prior to the end of the transfer, prepare blocking buffer and pour into blotting tub/container (*see* **Note 16**).
8. After transfer is finished, disassemble transfer sandwich and immediately place membrane into tub containing the blocking solution, such that the proteins will be on top of the membrane, not touching the bottom of the container. The benchmark molecular weight marker should be clearly visible on the membrane. Gel and filter paper can be discarded.
9. Block PVDF membrane for a minimum of 1 h at room temperature (*see* **Note 17**).
10. Immediately prior to the end of the blocking step, prepare the appropriate dilution of primary antibody in freshly made blocking solution (Table 1) (*see* **Note 16**).
11. Remove blocking buffer and discard, then immediately add the primary antibody solution, place a lid on the container and ensure airtight seal (so as to avoid buffer evaporation), and then place container on rocking platform at 4 °C to incubate overnight (approximately 16 h) (*see* **Note 18**).
12. Wash off primary antibody: Remove the primary antibody (*see* **Note 19**), rinse the membrane twice with 1× PBS.T, and then wash the membrane for a minimum of 30 min with a minimum of two changes of PBS.T. A "rinse" is defined as adding a small volume of 1× PBS.T to container and agitating briefly before discarding PBS.T, whereas a "wash" is adding at a larger volume of 1× PBS.T and placing container on a rocking platform at room temperature for a period of time (*see* **Note 20**).
13. After washing off the primary antibody, incubate the PVDF membrane in an appropriate dilution of secondary antibody, prepared in freshly made block buffer (Table 1), for 45 min to 1 h at room temperate on a rocking platform.
14. Wash off secondary antibody with 1× PBS.T, following the method outlined for washing off the primary antibody (*see* **step 12**).

15. Wash the membrane for approximately 5 min in 1× PBS to dilute out/remove any Tween20 detergent.
16. During the PBS wash, prepare ECL™ Prime chemiluminescence reagents by mixing equal parts solution A (luminol enhancer) and solution B (peroxide) and then expose the PVDF membrane to the ECL™ reagents for 5 min (*see* **Note 21**).
17. Secure the PVDF membrane between two sheets of overhead transparency so that it does not dry out.
18. Use a Fujifilm LAS-3000 digital image reader (or similar) to capture images at various exposure times as appropriate (*see* **Note 22**).

4 Notes

1. The cOmplete ULTRA mini tablets are stable when stored whole at 4 °C.
2. Appropriate personal protective equipment and care must be taken when preparing the 10% (w/v) SDS solution, as SDS solid/powder is a strong irritant. Alternatively, 10% SDS solution can be purchased directly from an appropriate supplier.
3. The SDS can sometimes precipitate out of solution. If this occurs, warm the solution or buffer at 37 °C until it has resolubilized.
4. Other percent acrylamide gels, for example, 12%, can be used, so long as there is adequate separation of small molecular weight proteins.
5. 1× MES running buffer can be reused approximately five times.
6. 1× transfer buffer can be reused approximately five times.
7. Cell pellets can be used immediately for experimentation or can be stored at either −20 °C (short term) or −80 °C (preferable) prior to use.
8. It is preferable to lyse cells in a relatively small volume, so as to make a concentrated (protein) lysate, which can be diluted further if required. Typically a 25 cm^2 (T25) flask yields a small pellet that is lysed in 100 μL or less, and a 75 cm^2 (T75) flask yields a pellet that is lysed in 150–200 μL.
9. The protocol is for a 15 μL final reaction volume, for ease of loading into wells, but can be adjusted for larger volumes where appropriate.
10. Generally there is better PNGaseF digest (i.e., less chance of "incomplete" digest) and more uniform heating of samples, if tubes are incubated in/surrounded by water. Therefore, if a

water bath is not available for use, a beaker or container of tap water can be placed inside a dry incubator/oven set at 37 °C to preheat, which effectively creates a 37 °C water bath.

11. The final total protein concentration can be increased to an approximate maximum of 100 μg in 10 μL, if PrP^C expression levels are low.
12. If needed, prepare an "untreated" control by omitting the PNGaseF and adding an extra 1 μL of lysis buffer or dH_2O. Untreated samples are otherwise treated identically to those containing PNGaseF.
13. Samples in sample buffer can be stored for longer periods at −20 °C until required for PAGE. In such cases, samples should be thawed at 4 °C prior to boiling (*see* Subheading 3.3, **step 4**).
14. Sometimes the running buffer can heat up considerably during electrophoresis, which may cause the gel to melt or warp. Approaches to minimize the chance of this happening:
    (a) Use chilled 1× running buffer, and if necessary replace buffer during electrophoresis with fresh/new chilled buffer.
    (b) Add a larger volume of (chilled) running buffer to the outer chamber (this chamber can be completely filled if necessary).
    (c) Resolve the gel slower at a lower voltage.
15. If the sponges are thin or "deflated," or if using a 1.0 mm instead of 1.5 mm gel, add more sheets of filter paper, to ensure good contact of each layer when the cassette is clamped closed.
16. For one mini gel, depending on the size of the blotting container, 15–20 mL of blocking buffer or antibody solution is sufficient.
17. Blocking can also take place overnight at 4 °C.
18. The primary antibody can be left on longer at 4 °C or for 2 or more hours at room temperature; however, this may increase background staining.
19. Primary antibodies, if precious or in limited supply, can often be kept (diluted in the block buffer, stored at −20 °C) and reused several times, although immunoreactivity typically reduces with each use.
20. Longer wash times generally reduce background and nonspecific staining and therefore make for a clearer image.
21. In order to minimize the volume of ECL reagents used, the following technique can be used:
    (a) Pipette approximately 1 mL per mini gel of prepared ECL solution onto a clean overhead transparency.

(b) Remove the membrane from the PBS and drain excess PBS by briefly holding one edge of the PVDF against a Kimwipe tissue.

(c) Gently lay the membrane face down on top of the ECL, allowing the solution to draw out under the whole membrane.

(d) After 5 min, place a second overhead transparency over the top of the membrane, pressing down to remove excess ECL reagents.

22. Instead of, or in addition to, using a digital imaging system to record and retain an image, the membrane can be exposed to film and developed/fixed through a standard photography film development protocol.

Acknowledgments

The author thanks Associate Professor Victoria Lawson for the 03R19 and 03R22 polyclonal antibodies, as well as the original RK13 cell line. This work was supported by an Australian Government National Health and Medical Research Council (NHMRC) program grant (#628946) and Training Fellowship (#567123) and a University of Melbourne Early Career Researcher Fellowship.

References

1. Lawson VA, Collins SJ, Masters CL et al (2005) Prion protein glycosylation. J Neurochem 93(4):793–801
2. Lewis V (2011) Proteolytic processing of the prion protein. In: Collins SJ, Lawson VA (eds) The cellular and molecular biology of prion disease. Research Signpost, Kerala, India
3. Ehlers MR, Riordan JF (1991) Membrane proteins with soluble counterparts: role of proteolysis in the release of transmembrane proteins. Biochemistry 30(42):10065–10074
4. Haigh CL, Collins SJ (2016) Endoproteolytic cleavage as a molecular switch regulating and diversifying prion protein function. Neural Regen Res 11(2):238–239
5. Lewis V, Hill AF, Haigh CL et al (2009) Increased proportions of C1 truncated prion protein protect against cellular M1000 prion infection. J Neuropathol Exp Neurol 68 (10):1125–1135
6. Jimenez-Huete A, Lievens PM, Vidal R et al (1998) Endogenous proteolytic cleavage of normal and disease-associated isoforms of the human prion protein in neural and non-neural tissues. Am J Pathol 153(5):1561–1572
7. Kuczius T, Koch R, Keyvani K et al (2007) Regional and phenotype heterogeneity of cellular prion proteins in the human brain. Eur J Neurosci 25(9):2649–2655
8. Lewis V, Johanssen VA, Crouch PJ et al (2016) Prion protein "gamma-cleavage": characterizing a novel endoproteolytic processing event. Cell Mol Life Sci 73(3):667–683
9. Chen SG, Teplow DB, Parchi P et al (1995) Truncated forms of the human prion protein in normal brain and in prion diseases. J Biol Chem 270(32):19173–19180
10. Vincent B, Paitel E, Saftig P et al (2001) The disintegrins ADAM10 and TACE contribute to the constitutive and phorbol ester-regulated normal cleavage of the cellular prion protein. J Biol Chem 276(41):37743–37746
11. Taylor DR, Parkin ET, Cocklin SL et al (2009) Role of ADAMs in the ectodomain shedding and conformational conversion of the prion protein. J Biol Chem 284(34):22590–22600
12. Cisse MA, Sunyach C, Lefranc-Jullien S et al (2005) The disintegrin ADAM9 indirectly contributes to the physiological processing of

cellular prion by modulating ADAM10 activity. J Biol Chem 280(49):40624–40631
13. Endres K, Mitteregger G, Kojro E et al (2009) Influence of ADAM10 on prion protein processing and scrapie infectiosity in vivo. Neurobiol Dis 36(2):233–241
14. McDonald AJ, Dibble JP, Evans EG et al (2014) A new paradigm for enzymatic control of alpha-cleavage and beta-cleavage of the prion protein. J Biol Chem 289(2):803–813
15. Pan T, Wong P, Chang B et al (2005) Biochemical fingerprints of prion infection: accumulations of aberrant full-length and N-terminally truncated PrP species are common features in mouse prion disease. J Virol 79(2):934–943
16. Mange A, Beranger F, Peoc'h K et al (2004) Alpha- and beta- cleavages of the amino-terminus of the cellular prion protein. Biol Cell 96(2):125–132
17. Pan T, Li R, Wong BS et al (2002) Heterogeneity of normal prion protein in two- dimensional immunoblot: presence of various glycosylated and truncated forms. J Neurochem 81(5):1092–1101
18. Watt NT, Taylor DR, Gillott A et al (2005) Reactive oxygen species-mediated beta-cleavage of the prion protein in the cellular response to oxidative stress. J Biol Chem 280 (43):35914–35921
19. McMahon HE, Mange A, Nishida N et al (2001) Cleavage of the amino terminus of the prion protein by reactive oxygen species. J Biol Chem 276(3):2286–2291
20. Pastore A, Zagari A (2007) A structural overview of the vertebrate prion proteins. Prion 1 (3):185–197
21. Kanaani J, Prusiner SB, Diacovo J et al (2005) Recombinant prion protein induces rapid polarization and development of synapses in embryonic rat hippocampal neurons in vitro. J Neurochem 95(5):1373–1386
22. Chen S, Yadav SP, Surewicz WK (2010) Interaction between human prion protein and amyloid-beta (Abeta) oligomers: role of N-terminal residues. J Biol Chem 285 (34):26377–26383
23. You H, Tsutsui S, Hameed S et al (2012) Abeta neurotoxicity depends on interactions between copper ions, prion protein, and N-methyl-D-aspartate receptors. Proc Natl Acad Sci U S A 109(5):1737–1742
24. Haigh CL, Drew SC, Boland MP et al (2009) Dominant roles of the polybasic proline motif and copper in the PrP23-89-mediated stress protection response. J Cell Sci 122(Pt 10):1518–1528
25. Haigh CL, Lewis VA, Vella LJ et al (2009) PrPC-related signal transduction is influenced by copper, membrane integrity and the alpha cleavage site. Cell Res 19(9):1062–1078
26. Guillot-Sestier MV, Sunyach C, Druon C et al (2009) The alpha-secretase-derived N-terminal product of cellular prion, N1, displays neuroprotective function in vitro and in vivo. J Biol Chem 284(51):35973–35986
27. Guillot-Sestier MV, Sunyach C, Ferreira ST et al (2012) α-Secretase-derived fragment of cellular prion, N1, protects against monomeric and oligomeric amyloid β (Aβ)-associated cell death. J Biol Chem 287(7):5021–5032
28. Sunyach C, Cisse MA, da Costa CA et al (2007) The C-terminal products of cellular prion protein processing, C1 and C2, exert distinct influence on p53-dependent staurosporine-induced caspase-3 activation. J Biol Chem 282(3):1956–1963
29. Shmerling D, Hegyi I, Fischer M et al (1998) Expression of amino-terminally truncated PrP in the mouse leading to ataxia and specific cerebellar lesions. Cell 93(2):203–214
30. Sakaguchi S, Katamine S, Shigematsu K et al (1995) Accumulation of proteinase K-resistant prion protein (PrP) is restricted by the expression level of normal PrP in mice inoculated with a mouse-adapted strain of the Creutzfeldt-Jakob disease agent. J Virol 69 (12):7586–7592
31. Bueler H, Aguzzi A, Sailer A et al (1993) Mice devoid of PrP are resistant to scrapie. Cell 73 (7):1339–1347
32. Mallucci G, Dickinson A, Linehan J et al (2003) Depleting neuronal PrP in prion infection prevents disease and reverses spongiosis. Science 302(5646):871–874
33. Fischer M, Rulicke T, Raeber A et al (1996) Prion protein (PrP) with amino-proximal deletions restoring susceptibility of PrP knockout mice to scrapie. EMBO J 15(6):1255–1264
34. Lawson VA, Priola SA, Wehrly K et al (2001) N-terminal truncation of prion protein affects both formation and conformation of abnormal protease-resistant prion protein generated in vitro. J Biol Chem 276(38):35265–35271
35. Westergard L, Turnbaugh JA, Harris DA (2011) A naturally occurring C-terminal fragment of the prion protein (PrP) delays disease

and acts as a dominant-negative inhibitor of PrPSc formation. J Biol Chem 286 (51):44234–44242

36. Johanssen VA, Johanssen T, Masters CL et al (2014) C-terminal peptides modelling constitutive PrPC processing demonstrate ameliorated toxicity predisposition consequent to alpha-cleavage. Biochem J 459(1):103–115
37. Turnbaugh JA, Unterberger U, Saa P et al (2012) The N-terminal, polybasic region of PrP(C) dictates the efficiency of prion propagation by binding to PrP(Sc). J Neurosci 32(26):8817–8830
38. Rutishauser D, Mertz KD, Moos R et al (2009) The comprehensive native interactome of a fully functional tagged prion protein. PLoS One 4(2):e4446
39. Lawson VA, Vella LJ, Stewart JD et al (2008) Mouse-adapted sporadic human Creutzfeldt-Jakob disease prions propagate in cell culture. Int J Biochem Cell Biol 40(12):2793–2801

Chapter 11

Cellular Analysis of Adult Neural Stem Cells for Investigating Prion Biology

Cathryn L. Haigh[#]

Abstract

Traditional primary and secondary cell cultures have been used for the investigation of prion biology and disease for many years. While both types of cultures produce highly valid and immensely valuable results, they also have their limitations; traditional cell lines are often derived from cancers, therefore subject to numerous DNA changes, and primary cultures are labor-intensive and expensive to produce requiring sacrifice of many animals. Neural stem cell (NSC) cultures are a relatively new technology to be used for the study of prion biology and disease. While NSCs are subject to their own limitations—they are generally cultured ex vivo in environments that artificially force their growth—they also have their own unique advantages. NSCs retain the ability for self-renewal and can therefore be propagated in culture similarly to secondary cultures without genetic manipulation. In addition, NSCs are multipotent; they can be induced to differentiate into mature cells of central nervous system (CNS) linage. The combination of self-renewal and multipotency allows NSCs to be used as a primary cell line over multiple generations saving time, costs, and animal harvests, thus providing a valuable addition to the existing cell culture repertoire used for investigation of prion biology and disease. Furthermore, NSC cultures can be generated from mice of any genotype, either by embryonic harvest or harvest from adult brain, allowing gene expression to be studied without further genetic manipulation. This chapter describes a standard method of culturing adult NSCs and assays for monitoring NSC growth, migration, and differentiation and revisits basic reactive oxygen species detection in the context of NSC cultures.

Key words Prion, Neural stem cell, Neurosphere, Colony-forming assay, Migration, Differentiation, ROS, DCFDA

1 Introduction

Neural stem cells (NSCs), by definition, demonstrate self-renewal and multipotency; they are able to differentiate into cells of CNS lineage, including neurons, astrocytes, and oligodendrocytes. It is readily possible to extract NSC cultures from knockout, wild-type, and PrP-overexpressing transgenic mice. Using these cultures it has

[#]Present address: Laboratory of Persistent Viral Diseases, Rocky Mountain Laboratories, National Institute of Allergy and Infectious Diseases, National Institutes of Health, 903 South 4th Street, Hamilton, MT 59840.

Victoria A. Lawson (ed.), *Prions: Methods and Protocols*, Methods in Molecular Biology, vol. 1658,
DOI 10.1007/978-1-4939-7244-9_11, © Springer Science+Business Media LLC 2017

been shown that PrP expression positively influences NSC renewal in cultures extracted from both murine embryos and adult mouse brain and increases the number of proliferating cells in vivo [1]. The effect of PrP expression on self-renewal and differentiation in murine embryonic stem cells was linked with transcriptional modulation of the pluripotency marker *Nanog* [2] and with the Notch signaling pathway [3]. PrP modulation of human embryonic stem cells has also been demonstrated, resulting in delayed differentiation in favor of increased proliferation and modulation of the lineage preference of spontaneously differentiating cells [4, 5]. Additionally, in the murine NSC cultures, a positive association was also found between PrP expression and neuronal differentiation [1]. Further, PrP expression increases when differentiation is induced [6], increased PrP expression favors more rapid maturation of astrocytes [7], and PrP expression is linked with increased neurogenesis following ischemic insult [8]. NSC cultures can also be infected with prions, although there is disparity in the literature as to whether proliferating NSCs or their differentiated progenies propagate infection [9–12]. We and others have found that both proliferating and differentiated cultures of NSCs display prion pathology [13–17] making them a valuable model for studying prion infection and toxicity. Furthermore, the potential for NSCs to be used in brain repair during prion disease has been investigated with initial promising results [18].

Once extracted from the brain, NSCs can be propagated in cytokine-rich media as balls of cells in suspension called neurospheres. The methods described herein include a selection of stem cell-specific protocols and routine cell-based ROS detection as carried out when using NSCs.

2 Materials

2.1 Neurosphere Routine Culture

1. Mouse PrP null, wild-type, and Tga20 adult NSCs extracted as described in Haigh et al. [13] and based on previously developed extraction protocols [19, 20].
2. Neural stem cell proliferation media.
   (a) NeuroCult mouse NSC basal medium (Stemcell Technologies, Melbourne, Australia).
   (b) NeuroCult proliferation supplements (Stemcell Technologies)—the proliferation supplements should be aliquoted into ten 50 ml centrifuge tubes of 5 ml each and stored at −20 °C to avoid freeze-thawing.
3. Phosphate-buffered saline (PBS), minus calcium and magnesium, pH 7.2.
4. Bovine serum albumin (BSA).
5. Glacial acetic acid.

6. 1.5 ml microcentrifuge tubes.
7. 15 and 50 ml conical centrifuge tubes.
8. Fibroblast growth factor (recombinant human FGF).
9. Epidermal growth factor (recombinant EGF).
10. 0.2% (w/v) heparin solution.
11. Uncoated tissue culture flasks (Nunc—Thermo Fisher, Melbourne, Australia).
12. Plugged glass Pasteur pipettes.
13. Cell culture incubator, humidified, set to 37 °C and 5% (v/v) CO_2.
14. Optional: penicillin/streptomycin solution.

2.2 Neural Colony-Forming Assay (NCFA)

1. 24-well cell culture plates.
2. Cell strainers.
3. Collagen solution.
4. Bright-field microscope with camera and software capable of measuring sphere diameter (we use Nikon, Coherent Scientific, Australia).

2.3 Neural Migration Assay

1. 6-well culture plates.
2. Laminin and/or poly-D-lysine matrix solutions.
3. Bright-field microscope with camera and software capable of measuring area.

2.4 Differentiation Assays

1. Neural stem cell differentiation media.
   (a) NeuroCult mouse NSC basal medium (Stemcell Technologies).
   (b) NeuroCult differentiation supplements (Stemcell Technologies)—the differentiation supplements should be aliquoted into ten 50 ml centrifuge tubes of 5 ml each and stored at −20 °C to avoid freeze-thawing.
2. 8-well chambered coverslips.
3. Laminin and/or poly-D-lysine matrix solutions as above.
4. PBS, minus calcium and magnesium, pH 7.2.
5. 16% (v/v) paraformaldehyde vials.
6. Triton X-100.
7. Fetal bovine serum.
8. BSA.
9. Primary antibodies: anti-neurofilament-L antibody, anti-glial fibrillary acidic protein (*see* **Note 1**).
10. Secondary antibodies – Alexa Fluor 488/647 (*see* **Note 2**) conjugated anti-mouse/anti-rabbit (Invitrogen).

11. Mounting medium with DAPI and anti-fade reagents.
12. Epifluorescence microscope with camera and software.

2.5 2′,7′-Dichlorodihydrofluorescein Diacetate (DCFDA) Assay for Reactive Oxygen Species

1. 96-well cell culture microtiter plate.
2. Low-fluorescence sealing tape.
3. Laminin and/or poly-D-lysine matrix solutions.
4. Opti-MEM I reduced serum medium (1×), liquid (without phenol red) (Invitrogen).
5. Proliferation supplements and cytokines/differentiation supplements as required.
6. 5-(and-6)-Chloromethyl-2′,7′-dichlorodihydrofluorescein diacetate, acetyl ester (CM-H_2DCFDA) (Invitrogen).
7. High-quality anhydrous DMSO.
8. Fluorescent microplate reader with light source and filter set suitable for detecting fluorescein isothiocyanate (FITC) (excitation, 488 nm; emission, 530 nm) and associated analysis software.

3 Methods

All routine culture and sterile cell experiments are carried out in a laminar flow cabinet, with sterilization of the cabinet and contents done using 80% (v/v) ethanol while in use and UV irradiation of the cabinet upon completion of experiments. This protocol was developed from the product information provided for the NeuroCult system from Stemcell Technologies, and it is recommended that it should be read in conjunction with the NeuroCult Technical Manual. Prion-specific observations are stated.

3.1 Neurosphere Routine Culture

3.1.1 Preparation

1. FGF.
   (a) Prepare a 0.1% (w/v) bovine serum albumin (BSA) solution in PBS and filter sterilize through a 0.22 μm filter.
   (b) Dilute one vial of 25 μg FGF into 2.5 ml of the sterile BSA-PBS solution to produce a final concentration of 10 μg/ml FGF.
   (c) Aliquot into 50 μl and 100 μl in sterile microfuge tubes and store at −20 °C. Do not freeze-thaw.
2. EGF.
   (a) Prepare a 10 mM acetic acid solution with 0.1% (w/v) BSA in sterile H_2O.
   (b) Dissolve one vial of 200 μg EGF in 100 μl of the acetic acid-BSA solution.

(c) Add 19.9 ml of basal medium containing proliferation supplements.

(d) Aliquot into 310 μl in sterile microcentrifuge tubes and store at −20 °C. Do not freeze-thaw—once thawed an aliquot can be stored at 4 °C until used.

3. Proliferation media.

(a) Thaw one 50 ml centrifuge tube containing 5 ml proliferation supplements.

(b) To this tube add 45 ml NSC basal media, 50 μl 0.2% (w/v) heparin solution, 50 μl FGF, 100 μl EGF.

(c) Invert gently to mix.

(d) Warm to 37 °C (*see* **Note 3**).

4. Fire polished pipettes.

(a) Using the plugged glass Pasteur pipettes, rotate the tip of the pipette in the hottest part of a Bunsen flame. The final pipette tip should be smooth all the way around without significant reduction of the aperture.

(b) Autoclave to sterilize.

3.1.2 Routine Culture

1. For routine expansion, neurospheres should be grown in tissue culture vessels that have not been pre-coated or treated to increase adherence. Cells should be kept in a humidified tissue culture incubator at 37 °C with 5% CO_2.

2. Neurospheres should be passaged at 100–150 μm in size; this usually corresponds to 5–8 days in culture depending upon genotype. Spheres should not be allowed to become too large as the centers become necrotic reducing viability (this is less apparent for Tga20 PrP-overexpressing NSCs which tend to break apart forming smaller spheres in liquid culture). Figure 1 shows example spheres in culture.

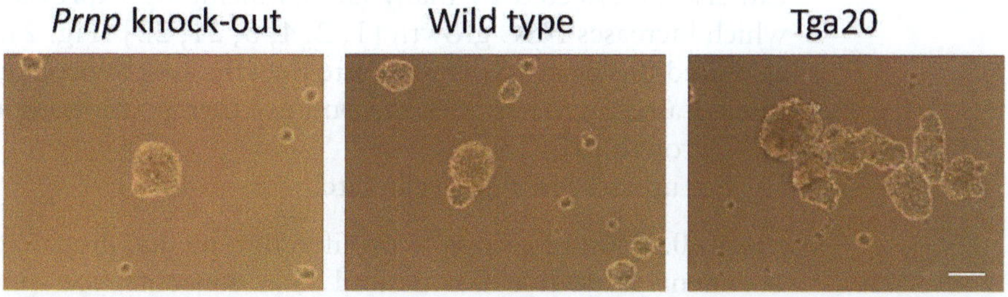

Fig. 1 Neurosphere morphology in suspension culture. Prnp knockout and wild-type spheres form balls of tightly adhered cells (with adhesion tightest for the knockout spheres) that show defined sphere boundaries. Tga20, PrP overexpressing, cells form less regular spheres with lower adhesion and cells showing a tendency to form satellite spheres on the edge of the main sphere resulting in less defined boundaries. Scale bar = 100 μm

3. Using 15 ml conical centrifuge tubes, collect neurospheres by centrifugation at 100 × *g* for 5 min (room temperature).
4. Discard media.
5. Using a fire-polished, plugged glass Pasteur pipette, transfer sufficient PBS (*see* **Note 4**) to resuspend the cell pellet and break up the neurospheres into a single-cell suspension by trituration (approximately 20–50 times depending on genotype; *prnp* knockout spheres tend to be most tightly adhered with Tga20 spheres easily broken apart). It is important to avoid introducing air bubbles to the suspension as much as possible while triturating as this will greatly reduce viability especially for the *prnp* knockout cells.
6. Wash cells 1× in PBS (*see* **Note 4**) and collect by centrifugation at 150 × *g* for 5 min (room temperature).
7. Resuspend in an appropriate amount of proliferation media (or PBS if the population is to be used for both proliferation and differentiation).
8. Count cells (*see* **Note 5**).
9. Cells should be seeded at the following densities in proliferation media for ongoing culture, 5×10^5/T25 flask (in 4 ml media) or 1.5×10^6/T75 flask (in 12 ml media), and returned to the incubator (*see* **Note 6**).
10. During culture cells can adhere to the flasks (this is particularly evident for the Tga20 cells); gentle agitation, by tapping the side of the flask, will dislodge the cells back into suspension.

3.2 Neural Colony-Forming Assay (NCFA)

The NCFA measures the self-renewal capacity of the NSCs. Counting the number of colonies formed indicates the number of single cells in the population that are able to divide and form new spheres, and sphere size is an indication of cell cycle time; larger spheres contain cells that are cycling more quickly than smaller spheres. Cell cycling is reduced as NSCs mature to be committed precursors and can also be affected by many factors including expression of PrP, which increases NSC growth [1, 2, 4, 5, 21, 22] (Fig. 2). We have also used this assay to demonstrate the effects of neurodegenerative disease-associated peptides and putative therapeutic compounds on NSC growth [22, 23].

Set up assay in a 24-well plate.

1. In 0.25 ml of complete proliferation media, prepare cell treatment at double the desired final concentration (include no treatment/carrier molecule controls).
2. Collect cells by centrifugation at 100 × *g* for 5 min.
3. Create a single-cell suspension using mechanical dissociation as per "neurosphere passage" protocol.

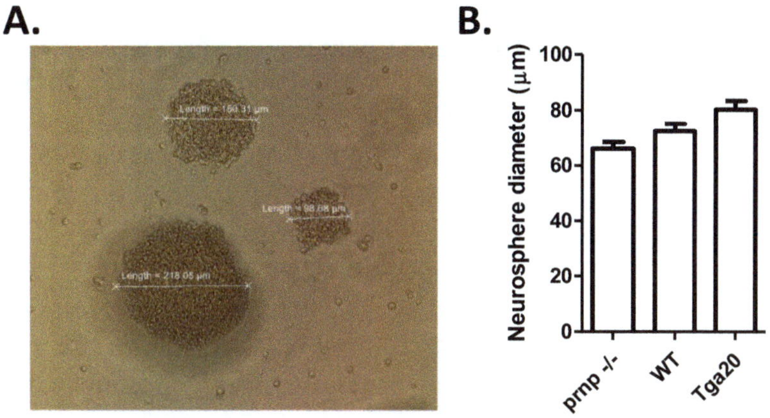

Fig. 2 Colony diameter measurement. (**a**) Bright-field image, captured at 10× magnification, showing example colonies formed after 21 days in culture. (**b**) Diameter quantification of prnp knockout (−/−), wild-type (WT), and overexpressing (Tga20) colonies, showing increased colony size with increased PrP expression

4. Place a cell strainer filter into the neck of a 50 ml conical centrifuge tube (*see* **Note 7**).
5. Gently drip the cell suspension onto the filter, and do not force the flow (as this will cause undissociated neurospheres to also pass through the filter).
6. Wash the filtered cells with PBS, and collect by centrifuging at 150 × *g* for 5 min.
7. Remove and discard PBS.
8. Resuspend cells in 1–2 ml of proliferation medium.
9. Count cells.
10. Add sufficient volume of cell suspension to contain 25,000 cells to the preprepared treatment medium and top up to a final volume of 0.365 ml with proliferation medium.
11. Add 0.135 ml of collagen solution (*see* **Note 8**) to each condition, mix gently by triturating a couple of times, and transfer the final 0.5 ml into one well of the 24-well plate.
12. Fill all leftover wells and well dividers with PBS to avoid plate dehydration.
13. Feed every 5 days with 60 μl complete proliferation media.
14. At 21 days count the number of colonies and measure colony diameters.

3.3 Neural Migration Assay

For integration of newly born neurons into the brain, the maturing cells must migrate to the site of the signals indicating their requirement. The below protocol describes a basic assay for observing migration.

3.3.1 Preparation

1. Laminin/PDL solutions.
   (a) 100 μg/ml poly-D-lysine in sterile water (e.g., 5 mg in 50 ml). Aliquot into appropriate sizes and store at −20 °C.
   (b) 10 μg/ml laminin in sterile water or PBS. Laminin stocks should be thawed in the fridge, aliquoted and stored at −20 °C until use.
2. Coat a 6-well culture plate with appropriate matrix.
   (a) Use 100 μg/ml PDL or 10 μg/ml laminin in sufficient volume to cover the well (*see* **Note 9**).
   (b) Incubate for at least 1 h at room temperature or overnight at 4 °C.
   (c) Wash once in PBS.
   (d) Leave PBS in well until ready to plate cells (do not allow plate to dry out).

3.3.2 Assay

1. Prepare cell treatments and controls at double the final concentration in 1 ml of proliferation medium.
2. Using neurospheres of approximately 100–150 μm in diameter (i.e., ready to be passaged), dilute into 1 ml of proliferation medium a sufficient volume of the neurosphere suspension to plate 20–50 spheres per well.
3. Add the 2 ml of spheres plus/minus treatment to each well.
4. Observe cells after 24 h and collect bright-field images at 1, 2, and 4 days. Measure the area migrated from the circumference of the sphere.
5. Optional: cells can be harvested post-assay for protein if desired.

3.4 NSC Differentiation Assays

There are a variety of assays that can be used to assess lineage distribution post differentiation. Here only indirect immunofluorescence (IIF) is described; flow cytometry and western blotting are viable alternatives; however, they will not provide information on any morphological changes displayed by the mature cells. A combination of techniques will provide the strongest data. Lineage preference during differentiation is influenced by both the environment stimulating differentiation and changes in gene expression caused by the environment to which the NSC is exposed before differentiation; as per the signals these cells would receive in their niche environment in the brain. Therefore, two variations on the differentiation assay are described: treatment during differentiation and treatment before beginning differentiation.

3.4.1 Preparation

1. Make a 4% (w/v) paraformaldehyde solution.
   (a) Dilute the 16% (w/v) solution 1 in 4 in PBS.
2. Differentiation media.
   (a) Mix 5 ml differentiation supplements to 45 ml basal media. Once made up differentiation media is stable at 4 °C for approximately 7–14 days.

3.4.2 Assay

1. Coat 8-well chambered coverslips with 100 μg/ml PDL or 10 μg/ml laminin as described for migration plates.
2. Prepare cell treatments and controls at double the final concentration in 0.25 ml of differentiation or proliferation medium (determined by whether you wish to do variation 1 or 2 or the assay as described in **steps** 7 and **8**).
3. Collect cells by centrifugation at 100 × *g* for 5 min.
4. Create a single-cell suspension using mechanical dissociation as per "neurosphere passage" protocol (*see* **Note 10**).
5. Wash cells in PBS, and collect cells by centrifuging at 150 × *g* for 5 min.
6. Remove and discard PBS.
7. Differentiation assay variation 1 (effect of test compound on lineage preference during differentiation). Resuspend cells in 1–2 ml of differentiation media.
8. Differentiation assay variation 2 (effect of test compound on lineage preference during growth prior to differentiation). Resuspend cells in 1–2 ml of proliferation media.
9. Count cells.
10. Dilute cells to be 3.2×10^5/ml in a sufficient volume of desired media for the number of wells required (1 ml per four wells).
11. Mix 0.25 ml cells with the 0.25 ml of differentiation medium (plus/minus treatment) and transfer the 0.5 ml into one well of the chambered coverslip. This results in a final concentration of 80,000 cells/well.
12. For cells treated during proliferation, change media to differentiation media after 24 h.
13. Incubate cells for 7 days in differentiation media.
14. Change media as it begins to turn acidic (repeated compound treatments can be done depending on the nature of the assay).
15. Fix cells by removing media, washing in PBS, and incubating in 4% (w/v) paraformaldehyde in PBS for 30 min in a fume hood.
16. Wash 3 × 3 min in PBS.
17. Permeabilize cells in 0.1% (v/v) Triton X-100 in PBS for 10 min.

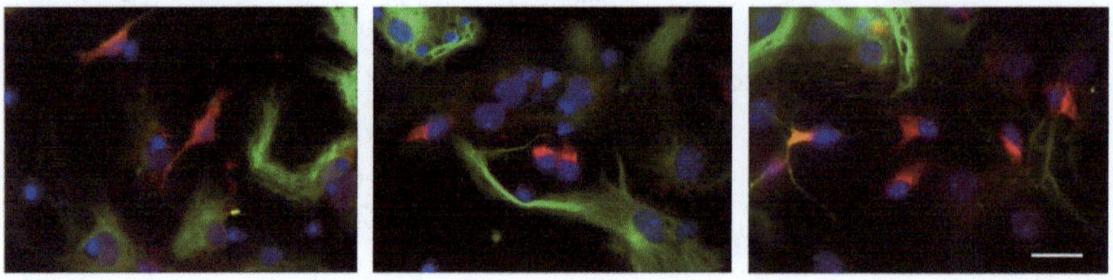

Fig. 3 Neuronal and astrocytic immunostaining of NSCs following 7 days in differentiation medium. Neurofilament-L neuronal staining is shown in red, GFAP astrocytic staining in green, and DAPI nuclei staining in blue. Scale bar = 20 μm

18. Wash 3 × 3 min in PBS.
19. Block in 10% (v/v) FBS and 0.5% (w/v) BSA in PBS for 30 min.
20. Wash once in PBS.
21. Incubate in primary antibody in 1% (v/v) FBS and 0.5% (w/v) BSA in PBS for a minimum of 2 h at room temperature (or 37 °C) or overnight at 4 °C.
22. Wash 3 × 5 min PBS.
23. Incubate in secondary antibody in 1% (v/v) FBS and 0.5% (w/v) BSA in PBS for 2 h at room temperature or 37 °C. Coverslips should be protected from light from this point on.
24. Wash 3 × 5 min in PBS.
25. Mount chambered coverslips by putting 2–3 drops mounting media with DAPI (*see* **Note 11**) into each chamber.
26. Allow to cure at room temperature for 24 h.
27. Store at 4 °C in the dark until viewing under an epifluorescence microscope equipped with appropriate filter sets for detection of the secondary antibody fluorophores and DAPI (Fig. 3).

3.5 DCFDA Assay

This assay was previously described in Haigh and Brown [24]; the below protocol has been modified for use with NSCs grown as adherent cultures.

1. Coat 96-well plates with a laminin/PDL matrix as described previously.
2. Seed cells at a density of 3.6 × 10^4 viable cells/well in proliferation medium. Allow cells to adhere to the matrix overnight. If differentiated cells are required, incubate for 7 days in differentiation medium, changing medium if it becomes acidic.
3. Prepare a 1 mM probe stock solution by diluting 50 μg (one vial) into 86.54 μl of high-quality anhydrous DMSO.
4. Prepare the 5 μM working solution by diluting 5 μl of probe stock per 1 ml of PBS.

5. Remove media from cells.
6. Add 50 μl of the 5 μM probe-PBS solution per well.
7. Incubate in the dark for 20 min at 37 °C.
8. Remove probe solution.
9. Add 100 μl of Opti-MEM plus proliferation supplements and cytokines as described for routine culture (or differentiation supplements if using differentiated cells) to each well.
10. Read the time 0 fluorescence intensity by using fluorescent microplate reader set to FITC wavelength excitation, 488 nm, and emission, 530 nm. Readings should be taken using a well-scanning mode to account for variation due to cell distribution.
11. Add control and test reagents to wells at least in triplicate.
12. After reagent addition, plates should be kept under incubator conditions, and fluorescence intensity can be measured as often as required; alternatively use a kinetic plate reader program to measure intensity over time using time 0/time of addition as the blank control.
13. At the conclusion of the assay, cells can be used for further assays or harvested as required.

4 Notes

1. The choice of primary antibody depends upon the lineage to be detected, and some antigens are expressed at different stages of differentiation. A guide to antibody choice depending upon experimental design can be found in [25]. The two antibodies used here are NF-L for mature neurons and GFAP, which is generally considered an astrocytic marker but is also expressed in NSCs.
2. Secondary antibodies should have fluorophores of sufficiently separated fluorescence emission spectra as to not overlap and can be chosen based upon filters available.
3. Once made up proliferation media is only kept for up to 7 days stored at 4 °C.
4. Penicillin/streptomycin can be included in the PBS to reduce the risk of introducing an infection during passage but is not included in normal media due to genetic pressure on the cells.
5. We count cells using trypan blue exclusion by either an automated count using a Countess cytometer (Invitrogen) or manually using a hemocytometer.
6. During culture, media should not be allowed to become too acidic. *Prnp* knockout cells turn media acidic faster than

wild-type or Tga20 cells. If the media is beginning to yellow in color, top up with fresh media or perform a half media exchange. Note that flasks should not be overfilled as this increases the risk of microbial contamination.

7. Oversized spheres trapped on the filter can be collected and transferred into fresh media for continued culture or passage.
8. Collagen solution is liquid at 4 °C and solidifies at 37 °C; keep this solution cool while using or in the fridge until just prior to use to avoid it solidifying.
9. NSC migration (and the morphology of migrating cells) will vary greatly depending upon which matrix is used [26] and other matrices to those mentioned are available. Researchers may wish to consider trying several before ruling out migration changes.
10. Depending on the nature of the differentiation assay, cells can be induced to differentiate directly from neurospheres or from a single-cell suspension (as described). Each experimental setup will provide different information as the position of a cell in the neurosphere will limit its exposure to the surrounding environment.
11. If mounting media containing DAPI is not used, DAPI can be included in the final PBS wash at a concentration of 1 μg/ml.

References

1. Steele AD, Emsley JG, Ozdinler PH et al (2006) Prion protein (PrPc) positively regulates neural precursor proliferation during developmental and adult mammalian neurogenesis. Proc Natl Acad Sci U S A 103:3416–3421
2. Miranda A, Pericuesta E, Ramirez MA, Gutierrez-Adan A (2011) Prion protein expression regulates embryonic stem cell pluripotency and differentiation. PLoS One 6: e18422
3. Martin-Lanneree S, Halliez S, Hirsch TZ et al (2016) The cellular prion protein controls notch signaling in neural stem/progenitor cells. Stem Cells 35(3):754–765. doi:10.1002/stem.2501
4. Lee YJ, Baskakov IV (2013) The cellular form of the prion protein is involved in controlling cell cycle dynamics, self-renewal, and the fate of human embryonic stem cell differentiation. J Neurochem 124:310–322
5. Lee YJ, Baskakov IV (2010) Treatment with normal prion protein delays differentiation and helps to maintain high proliferation activity in human embryonic stem cells. J Neurochem 114:362–373
6. Peralta OA, Huckle WR, Eyestone WH (2011) Expression and knockdown of cellular prion protein (PrPC) in differentiating mouse embryonic stem cells. Differentiation 81:68–77
7. Hartmann CA, Martins VR, Lima FR (2013) High levels of cellular prion protein improve astrocyte development. FEBS Lett 587:238–244
8. Doeppner TR, Kaltwasser B, Schlechter J et al (2015) Cellular prion protein promotes post-ischemic neuronal survival, angioneurogenesis and enhances neural progenitor cell homing via proteasome inhibition. Cell Death Dis 6:e2024
9. Herva ME, Relano-Gines A, Villa A, Torres JM (2010) Prion infection of differentiated neurospheres. J Neurosci Methods 188:270–275
10. Giri RK, Young R, Pitstick R et al (2006) Prion infection of mouse neurospheres. Proc Natl Acad Sci U S A 103:3875–3880
11. Milhavet O, Casanova D, Chevallier N et al (2006) Neural stem cell model for prion propagation. Stem Cells 24:2284–2291

12. Relano-Gines A, Gabelle A, Hamela C et al (2013) Prion replication occurs in endogenous adult neural stem cells and alters their neuronal fate: involvement of endogenous neural stem cells in prion diseases. PLoS Pathog 9: e1003485
13. Haigh CL, McGlade AR, Lewis V et al (2011) Acute exposure to prion infection induces transient oxidative stress progressing to be cumulatively deleterious with chronic propagation in vitro. Free Radic Biol Med 51:594–608
14. Sinclair L, Lewis V, Collins SJ, Haigh CL (2013) Cytosolic caspases mediate mislocalised SOD2 depletion in an in vitro model of chronic prion infection. Dis Model Mech 6:952–963
15. Iwamaru Y, Takenouchi T, Imamura M et al (2013) Prion replication elicits cytopathic changes in differentiated neurosphere cultures. J Virol 87:8745–8755
16. Biasini E, Turnbaugh JA, Massignan T et al (2012) The toxicity of a mutant prion protein is cell-autonomous, and can be suppressed by wild-type prion protein on adjacent cells. PLoS One 7:e33472
17. Collins SJ, Haigh CL (2016) Simplified murine 3D neuronal cultures for investigating neuronal activity and neurodegeneration. Cell Biochem Biophys 75(1):3–13. doi:10.1007/s12013-016-0768-z
18. Relano-Gines A, Lehmann S, Bencsik A et al (2011) Stem cell therapy extends incubation and survival time in prion-infected mice in a time window-dependant manner. J Infect Dis 204:1038–1045
19. Chojnacki A, Weiss S (2008) Production of neurons, astrocytes and oligodendrocytes from mammalian CNS stem cells. Nat Protoc 3:935–940
20. Ahlenius H, Kokaia Z (2010) Isolation and generation of neurosphere cultures from embryonic and adult mouse brain. Methods Mol Biol 633:241–252
21. Santos TG, Silva IR, Costa-Silva B et al (2011) Enhanced neural progenitor/stem cells self-renewal via the interaction of stress-inducible protein 1 with the prion protein. Stem Cells 29:1126–1136
22. Collins SJ, Tumpach C, Li QX et al (2015) The prion protein regulates beta-amyloid mediated self-renewal of neural stem cells in vitro. Stem Cell Res Ther 6:60
23. Haigh CL, Tumpach C, Collins SJ, Drew SC (2016) A 2-substituted 8-hydroxyquinoline stimulates neural stem cell proliferation by modulating ROS signalling. Cell Biochem Biophys 74:297–306
24. Haigh CL, Brown DR (2008) Investigation of PrPC metabolism and function in live cells: methods for studying individual cells and cell populations. Methods Mol Biol 459:21–34
25. García Gil-Perotin S, Alvarez A, Garcia-Verdugo JM (2009) Identification and characterization of neural progenitor cells in the adult mammalian brain. Adv Anat Embryol Cell Biol 203:1–101. Springer
26. Kearns SM, Laywell ED, Kukekov VK, Steindler DA (2003) Extracellular matrix effects on neurosphere cell motility. Exp Neurol 182:240–244

Chapter 12

Neurotoxicity of Prion Peptides on Cultured Cerebellar Neurons

Giuseppe D. Ciccotosto, Metta Jana, and Roberto Cappai

Abstract

Prion neurotoxicity has been modeled in vitro using synthetic peptides derived from the PrP^C sequence. The major region of neurotoxicity has been localized to the hydrophobic domain located in the middle of the PrP protein. Neurotoxicity assays are typically performed on cultured mouse cerebellar neurons derived from neonatal pups, and cell viability can be monitored by assays including MTT or MTS, cell death by LDH release, or apoptosis by caspase cleavage assays. These neurotoxicity studies have been useful in identifying cofactors, such as PrP^C and metals, as modulators of PrP peptide-mediated neurotoxicity. Given the biosafety issues associated with handling and purifying infectious prions, the use of synthetic peptides, which display a dependence upon PrP^C expression for toxicity, as per the PrP^{Sc} agent for infectivity, supports the relevance of using these synthetic peptides for understanding PrP-mediated neurotoxicity.

Key words Prion protein (PrP), $PrP_{106–126}$, Neurotoxicity, Cultured cerebellar neurons, Cell viability, Lipid peroxidation, immunofluorescence

1 Introduction

Gerstmann-Sträussler-Scheinker (GSS) disease is a prion-related encephalopathy that is characterized by the deposition of the prion protein in the central nervous system. Tagliavini et al. (1993) purified the amyloid deposits from a GSS kindred and identified a PrP fragment corresponding to residues 58–150 as a major component. They tested a series of synthetic peptides overlapping this region and identified peptides corresponding to residues 127–147 and 106–126 as having distinct fibril-forming properties [1]. $PrP_{127–147}$ formed twisted fibrils similar to scrapie-associated fibrils. $PrP_{106–126}$ formed straight fibrils similar to those extracted from GSS brains. They surmised that the region 106–147 encompassed the amyloid-forming domain of PrP. In relation to toxicity, Forloni et al. (1993) demonstrated that the synthetic $PrP_{106–126}$ peptide was neurotoxic in a neuronal cell culture model [2]. A range of synthetic peptides derived from the prion protein (PrP^C)

Victoria A. Lawson (ed.), *Prions: Methods and Protocols*, Methods in Molecular Biology, vol. 1658,
DOI 10.1007/978-1-4939-7244-9_12, © Springer Science+Business Media LLC 2017

display toxicity against cultured neurons in vitro[2–4] and in vivo [5]. $PrP_{106-126}$ peptide is neurotoxic and will induce apoptosis in several primary cell culture models including cortical [6–8], hippocampal [2, 9], microglial [7], and cerebellar neurons [10–13]. Further, cerebellar granule neurons from Prnp gene knockout mice were resistant to $PrP_{106-126}$ neurotoxic activity [10, 11] suggesting that the cellular prion protein, PrP^c, is mediating this neurotoxicity. In addition, $PrP_{106-126}$ toxicity activates pro-apoptotic pathways with upregulation of annexin V binding and caspase-3-, caspase-6-, and caspase-8-like activity [8, 9] as well as activating arachidonic acid metabolism through the 5-lipoxygenase pathway to mediate toxicity [13]. The nature of the $PrP_{106-126}$ toxic species indicates neurotoxicity appears to correlate with $PrP_{106-126}$ being in a non-amyloid/non-fibrillar state [14]. Our own sequence-activity studies have shown that the hydrophobic residues in the C-terminal hydrophobic region of $PrP_{106-126}$ are necessary for neurotoxicity [15]. Moreover, $PrP_{106-126}$ neurotoxic activity is dependent on the binding of copper and zinc metal ions to the peptide, while mutating either His_{111}, Met_{109}, or Met_{112} diminished the metal binding and subsequent toxicity [12].

2 Materials

Use rats at 6–8 days postnatal or mice at 5–7 days postnatal (the use of animals for experimental models must be approved by your local animal ethics committee).

2.1 Cell Culture Stock and Media Solutions (See Notes 1–4)

1. $MgSO_4$ (3.85 w/v % stock): Dissolve 1.54 g $MgSO_4·7H_2O$ in 50 mL dH_2O. Filter sterilize solution. Store at 4 °C.
2. $CaCl_2$ (1.2 w/v % stock): Dissolve 120 mg $CaCl_2·2H_2O$ in 10 mL of dH_2O. Filter sterilize solution. Store at RT.
3. Poly-D-lysine (PDL, 100× stock solution): Dissolve 5 mg in 10 mL dH_2O (0.5 mg/mL). Filter sterilize solution. Store in the dark at 4 °C (*see* **Note 5**).
4. Poly-D-lysine (1X, working solution): Add 0.5 mL of 0.5 mg/mL PDL to 49.5 mL dH_2O (5ng/mL). Filter sterilize solution. Store at 4 °C.
5. Trypsin (20× stock): Dissolve 25 mg in 10 mL Krebs buffer (~7500 units/mg). Filter sterilize solution. Store in 0.75 mL aliquots at −20 °C.
6. DNase/soy bean trypsin inhibitor (SBTI) (20× stock): Dissolve 8 mg DNase and 26 mg SBTI in 10 mL Krebs buffer. Filter sterilize. Store in 0.5 mL aliquots at −20 °C (*see* **Note 6**).

7. Cytosine β-D arabinofuranoside (AraC) (1000× stock): Dissolve 4.8 mg in 2 mL of dH_2O. Filter sterilize. Store in 50 μL aliquots at −20 °C.
8. Krebs buffer (10× stock): Dissolve 36.25 g of NaCl, 2.0 g KCl, 0.7 g $NaH_2PO_4.H_2O$, 13.0 g of D-glucose, 0.05 g of phenol red, and 29.7 g of HEPES acid in 450 mL of dH_2O. Adjust to pH 7.4. Bring volume to 500 mL. Filter sterilize. Store at 4 °C.
9. Krebs buffer: Dissolve 1.5 g of Bovine serum albumin (Fraction V) in 446 mL of dH_2O. Add 50 mL 10× Krebs stock. Add 4 mL of 3.85% $MgSO_4$. Filter sterilize. Store at 4 °C.
10. Dulbecco's Phosphate-Buffered Saline (D-PBS) (10× stock): Liquid contains no calcium or magnesium. Dissolve 2 g of KCl, 2 g of KH_2PO_4, 80 g of NaCl, and 21.6 g of $Na_2HPO_4{\cdot}7H_2O$ in 1 L of dH_2O. Adjust to pH 7.0.
11. D-PBS buffer: Add 50 mL of 10× D-PBS to 450 mL of dH_2O. Adjust to pH 7.1. Filter sterilize. Store at 4 °C.
12. BME plating media: To 500 mL bottle of Basal medium Eagle with Earle's salts (BME) (Invitrogen), add 50 mL of heat inactivated fetal calf serum (FCS) (*see* **Note** 7), 5 mL of 200 mM glutaMAX (Invitrogen), and 0.5 mL of 10 mg/mL gentamicin (Invitrogen). Store at 4 °C.
13. Primary neuronal growth medium: Add 1 mL of B27 supplements (Invitrogen) to 50 mL of Neurobasal (NB) media (Invitrogen), 50 μL of 10 mg/mL gentamicin, and 125 μL of 200 mM glutaMAX. Store unused media at 4 °C. Discard unused media after 3 to 5 days of storage.
14. Primary neuronal experimental medium: Add 1 mL of B27 supplements minus antioxidants (Invitrogen) to 50 mL of NB, 50 μL of 10 mg/mL gentamicin. Store unused media at 4 °C. Discard unused media after 3–5 days of storage.

2.2 Isolating Cerebellar Neurons

1. Dissecting instruments: one pair of small dissecting scissors, straight, 115 mm. One pair of large dissecting scissors straight, 170 mm. One tissue forceps with teeth, 125 mm. One pointed No.5 tweezers, 114 mm. One scalpel blade (profile 24). Dissecting instruments should be sterilized by autoclaving cleaned utensils or by soaking instruments in 80% ethanol (v/v) before using them.
2. Cell counting: hemacytometer (Neubauer, 0.1 mm depth). Coverglass for hemacytometer. Trypan blue solution (0.4%). Tally counter. Inverted light microscope.
3. Disposable plasticware and glassware: round dishes (30 and 90 mm diameter), polystyrene tubes (50 and 15 mL), and microfuge tubes (1.5 mL). Pipette tips, filter plugged, and

unplugged. Standard long glass Pasteur pipettes (cotton plugged and unplugged). Fire polished cotton plugged long Pasteur pipettes (*see* **Note 8**). 48-well plates (for toxicity or caspase assays). 12-well plates (for protein extractions). For immunofluorescence staining of cells, use 4- or 12-well plates and 12 mm glass coverslips #1.5 (glass chamber slides are also suitable). Plasticware and glassware must be clean and sterile before use.

2.3 Toxicity Assays

1. High-quality synthetic peptide $PrP_{106-126}$.
2. MTS assay (we use CellTiter96 Aqueous one solution cell proliferation assay by Promega) supplied as a 10× stock solution. Store unused 5–10 mL aliquots at −20 °C, and keep 1 aliquot at 4 °C with the tube wrapped in foil.
3. MTT assay (4,5-dimethylthiazol-2-yl)-2,5-diphenyltetrazolium). MTT comes in a yellow powder. Make a 10× stock by dissolving 5 mg/mL in D-PBS, and store at 4 °C with the tube wrapped in foil. If you plan to use MTT, you will also need DMSO to dissolve the formazan crystal deposits on the cells (after aspirating off the conditioned media).
4. LDH cytotoxicity assay (we use the LDH cytotoxicity assay kit II from BioVision).
5. Lipid peroxidation assay (we use LPO 586 lipid peroxidation kit from Oxis International).
6. Lipid peroxidation assay detection of MDA with thiobarbituric acid reactive substances (TBARS) assay uses TBA solution: 15% trichloroacetic acid, 1.25% TBA, and 5.5% hydrochloric acid.
7. Glutathione peroxidase (**GPx**) and reductase assays (we use GPx-340 and GR-340, respectively, Oxis International Inc., Foster City, CA).
8. BCA assay kit.
9. Plate reader which has a filter setting at 490 nm absorbance (MTS and MTT), 450 nm (LDH assay), 586 nm (lipid peroxidation assay), and 340 nm (glutathione peroxidase and reductase assays).
10. Spectrophotometer cuvette and read absorbance at 532 nm (TBARS assay).
11. Additional reagents: Tris–HCl buffer pH 7.4; 500 mM butylated hydroxytoluene; 50 mM Tris–HCl pH 7.5 containing 1 mM 2β-mercaptoethanol.
12. Polytron homogenizer.

2.4 Cell Binding Assay

1. Carbonate buffer; 0.1 M $NaHCO_3$.
2. Lysis buffer (20 mM Tris, pH 7.4, 1 mM EDTA, 0.25 M sucrose; 0.1% Triton X-100) containing a protease inhibitor mix.
3. BCA kit.
4. Electrophoresis equipment.
5. SDS-PAGE gel (e.g., NuPAGE® 12% Bis-Tris gel, Invitrogen, Australia).
6. Nitrocellulose membrane.
7. Peptide-specific and loading control antibodies.
8. Species-specific secondary antibodies.
9. Chemiluminescent developer.

2.5 Immuno-fluorescence Staining of Cells

1. Carbonate buffer ; 0.1 M $NaHCO_3$.
2. Blocking buffer: Add 10 mL of normal goat serum to 90 mL of D-PBS. Make aliquots and store at −20 °C for long-term storage. Store at 4 °C for short-term storage.
3. Preparation of 4% paraformaldehyde (PF) in D-PBS: Add 4 g of paraformaldehyde to 50 mL dH_2O in a glass bottle. Place a clean magnetic stirrer in solution, and place bottle on top of a heated stirrer. Turn on heat to high and stirrer. Monitor temperature of solution. Once it reaches 60 °C (this takes only a few minutes), turn off heating element. Add one drop of 1 N NaOH (only). The cloudy solution will become clear after a few minutes. Allow the mixture to cool to room temperature. Add 50 mL of 2× D-PBS. Check that pH is less than 8 using litmus stick (do not use pH meter). Store at 4 °C for less than 10 days or make aliquots and store at −20 °C. Note: PF is toxic. Prepare reagent in laminar flow hood and use PPE. Use dedicated glassware for the preparation of this reagent.
4. Permeabilization buffer 0.075% (w/v): Add 7.5 μL of Triton X-100 to 100 mL of block buffer. Mix and make 10 mL aliquots. Store at −20 °C.
5. Mounting media (we use ProLong Gold antifade mounting media, Life Technologies Invitrogen).

3 Methods

3.1 Setup Procedures

1. Coating with PDL: Coat plates, dishes, and glass coverslips with PDL overnight for best results (or for at least 2 h at 37 °C). Add enough PDL solution to totally cover the surface area of vessel. Before plating neurons, aspirate off solution and

let plates air dry with lids partially removed. Importantly, for the glass coverslips, after aspirating off the PDL solution, wash the wells using sterile water and then aspirate off the water.

2. Thaw out frozen aliquots of trypsin and DNase/SBTI.
3. Place the BME media in a 37 °C water bath.
4. Ensure that the dissecting instruments are clean, spray with 80% ethanol, and place in the laminar flow hood ready for dissection.
5. Label the following tubes and add the following solutions.

   Tube #1 (50 mL tube): Add 15 mL Krebs and 750 μL trypsin.

   Tube #2 (15 mL tube): Add 10 mL Krebs and 500 μL DNase/SBTI.

   Tube #3 (15 mL tube): Take 1.5 mL from tube #2.

   Tube #4 (15 mL tube): Add 6 mL Krebs and 7.5 μL $CaCl_2$.

   Tube "CELLS" (15 mL tube): Leave blank for now.

3.2 Cerebellar Granule Neuron Isolation Procedures

3.2.1 Isolating Cerebellum

1. The postnatal pups are separated from the mother and placed in a separate box. Keep warm (e.g., shine light directly into box but shielded from pups). Pups should be used within 30 min of being separated from the mother.
2. Place a pup on a clean 90 mm dish, and hold it firmly in the trunk region. Briefly spray 80% ethanol over the head region; then using the larger pair of sharp scissors, decapitate just above the shoulder region. Transfer the head falling into a sterile 90 mm dish, and discard the remaining carcass into a biohazard bag.
3. The whole brain is initially removed. This is done using the pair of smaller pointed scissors and by inserting one side of the scissors into the spinal cord canal opening, and move it toward the left ear (3–5 mm) while at the base of the head and cut. Repeat on the other side.
4. Using forceps, gently lift skull and skin up and tear away the tissue. Repeat until the brain is totally exposed.
5. Using fine forceps, gently tease the brain out of the base of the skull and onto the dish. Using the scalpel blade, cut away the cerebellum region from the brain (Fig. 1), and transfer it to the 30 mm dish containing Krebs buffer solution.
6. Using the fine forceps, gently tease off the meninges from the cerebellum.
7. Transfer cleaned cerebellum to new 60 mm dish.
8. Repeat procedure on other pups.

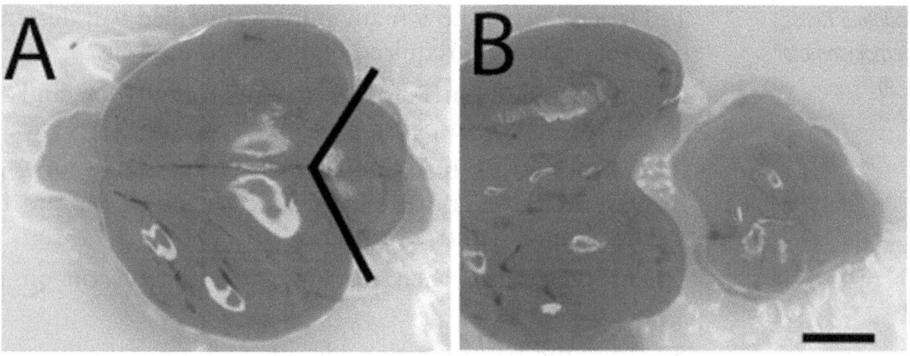

Fig. 1 The whole brain was removed from the skull of the mouse with thick black lines indicating dissection required to isolate the cerebellum **(A)**. The isolated cerebellum cut away from the whole brain **(B)**. *Scale bar* = 2 mm

3.2.2 Preparing a Single Cell Suspension

1. Using a scalpel blade, chop the cerebellum tissue into small <1 mm pieces. Add 1 mL of media from ***tube 1*** to chopped tissue. Collect the chopped tissue suspended in media and transfer into ***tube 1.***
2. Place ***tube 1*** into a gently shaking water bath set at 37 °C and incubate for 20 min.
3. Remove ***tube 1*** from water bath and spray with 80% ethanol, wipe tube down, and return tube to rack in laminar flow hood.
4. Pour the contents of ***tube 2*** into ***tube 1***, and mix gently by inverting tube several times. The DNA is digested by the DNase resulting in tissue pieces not sticking together.
5. Place ***tube 1*** into bench top centrifuge and spin down cells at 200 × *g* for 3 min. Spray tube with ethanol and return ***tube 1*** to hood. Aspirate off the supernatant without disturbing pellet. Do this step manually if you are concerned about aspirating the pellet.
6. Add contents of ***tube 3*** to ***tube 1***. Using the plugged fire polished Pasteur pipette, carefully triturate tissue suspension (i.e., pass up and down a pipette slowly) 20 times avoiding introducing bubbles in the suspension.
7. Add the contents of ***tube 4*** to ***tube 1*** and let the tube stand for 10 min at RT. Any undigested clumps should settle to the bottom of tube during this time. Carefully transfer the upper single cell suspension (avoiding any potential tissue pieces) to tube labeled "CELLS."
8. Centrifuge cell suspension at 200 × *g* for 5 min at RT. Spray tube with 80% ethanol, wipe down tube, and return to hood.
9. Aspirate off supernatant. Resuspend cell pellet initially in 1 mL of BME media (i.e., triturate three times); then dilute to 10 mL (suspension volume) with BME plating medium.

3.2.3 Counting Cells Using a Hemocytometer (See Note 9)

1. Depending on how many pups were killed will determine the level of dilution required before counting cells.
2. Firstly, add 0.4 mL of BME to a microfuge tube, and then add 0.1 mL of cell suspension to tube.
3. Transfer 50 μL to diluted cells to a microfuge tube, and mix with 50 μL of Trypan blue. Mix cells and slowly add 13 μL to the cleaned hemocytometer.
4. Let cells settle then place hemocytometer on an inverted microscope.
5. Using the 10× objective, identify and count the live cells which appear round and opaque colored in each outer four corner squares (dead cells will have absorbed the trypan blue and be blue colored).
6. Calculate the "average cell number" per corner.
7. Total cell number = average cell number × dilution × suspension volume × 10,000.
8. Resuspend cells to a density of 1.5 million cells/mL in BME media.

3.2.4 Seeding Cells for Various Experiments (See Notes 10 and 11)

1. Ensure that all vessels are PDL coated.

   For **cell viability or cell toxicity assay**, use the 48-well plates.

   Seed cells at a density of 375,000 cells/well.

   Add 250 μL/well of a 1.5 million cells/mL suspension.
2. For **lipid peroxidation measurement**, use 6-well plates.

   Seed cells at a density of 6,000,000 cells/well.

   Add 4 mL/well of 1.5 million cells/mL suspension.
3. For **TBARS assay**, use 24-well plates.

   Seed cells at a density of 2,100,000 cells/well.

   Add 1.4 mL/well of 1.5 million cells/mL suspension.
4. For **glutathione assays**, use 24-well plates.

   Seed cells at a density of 600,000 cells/well.

   Add 500 μL/well of 1.5 million cells/mL suspension.
5. For **cell binding assays**, use the 48-well plates.

   Seed cells at a density of 375,000 cells/well.

   Add 250 μL/well of a 1.5 million cells/mL suspension.
6. For **immunofluorescence staining of cells,** use the 4- or 12-well plates (containing PDL coated 12 mm glass coverslips). Seed cells at a density of 50,000 cells/well.

   Add 500 μL/well of a 100,000 cells/mL suspension.

3.2.5 Growth and Maintenance of Cell Cultures

1. Place neuronal cell cultures in a humidified 37 °C incubator with 5% CO_2.
2. After 2 h, remove off 80% v/v of the BME plating media carefully (trying to avoid dislodging attached cells).
3. Add back 60% v/v of freshly prepared NB/B27 media and return to incubator.
4. After 24 h, add AraC to help minimize astrocyte and microglial cell growth.
5. Add 20% v/v AraC to cells growing in plates.
6. Cells are ready for experiments after growing for 6 days in the incubator.
7. Toxicity assays (*see* **Note 12**).

3.2.6 Aggregation of $PrP_{106–126}$ Peptide

1. Prepare a 2 mM stock solution by dissolving the peptide $PrP_{106–126}$ directly in Neurobasal medium supplemented with B27 lacking antioxidant supplements (NB/AO).
2. Aggregate peptide by incubating overnight at 37 °C.
3. The synthetic $PrP_{106–126}$ peptide aggregates rapidly into fully formed fibrils.

3.2.7 Adding $PrP_{106–126}$ to Cultures

1. Dilute aggregated $PrP_{106–126}$ to a final concentration of 20 μM in NB/AO media, and treat for at least 3 days to induce neurotoxicity of cultured cells.
2. For cell viability/toxicity assays, perform experiments in triplicates in a final volume of 200 μL/well (i.e., in a 48-well plate). Up to seven treatments plus a vehicle experiment can be performed/plate. Include a vehicle on each separate plate used.
3. Prepare a working plate (i.e., use a sterile 48-well plate if testing lots of samples). Number wells on the a 48-well plate in an 8 × 6 well format whereby each set of 8-well treatments will be transferred to one plate containing neurons.
4. For PrP treatment, add 6.4 μL of 2 mM $PrP_{106–126}$, and make the volume up to 640 μL with freshly prepared NB/AO. Repeat procedure by adding $PrP_{106–126}$ and/or other reagents and bring volume to 640 μL. If using DMSO, use a maximum volume of 3.2 μL (D-PBS up to a volume of 48 μL), and include DMSO (or D-PBS) in your vehicle sample too. Add $PrP_{106–126}$ and reagents and top up with vehicle buffer in working plate first; then top up with prewarmed NB/AO media and mix.
5. Remove plates coated with cells from incubator and bring to hood. Remove media from each well (do one plate at a time), and transfer experimental media to cells on plate; then return to incubator.

3.3 Cell Viability and Cell Death Assays

3.3.1 MTS and MTT Assays

Cell viability assays are used to measure how healthy cells are after treatment, while a cell death assay measures the opposite. The 3-(4,5-dimethylthiazol-2-yl)-2,5-diphenyltetrazolium (MTT) and 3-(4,5-dimethylthiazol-2-yl)-5-(3-carboxymethoxyphenyl)-2-(4-sulfophenyl)-2H-tetrazolium, inner salt (MTS) assays are used to measure the redox potential (electron transfer activity) of the cell which is a good indicator of how healthy the cell is. The MTT and MTS tetrazolium compounds are reduced by live healthy cells into a colored formazan by-products. MTT and MTS differ substantially in that the MTS formed by-product is soluble, while the MTT formazan product deposits on the cells as a crystalline precipitate that requires an additional step to dissolve the crystals before recording the absorbance readings.

MTS assay

1. For each 48-well plate, you need to prepare a working volume of 150 μL × 24 wells = 3.60 mL of fresh NB/AO media plus 10% MTS (i.e., 3.24 mL of NB/AO media plus 0.36 mL of 10× MTS).
2. Aspirate off the experimental media from the cells.
3. Add 150 μL of freshly prepared NB/AO/MTS reagent to the cells in culture.
4. Return the cells to the humidified incubator for 1–4 h depending cell density. For the amount of cerebellar cells plated, 1–2 h is generally sufficient to see enough color change to get a reasonable data set.
5. When media are ready to be read, transfer 130 μL to a clean flat-bottomed 96-well plate. Ensure that background media (NB/AO/MTS) not exposed to cells are included on the plate to represent the background control value (the background value is subtracted from vehicle and treated absorbance readings).
6. Ensure that no bubbles are present in the wells while reading the samples. The plates can be read on any plate reader which has a filter setting at 490 nm absorbance.
7. Control wells are the vehicle-treated cells, and they should have maximum absorbance readings. The data set is now normalized to the vehicle-treated cells which is set to 100% and the absorbance values for treated cells adjusted accordingly.

MTT assay

8. For each 48-well plate, you need to prepare a working volume of 150 μL × 24 wells = 3.60 mL of fresh NB/AO media plus 10% MTT (i.e., 3.24 mL of NB/AO media plus 0.36 mL of 10× MTT).

9. Aspirate off the experimental media from the cells.
10. Add 150 μL of freshly prepared NB/AO/MTT reagent to the cells in culture.
11. Return the cells to the humidified incubator for 1–4 h depending on cell density. For the amount of cerebellar cells plated, 1–2 h is generally sufficient to see enough color change to get a reasonable data set.
12. Aspirate off the culture media, and add 150 μL of DMSO to wells to fully dissolve formazan crystals. Once fully dissolved, transfer 130 μL to a clean flat-bottomed 96-well plate. Ensure that no bubbles are present in the wells before reading the plate. The plates can be read on any plate reader which has a filter setting at 570 nm absorbance.
13. Control wells are the vehicle-treated cells, and they should have maximum absorbance readings. The data set is now normalized to the vehicle-treated cells which is set to 100% and the absorbance values for treated cells adjusted accordingly.

3.3.2 LDH Cytotoxicity Assay

The lactate dehydrogenase (LDH) assay is used to measure and quantitate cell death. LDH is a stable cytoplasmic enzyme present in all cells. It is rapidly released into the cell culture media upon damage of the plasma membrane. The increase in the amount of enzyme activity indirectly correlates to the amount of formazan formed during a limited time period. This formazan dye is water soluble and can be detected using a plate reader.

1. The LDH cytotoxicity assay kit II from BioVision (Mountain View, CA) utilizes WST reagent for a fast and more sensitive detection of LDH released from damaged cells. The manufacturer's instructions are excellent, and I have outlined the major steps required to successfully complete this assay.
2. The LDH assay requires collection of the treated culture medium for assay.
3. The following three controls have to be performed in each experimental setup.
4. **Background**: The background control will measure reagents and LDH background from culture medium. The background value has to be subtracted from all other values.
5. **Low control** provides information on LDH activity released from the untreated normal cells.
6. **High control** provides information on max releasable LDH activity following treatment of cells with cell lysis.
7. All controls and experimental wells should be done in triplicate.
8. **Background control wells:** These wells have only assay medium in them.

9. **Low control wells:** These wells will have untreated cells growing in assay medium.
10. **High control wells :** Add 10 μL of cell lysis solution to each well and mix. Incubate at 37 °C for at least 1 h.
11. Transfer 10 μL of culture media into an optically clear 96-well plate (flat bottom). Repeat for all samples.
12. Add 100 μL LDH reaction mix to each well, mix, and incubate for 30 min at RT.
13. Measure the absorbance of all controls and samples with a plate reader using a 450 nm filter.
14. Calculation of the percentage cytotoxicity (%) = 100 × (test sample−low control)/(high control−low control).

3.3.3 Lipid Peroxidation Assay (See Note 13)

Lipid peroxidation refers to the oxidative degradation of lipids, and it is a well-established mechanism of cellular injury, and it is used as an indicator of oxidative stress in cells and tissues. Polyunsaturated fatty acids are the most vulnerable source of decomposition by the free radicals resulting in the generation of secondary products of malondialdehyde (MDA) and 4-hydroxyalkenals (HNE) by-products. Therefore, MDA and HNE can be measured using the LPO 586 lipid peroxidation kit (Oxis international). Another common method for measuring MDA, referred to as the thiobarbituric acid reactive substances (TBARS) assay, is to react it with thiobarbituric acid (TBA) to form fluorescent red adducts and record the absorbance at 532 nm using a plate reader.

1. This assay requires a homogenate concentration of 5 million cells/0.1 mL.
2. Treat cells with $PrP_{106–126}$ for desired time.
3. At the end of the experiment, briefly wash cells three times with D-PBS.
4. Add 1 mL of cold D-PBS to wells and scrape cells off dish and collect dissociated cells into a microcentrifuge tube. Centrifuge tube at 5000 × *g* for 5 min to pellet cells.
5. Remove D-PBS buffer leaving cell pellet in tube.
6. Add 120 μL of cold 20 mM Tris buffer pH 7.4 to cell pellet.
7. Add 1.2 μL cold 500 mM butylated hydroxytoluene (to prevent oxidation).
8. Gently mix homogenate. Centrifuge at 3000 × *g* at 4 °C for 10 min.
9. Take a 10 μL sample aliquot for BCA protein assay measurement.
10. Transfer the remaining supernatant to a clean tube and freeze supernate sample and store at −70 °C or use immediately (keep on ice until use).

11. Assay procedure for MDA and HNE assay.
12. Transfer 100 μL of sample to a new microfuge tube (A).
13. Prepare standards and background samples- 4-HNE (S1), TMOP (S2), sample blank (Asb), and reagent blank (Ao), and bring volume to 100 μL.
14. Add 5 μL 500 mM butylated hydroxytoluene.
15. Add 325 μL 10.3 mM N-methyl-2-phenylindole, in acetonitrile (R1).
16. Gently vortex samples for a few seconds.
17. Add 75 μL 15.4 M methansulfonic acid (R2).
18. Mix well by vortexing. Incubate sample mix at 45 °C for 45 min. Centrifuge samples at 15,000 × *g* for 10 min.
19. Transfer 200 μL of clear supernatant to a single well in a 96-well plate that is resistant to acetonitrile and strong acid.
20. Read samples using a plate reader set at 586 nm (or any wavelength between 580 and 590 nm).
21. Calculations for MDA and HNE assay: The calculated molar extinction coefficient (ε) at 586 nm is approximately 110,000 for MDA and HNE. To calculate, the concentration of your unknown sample analyte, the following formula [A–Asb–Ao] × 5/ε is used.

3.3.4 TBARS Assay

1. To cells treated with peptides or drugs, bring media volume to 600 μL.
2. Add 400 μL of TBA solution. Media may go cloudy due to protein precipitation.
3. Incubate the plate for 10 min at RT on a plate shaker.
4. Transfer supernatant from each well to a clean 15 mL tube.
5. Heat at 95 °C for 20 min. Cool to RT and then centrifuge at 3000 × *g* for 5 min.
6. Transfer the clarified supernatant to a spectrophotometer cuvette, and read absorbance at 532 nm.
7. Calculations for TBARS assay: Cell-free medium alone is incubated with the TBA solution as above and subtracted from test readings. The TBARS values are given as optical density (OD) units (× 10^{-3})/well. Cell numbers are determined by cell viability and total protein (BCA) assays.

3.3.5 Glutathione Peroxidase and Reductase Assays (See Note 13)

Glutathione peroxidase (**GPx**) is important for inhibiting lipid peroxidation of the cell membrane. The function of GPx is to reduce lipid hydroperoxides to their corresponding alcohols and to reduce free hydrogen peroxide to water.

For example, GPx catalyzes the following reaction:

$$2GSH + H_2O_2 \rightarrow GSSG + 2H_2O,$$

reduced monomeric glutathione (**GSH**) and glutathione disulfide (**GSSG**).

Glutathione reductase (**GR**) then reduces the oxidized glutathione in the presence of nicotinamide adenine dinucleotide phosphate (**NADPH**) oxidase and hydrogen (**H**) to complete the cycle.

$$GSSG + NADPH + H^+ \rightarrow 2GSH + NADP^+.$$

Therefore the production of GSH is essential since it acts as an antioxidant protecting the neurons from cellular oxidative stress events. Determining GPx and GR levels in cerebellum cultures is performed using the colorimetric assay kit as per the manufacturer's instructions (GPx-340 and GR-340, respectively, Oxis International Inc., Foster City, CA). Below is a brief summary of the methods and calculations used to determine GPx and GR levels.

Harvesting cells for GPx assay:

1. To cells treated with peptides or drugs, washed one time with D-PBS.
2. Add 500 μL of ice-cold 50 mM Tris–HCl pH 7.5 containing 1 mM 2β-mercaptoethanol to wells, and scrape cells off, and transfer to microfuge tube.
3. Homogenize sample using a polytron homogenizer for 10 s.
4. Centrifuge tube at 8500 × *g* for 10 min at 4 °C.
5. Transfer supernatants to new tube, and store at −80 °C until ready for assay.

Assay procedure for GPx assay:

1. Immediately prior to assay, warm all reagents and sample to room temperature.
2. Place 1 mL cuvette on bench.
3. Add 70 μL of sample, 350 μL of assay buffer, and 350 μL of NADH reagent.
4. Place cuvette in spectrophotometer.
5. Add 350 μL of 0.007% tert-Butyl hydroperoxide.
6. Mix well and record absorbance value at 340 nm (A_{340}) for 3 min at 30 s intervals.

Calculations for glutathione peroxidase (GPx) assay:

1. GPx activity is calculated by measuring the rate of decrease in the A_{340} per min.
2. This can be done by performing a linear regression of the A_{340} as a function of time.

3. The molar extinction coefficient (ε) for NADPH is 6220/M/cm.
4. Then mU/mL = (A_{340}/min)/ε.

Harvesting cells for measuring GR activity:

1. To cells treated with peptides or drugs, washed one time with D-PBS.
2. Add 500 μL of ice-cold 60 mM KPO_4 buffer pH 7.5.
3. Homogenize sample using a polytron homogenizer for 10 s.
4. Centrifuge tube at 8500 × *g* for 10 min at 4 °C.
5. Transfer supernatants to new tube, and store at −80 °C until ready for assay.

Assay procedure for measuring GR activity:

1. Immediately prior to assay, warm all reagents and sample to room temperature.
2. Place 1 mL cuvette on bench.
3. Add 200 μL of sample and 400 μL of GSSG.
4. Place cuvette in spectrophotometer.
5. Add 400 μL of NADH.
6. Mix well and record absorbance value at 340 nm (A_{340}) for 5 min at 1 min intervals.

GR activity calculations:

1. Need to calculate the rate of decrease in the A_{340} per min which is a measure of GR activity.
2. This can be done by performing a linear regression of the A_{340} as a function of time. The molar extinction coefficient (ε) for NADPH is 6220 M/cm.
3. Then mU/mL = (A_{340}/min)/ε.

GPx and GR data calculation adjustments for protein content:

1. Measure protein concentration (mg/mL) of each sample (from homogenized cultures before centrifugation) using the BCA protein assay kit (Pierce, Rockford, IL).
2. GPx or GR activity data result (mU/mL) divided protein concentration (mg/mL) to give the final data as mU/mg protein.
3. One GR unit reduces 1 μmol of GSSG/min at 25 °C, pH 7.6.
4. One GPx unit consumes 1 μmol NADPH/min at 25 °C, pH 7.6.

3.4 Cell Binding Assay

1. Keep a 10 μL aliquot of peptide aggregated material for standard curve.
2. At the end of treatment, conditioned media is removed from wells.
3. Wash cells with 500 μL of ice-cold carbonate buffer (0.1 M $NaHCO_3$) for 3 min. Repeat two more times.
4. Rinse with ice-cold D-PBS.
5. Add 500 μL ice-cold D-PBS to wells, and scrape cells off plate, and transfer cell suspension to a microfuge tube.
6. Centrifuge tubes at 16,000 × *g* for 5 min at 4 °C. Aspirate off D-PBS buffer.
7. Add 50 μL of lysis buffer (with protease inhibitor) to cell pellet.
8. Sonicate in an ice water bath for 10 min.
9. Centrifuge tubes at 16,000 × *g* for 5 min at 4 °C. Transfer supernatant to new tubes.
10. Determine total protein concentration of each cell lysate using BCA kit.
11. Load 10 μg of protein on an SDS-PAGE gel (e.g., NuPAGE® 12% Bis-Tris gel, Invitrogen, Australia) for western blot analysis.
12. Transfer gels onto a nitrocellulose membrane.
13. Load three to four different concentrations of peptide for a standard curve.
14. Use specific antibody to detect peptide added to cells for treatment on western blot.
15. Measure protein loading controls by detecting GAPDH or beta-actin.
16. Use software program such as ImageJ (NIH v1.48) to quantitate band intensity levels of standard curve and samples.
17. Plot band intensity vs. peptide concentration.
18. Measure band intensity of peptide-treated cultures. Calculate concentration of peptide bound to cells from standard curve.
19. Normalize values according to loading controls for experiment.

3.5 Immunofluorescence Staining of Cells (See Note 14)

1. Immunofluorescence studies can be done on neuronal cells grown on either glass coverslips or chambered glass slides. For glass coverslips, use 12 mm diameter circle glass coverslips which fit in 24-well tissue culture dishes or 4-well dishes. Larger rectangular or square coverslips up to ~24 mm will fit into 6-well dishes.
2. To treated neuronal cells (on 12 mm coverslips in 4-well plates), aspirate off media and wash 1× with D-PBS.

3. Aspirate off D-PBS. Add 0.5 mL of 4% paraformaldehyde and incubate for 20 min at RT.
4. Aspirate off 4% paraformaldehyde. Wash with 0.5 mL D-PBS.
5. If cells are treated with $PrP_{106-126}$, wash cells with 0.5 mL cold carbonate wash buffer for 5 min at RT. Aspirate off buffer.
6. Add 0.5 mL permeabilize buffer and incubate for 20 min at RT. Aspirate off buffer.
7. Wash cells with 0.5 mL D-PBS. Aspirate off buffer.
8. Add 0.5 mL block buffer and incubate for 60 min at RT.
9. Prepare antibody by diluting in block buffer to desired concentration.
10. Incubate cells in 500 μL of primary antibody and incubate overnight at 4 °C (or for 2 h at RT).
11. Alternatively, to minimize antibody usage, use a moisture box (e.g., 20 × 40 cm box which has a lid and flat base). Cut a piece of parafilm and number one side (e.g., 1–12) 3 cm apart to indicate where coverslips are to be placed. Spray some water on the dish, and place the parafilm on the base of the dish with the numbers facing down. Place 50 μL of diluted antibody on the parafilm at each number according to your protocol. Gently pick up the coverslips using fine forceps, and place the coverslip on top of the droplet with the CELLS coming into contact with the antibody. Repeat procedure for all coverslips. Seal the container and incubate for the desired time period.
12. Aspirate off antibody from wells (or transfer coverslips from moisture box back into plates with cells facing UP). Wash cells with 0.5 mL D-PBS three times at 5 min intervals.
13. To cells, add appropriate secondary antibody in block (fluorescently tagged), and incubate for 60 min at RT in the dark (to conserve antibody, you can repeat **step 11** above).
14. Aspirate off antibody (or transfer coverslips back into plate cells facing UP).
15. Wash cells with 0.5 mL D-PBS three times at 5 min intervals.
16. Label glass slides with relevant experimental details. Place a drop (10 μL) of ProLong Gold antifade mounting media (Life Technologies Invitrogen) on the glass slide. Pick up coverslip, dip into clean dH_2O, remove excess water by wicking on a piece of lint free tissue paper, and then gently place coverslip (cells facing down) on the mounting media.
17. Repeat procedure for other coverslips. Add only three to four coverslips per glass slide.
18. Allow mounting media to dry for 24 h at RT, and then store at 4 °C for longer-term storage.

4 Notes

1. All chemicals were purchased from Sigma-Aldrich (St. Louis, MO) unless otherwise specified.
2. Unless stated otherwise, all solutions should be prepared in water that has a resistivity of 18.2 MOhm/cm and total organic content of less than five parts per billion. This standard is referred to as "dH_2O" in this text.
3. To filter sterilize solutions, use a 0.22 micron membrane filter (Millipore).
4. Use aseptic procedures in the preparation of cell culture solutions and in the preparation of cell cultures for experiments to eliminate potential contamination of cell cultures.
5. Store solutions in tubes or bottles made of polystyrene and not glass.
6. Do not add concentrated DNase directly to cell suspension as this will result in decreased cell recovery.
7. Heat inactivate FCS by heating it to 55 °C in a water bath for 30 min only. Longer heating will cause serum protein to precipitate.
8. To make fire polished pipettes, put the glass tip into a naked flame (i.e., use a Bunsen burner), and rotate briefly to remove sharp edges. Using a microscope, check if sharp edges are gone and that you have not sealed the end by flaming it too long.
9. The ideal number of cells to count per corner is between 50 and 100 cells. If there are more cells, then dilute out accordingly to minimize on counting errors.
10. For 48-well plates, add cells only to the inner 24 wells and 200 μL of dH_2O the outer 24 wells as this helps minimize evaporation in the wells containing cells.
11. These are the actual surface area (not calculated) for various vessels:

    For multiwell plates 6, 12, 24, 48, and 96 wells, area is 10, 4, 2, 1, and 0.3 cm^2, respectively.
    For round dishes 35, 60, and 100 mm diameter, area is 10, 20, and 60 cm^2, respectively.
12. We purchased our peptides from The ERI Amyloid Laboratory (Oxford, CT).
13. The methods described for measuring lipid peroxidation and glutathione peroxidase and glutathione reductase levels are not comprehensive, and it is advisable to read the manufacturer's instructions for details.

14. For control nuclear staining of cells, DAPI, Hoechst 33,258, and ToPro3 (Invitrogen) are great markers. Of the three dyes, DAPI fades the least, and it is relatively stable, and neurons will fluorescence for many months and years. ToPro3 is ideal if your imaging setup does not have the capacity to take images using UV but does have far red filters. For DAPI and Hoechst, it is also just as easy to dilute these dyes to final working concentrations in the mounting media rather than having to add it to the cells in block .

References

1. Tagliavini F, Prelli F, Verga L et al (1993) Synthetic peptides homologous to prion protein residues 106-147 form amyloid-like fibrils in vitro. Proc Natl Acad Sci U S A 90:9678–9682
2. Forloni G, Angeretti N, Chiesa R et al (1993) Neurotoxicity of a prion protein fragment. Nature 362:543–546
3. Thompson A, White AR, McLean C et al (2000) Amyloidogenicity and neurotoxicity of peptides corresponding to the helical regions of PrP(C). J Neurosci Res 62:293–301
4. Brown DR (2000) Prion protein peptides: optimal toxicity and peptide blockade of toxicity. Mol Cell Neurosci 15:66–78
5. Bergstrom AL, Cordes H, Zsurger N et al (2005) Amidation and structure relaxation abolish the neurotoxicity of the prion peptide PrP106-126 in vivo and in vitro. J Biol Chem 280:23114–23121
6. Thellung S, Florio T, Corsaro A et al (2000) Intracellular mechanisms mediating the neuronal death and astrogliosis induced by the prion protein fragment 106-126. Int J Dev Neurosci 18:481–492
7. White AR, Maher F, Brazier MW et al (2003) Diverse fibrillar peptides directly bind the Alzheimer's amyloid precursor protein and amyloid precursor-like protein 2 resulting in cellular accumulation. Brain Res 966:231–244
8. White AR, Guirguis R, Brazier MW et al (2001) Sublethal concentrations of prion peptide PrP106-126 or the amyloid beta peptide of Alzheimer's disease activates expression of proapoptotic markers in primary cortical neurons. Neurobiol Dis 8:299–316
9. Forloni G, Bugiani O, Tagliavini F, Salmona M (1996) Apoptosis-mediated neurotoxicity induced by beta-amyloid and PrP fragments. Mol Chem Neuropathol 28:163–171
10. Brown DR, Herms J, Kretzschmar HA (1994) Mouse cortical cells lacking cellular PrP survive in culture with a neurotoxic PrP fragment. Neuroreport 5:2057–2060
11. White AR, Collins SJ, Maher F et al (1999) Prion protein-deficient neurons reveal lower glutathione reductase activity and increased susceptibility to hydrogen peroxide toxicity. Am J Pathol 155:1723–1730
12. Jobling MF, Huang X, Stewart LR et al (2001) Copper and zinc binding modulates the aggregation and neurotoxic properties of the prion peptide PrP106-126. Biochemistry 40:8073–8084
13. Stewart LR, White AR, Jobling MF et al (2001) Involvement of the 5-lipoxygenase pathway in the neurotoxicity of the prion peptide PrP106-126. J Neurosci Res 65:565–572
14. Salmona M, Malesani P, De Gioia L et al (1999) Molecular determinants of the physicochemical properties of a critical prion protein region comprising residues 106-126. Biochem J 342:207–214
15. Jobling MF, Stewart LR, White AR et al (1999) The hydrophobic core sequence modulates the neurotoxic and secondary structure properties of the prion peptide 106-126. J Neurochem 73:1557–1565

Part III

Methods for Cell Free Propagation of Prions

Chapter 13

Methods of Protein Misfolding Cyclic Amplification

Natallia Makarava, Regina Savtchenko, and Ilia V. Baskakov

Abstract

Protein misfolding cyclic amplification (PMCA) amplifies infectious prions in vitro. Over the past decade, PMCA has become an essential tool in prion research. The current chapter describes in detail the PMCA format with beads (PMCAb) and several methods that rely on PMCAb for assessing strain-specific prion amplification rates, for selective amplification of subtypes of PrP^{Sc} from a mixture, and a PMCAb approach that can replace animal titration of scrapie material. Development of PMCAb-based methodology is important for addressing a number of research topics including prion strain evolution, selection and adaptation, strain-typing, prion detection, and biochemical requirements for prion replication.

Key words Prion protein, Prion diseases, Protein misfolding cyclic amplification, Sonication, Self-propagating protein states, Prion detection, Prion strains, Replication rate

1 Introduction

Protein misfolding cyclic amplification (PMCA) is the sole technique that can successfully amplify the disease-associated, infectious form of the prion protein (PrP^{Sc}) in vitro [1]. PMCA reactions consist of two alternating steps: sonication and incubation (Fig. 1). Sonication fragments PrP^{Sc} particles or fibrils into smaller pieces, a process that is believed to result in the multiplication of active centers of PrP^{Sc} growth. During the incubation step, small PrP^{Sc} particles grow by recruiting and converting the normal, cellular form of the prion protein (PrP^{C}) into PrP^{Sc}. Since its introduction in 2001 [2], PMCA has become an important tool in prion research. To date, amplification of PrP^{Sc} in PMCA can be achieved by using PrP^{C} in crude brain homogenate [1], PrP^{C} purified from normal mammalian brains [3], or recombinant PrP (rPrP) produced in *E. coli* [4, 5] as substrates. In recent years, PMCA was used for identifying cofactors that are involved in prion replication [4–6], assessing the impact of glycosylation on replication of prion strains [7, 8], assessing the ability of prion strains to replicate in heterologous substrate and to cross species barriers [9, 10], and

Victoria A. Lawson (ed.), *Prions: Methods and Protocols*, Methods in Molecular Biology, vol. 1658,
DOI 10.1007/978-1-4939-7244-9_13, © Springer Science+Business Media LLC 2017

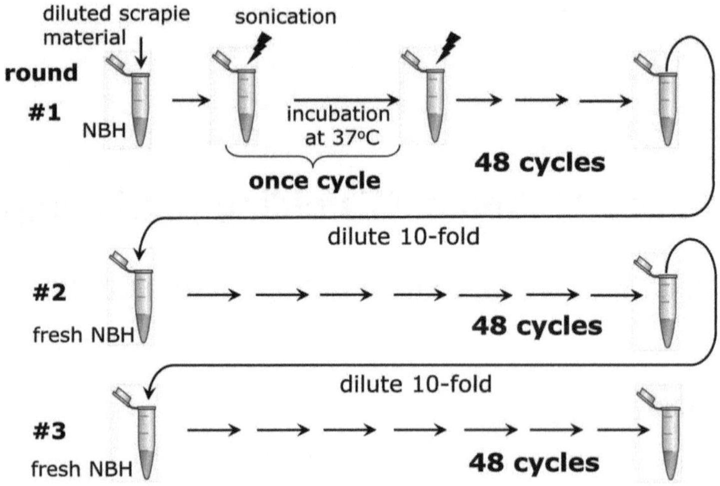

Fig. 1 Schematic diagram that illustrates the principles of PMCA. Prion amplification is initiated by mixing diluted scrapie material with normal brain homogenate (NBH) used as a source of PrP^C. A single PMCA round consists of multiple repetitive cycles of sonication and incubation at 37 °C. A serial PMCA consists of multiple rounds, where each subsequent round is initiated by diluting PMCA products from the previous round into fresh NBH

elucidating prion interference and prion adaptation to new hosts or new replication environments [11–13]. Due to its capacity to amplify a single PrP^{Sc} particle [14, 15], PMCA provides one of the most sensitive approaches for detecting miniscule amounts of prion infectivity [14, 16, 17], including detection of prions in blood or peripheral tissues at preclinical stages of the disease [18, 19].

While PMCA-based methods offer tremendous opportunities for the development of new tools and assays in the field, a number of limitations have surfaced over the years. The efficiency of amplification depends to a large extent on the prion strain or type of Creutzfeldt-Jakob disease (if human prions are assayed). Amplification efficiency in PMCA could also vary significantly depending on minor variations in experimental parameters, including those that are difficult to control, such as the age of the sonicator's horn and individual patterns of horn corrosion. In addition, cross-contamination has been reported as one of the major issues that could undermine interpretation of the results [20].

Previously, our laboratory introduced a new PMCA format with beads (PMCAb) [21], an important modification that improved PMCA efficiency, sensitivity, and robustness. The current chapter describes in detail PMCAb and several methods that rely on PMCAb for assessing strain-specific PrP^{Sc} amplification rates, for selective amplification of PrP^{Sc} from a mixture, and a PMCAb approach that can replace animal titration of scrapie material. Further development of PMCA methodology is important for addressing broader scientific topics including prion evolution, behavior of prion strains, strain-typing, and prion amplification requirements.

2 Materials

All solutions are prepared with deionized water purified using a Barnstead MicroPure UV water system (Thermo Scientific), unless noted otherwise.

2.1 Preparation of 10% Normal Brain Homogenate for Protein Misfolding Cyclic Amplification

1. Conversion buffer: to prepare 50 ml, use 5 ml of Ca^{2+}-free and Mg^{2+}-free 10× phosphate-buffered saline, pH 7.4 (10×PBS), 1.5 ml of 5 M sodium chloride, 0.1 ml of 0.5 M ethylenediaminetetraacetic acid, pH 8.0 (EDTA), 1 tablet of protease inhibitors (cOmplete, EDTA-free), 5 ml 10% (w/v) Triton X-100, and RNase-free water. Sterilize by filtration through 22 μm syringe-driven filter unit (SLGV033RB); keep on ice.
2. Brains from healthy weanling hamsters, perfused immediately after euthanasia with PBS, pH 7.4, supplemented with 5 mM EDTA (*see* **Note 1**). Three brains will produce about 30 ml of PMCA substrate (*see* **Note 2**).
3. 30 ml Potter-Elvehjem tissue grinders with PTFE pestle (358049, Wheaton Science Products, Millville, NJ) (*see* **Note 3**).
4. Laboratory balance to weigh brains.
5. Refrigerated microcentrifuge.
6. Motor-driven homogenizer (Heidolph, Elk Grove Village, IL).
7. RNase-free tubes and RNase-free aerosol barrier pipette tips.

2.2 Protein Misfolding Cyclic Amplification with Beads (PMCAb)

1. RNase-free 0.2 ml thin-wall tubes with domed caps (14230205, Fisher Scientific).
2. Polytetrafluoroethylene (PTFE) spheres, 3/32″ diameter (B000FMUEXG, www.amazonsupply.com).
3. Brain homogenate prepared as in Subheading 3.1.
4. Seeds: the source of scrapie material depends on the experimental design. We usually use diluted 10% brain homogenate from terminally ill animals, which is prepared the same way as NBH (*see* Subheading 3.1, **steps 1–7**).
5. Sonicator with a microplate horn (Q700 with 431MPX, Qsonica, Newtown, CT) at 37 °C (*see* **Note 4**).
6. Floating rack (*see* **Note 5**).
7. Proteinase K (P8102S, New England BioLabs, Ipswich, MA).
8. 10% sodium dodecyl sulfate (SDS, Fisher Scientific).

2.3 Protein Misfolding Cyclic Amplification with Beads Using Partially Deglycosylated Substrate (dgPMCAb)

1. All materials as mentioned for PMCAb (*see* Subheading 2.2).
2. Peptide-N-Glycosidase F (PNGase F, glycerol-free, New England BioLabs).
3. Laboratory rotator.
4. Incubator.

2.4 Protein Misfolding Cyclic Amplification with Beads in RNA-Depleted Environment

1. All materials as mentioned for PMCAb (*see* Subheading 2.2).
2. RNase A (R4875, Sigma-Aldrich, St. Louis, MO): dissolve in water to prepare a 10 mg/ml solution; store at −20 °C in small aliquots.
3. Laboratory rotator.
4. Incubator.

2.5 PMCA-Based Titration

1. All materials as mentioned for PMCAb (*see* Subheading 2.2).

2.6 Estimation of Amplification Rate in PMCA

1. All materials as mentioned for PMCAb (*see* Subheading 2.2).

3 Methods

Low efficiency of protein misfolding cyclic amplification can be caused by several factors:

1. Presence of blood in NBH, when using nonperfused brains to prepare the substrate.
2. Degradation of PrP^{C} in NBH due to prolonged storage and multiple freeze/thaw cycles.
3. Inefficient sonication due to a very old or damaged horn (*see* **Note 6**).
4. Inadequate sonication protocol (*see* **Note 7**).

Once sonication protocol is established, prions can by amplified very efficiently. All PMCA reactions should be set up and performed in a way that minimizes cross-contamination of the samples (*see* **Note 8**), since even a miniscule amount of prions can generate false-positives [20].

3.1 Preparation of 10% Normal Brain Homogenate for Protein Misfolding Cyclic Amplification

1. Keep tissue grinders, conversion buffer, and fresh brains on ice (*see* **Notes 1** and **3**). Place on ice a 50 ml tube for collection of the homogenate.
2. Tare a Petri dish on a balance, transfer brains into the Petri dish, record the weight quickly, and place the Petri dish with brains on ice to keep brains cold.

3. Calculate the amount of conversion buffer needed to prepare a 10% homogenate. Each gram of brains requires 9 ml of conversion buffer. If the calculated volume of conversion buffer exceeds 15 ml, work in batches in order not to overfill the grinder.
4. Attach the pestle of the tissue grinder to the motor-driven homogenizer.
5. Transfer the brains into the glass tube of the grinder, and add the needed volume of the ice-cold conversion buffer.
6. Turn on the homogenizer and operate on a speed that is easy to control. Run the rotating pestle up and down through the tube, and process the tissue until homogeneous. Make sure the grinder remains cold. Cool on ice, if needed.
7. Pour the prepared 10% brain homogenate into a precooled 50 ml tube. Keep on ice. If working in batches, combine all homogenate to produce a homogenous pool, and then aliquot for storage as desired (*see* **Note 2**).
8. Before assembling a protein misfolding cyclic amplification reaction, thaw the 10% normal brain homogenate, aliquot into RNase-free 1.5 ml microcentrifuge tubes, and centrifuge 2 min at 500 × *g*. Carefully collect the supernatant and use it as a substrate for PMCA (10% NBH). Leftovers of the precleared NBH can be stored at −80 °C for 3–6 months.

3.2 Protein Misfolding Cyclic Amplification with Beads (PMCAb)

A single PMCAb round consists of multiple cycles, each of which consists of brief sonication and incubation at 37 °C (Fig. 1). In a serial PMCA, the end products of a single PMCA round are diluted to fresh NBH to seed the next PMCAb round (Fig. 1).

1. Fill the microplate horn of the sonicator with 350 ml of water and make sure it reaches 37 °C before starting PMCA (*see* **Note 4**).
2. Thaw a sufficient amount of prepared brain homogenate (*see* Subheading 3.1). To distinguish between amplified material and the residual amount of seeds, include round 0 in the experiment. Round 0 should contain the same amount of seeds as the round 1, but not subjected to sonication and incubation cycles.
3. Using clean forceps, distribute 1–3 PTFE spheres (beads) into each PMCA tube [21].
4. Aliquot precleared NBH: 90 μl per tube, if using tenfold dilution between rounds (*see* **Note 7**).
5. If several PMCAb rounds are planned, keep rounds 0 and 1 on ice, and freeze the rest at −80 °C.

6. If using 10% scrapie brain homogenate (BH) as a source of seeds, prepare corresponding dilution: sonicate 80–100 μl of 10% BH in a 0.2 ml PMCA tube for 30 s using Qsonica microplate horn at 150–200 Wt sonication power. Centrifuge the tube briefly to bring all liquid down before opening the tube. Dilute sonicated 10% brain homogenate into conversion buffer to achieve the desired dilution. Usually, the final dilution of scrapie brain 10^{-4} (for 263 K) provides sufficient amount of seeds for amplification in the first round, leaving the residual amount of the original seeds undetectable by standard immunoblotting.
7. Add 10 μl of the diluted seeds to 90 μl of NBH in rounds 0 and 1. Close the tubes of round 0 and freeze them at −20 °C until ready to analyze.
8. Close the tubes of round 1 and place in a floating rack (*see* **Note 5**). Make sure that the tubes do not touch the bottom of the horn. To achieve minimal variations in sonication efficiency, position the tubes equidistantly around the center of the horn and horn surface. Cover the horn with foil to reduce evaporation.
9. Program the sonicator: as a rule, we use 30 s sonication pulses at 150–200 Wt delivered every 30 min during a 24-hour period. This gives 48 cycles within one round of PMCA (*see* **Note 7**). Start sonication.
10. The next day, after one round of PMCAb is completed, centrifuge the tubes briefly before opening. Thaw the aliquots for the next round, and add 10 μl from completed round to the 90 μl of the fresh substrate of the next round tubes (*see* **Note 7**).
11. Sonicate the tubes by the same protocol as during the previous round.
12. Store completed PMCAb rounds frozen at −20 °C until ready to analyze. PrP^{Sc} produced in PMCAb and stored at −20 °C retains its seeding abilities for at least 6 months.
13. To visualize amplification of PrP^{Sc} after PMCAb, we use immunodetection after proteinase K (PK) digestion: combine 10 μl of the PMCAb reaction with 5 μl of 1% sodium dodecyl sulfate (SDS) and 5 μl of diluted PK, to achieve 20–50 μg/ml final concentration of PK in the reaction, and incubate at 37 °C for 1 h (*see* **Note 9**). Stop the reaction by adding sample buffer of your choice and denaturing the samples in a boiling water bath for 10 min. If round 0 without PK treatment is used as a weight marker in a gel and as a control for PK digestion, subject the tube with round 0 to a single 30 s sonication pulse to reduce sample viscosity, then add sample buffer, and boil with the rest of the samples. Proceed with the electrophoresis and Western blot as described elsewhere.

3.3 Protein Misfolding Cyclic Amplification with Beads Using Partially Deglycosylated Substrate (dgPMCAb)

Treatment of NBH with PNGase partially deglycosylates PrP^C and shifts the ratio of the three PrP^C glycoforms making mono- and unglycosylated forms more abundant. A shift in glycoform ratio changes the selectivity of amplification and can be used for selective amplification of classical or atypical PrP^{Sc} or for strain typing [8] (Fig. 2a).

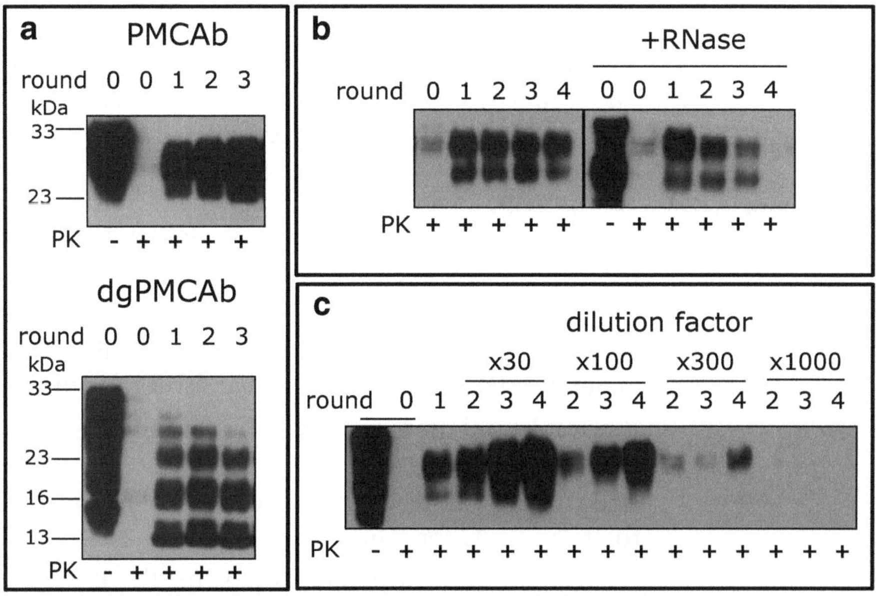

Fig. 2 Representative results from the experiments on PMCAb, dgPMCAb, and PMCAb in an RNA-depleted environment and an estimation of amplification rate. Each reaction was done in the presence of three PTFE spheres. The sonication protocol consisted of 30 s sonication pulses at 50% power performed every 30 min during 24 h. (**a**) PMCAb and dgPMCAb reactions were seeded with 10^{-2} dilution of brain material from the first passage of a synthetic strain LOTSS [22]. Ten microliters of reaction products from a completed round were used to seed 90 μl of the subsequent round. In the first passage, brain-derived LOTSS represents a mixture of authentic PrP^{Sc} and atypical PrPres, which are selectively amplified in PMCAb (*upper panel*) and dgPMCAb (*lower panel*), respectively. Amplification of PrP^{Sc} was visualized with 3F4 antibody, while atypical PrPres was detected with SAF-84. (**b**) PMCAb in an RNA-depleted environment illustrates RNA dependency of amplification of 263 K scrapie material. The reactions were seeded with 10^{-3} diluted brain material from terminally ill hamsters. Ten microliters of reaction products from a completed round were used to seed 90 μl of the subsequent round. Western blot was stained with SAF-84. (**c**). Estimation of an amplification rate for LOTSS. PMCAb reactions were seeded with 10^{-4} diluted brain material from the second passage of LOTSS. For the dilution factor ×30, 10 μl of a completed reaction were mixed with 20 μl of conversion buffer, and 10 μl of this dilution was used to seed subsequent PMCAb rounds. For the dilution factor ×100, 10 μl of a completed reaction were mixed with 90 μl of conversion buffer, and 10 μl of this dilution was used to seed subsequent PMCAb rounds. For the dilution factors ×300 and ×1000, 10 μl of a completed reaction was diluted threefold and tenfold, respectively, into conversion buffer and then diluted further by mixing 10 μl with 90 μl of conversion buffer each, and 10 μl out of these dilutions were used to seed subsequent PMCAb rounds. An amplification rate of LOTTS was estimated to be 300-fold per PMCAb round. An increase in intensity of the signal from round to round illustrates the ability of synthetic strains to quickly adapt to an PMCAb environment and increase its amplification efficiency as previously described [23]. Western blot was stained with SAF-84

1. Thaw sufficient amount of prepared brain homogenate (*see* Subheading 3.1).
2. Preclear brain homogenate as in Subheading 3.1, **step 8**, if it is not already done.
3. Add 1500 U/ml PNGase F to an aliquot in a RNase-free sterile tube and rotate at 37 °C for 5 h.
4. Proceed using the same method as for the PMCAb (*see* Subheading 3.2).

3.4 Protein Misfolding Cyclic Amplification with Beads in RNA-Depleted Environment

Prion strains from different species or certain PrP^{Sc} subtypes differ with respect to RNA dependency of amplification [8, 24, 25]. PMCAb amplification in an RNA-depleted environment can be used to assess their RNA dependency. If amplification of a strain or PrP^{Sc} subtype is found to be RNA independent, removing all RNA from NBH by treatment with RNases improves selectivity of amplification for such strain or PrP^{Sc} subtype (Fig. 2b).

1. Thaw a sufficient amount of prepared brain homogenate (*see* Subheading 3.1). Partially deglycosylated brain homogenate as in Subheading 3.3 can also be used for RNA depleting.
2. Preclear brain homogenate as in Subheading 3.1, **step 8**, if it is not already done.
3. Add 100 μg/ml RNase A to an aliquot in a sterile tube and rotate at 37 °C for 1 h.
4. Proceed using the same method as for the PMCAb (*see* Subheading 3.2).

3.5 PMCAb-Based Titration of Scrapie Material

To determine the titer of PMCAb-active particles in infectious material, serial tenfold dilutions are prepared and used to seed multiple rounds of PMCAb. As a result, infectious material diluted up to a single PMCAb-active particle can be amplified and detected by immunoblot. This method offers a fast and accurate way of quantitating prion titer and can replace prion titration by animal bioassay [15]. Perform titration in several independent repetitions to allow statistical analysis of the data (*see* **Note 10**).

1. Label PMCA tubes for all repetitions and rounds in the titration experiment, place PTFE spheres in all the tubes, and dispense NBH (*see* Subheading 3.2).
2. Prepare tubes for dilutions by dispensing 90 μl of conversion buffer into each tube (*see* **Note 11**). Keep the tubes open under the hood on the rack to eliminate handling, which might increase the risk of cross-contamination.

3. Sonicate 10% homogenate of scrapie material as in Subheading 3.2, **step 6**. Centrifuge the tube briefly to bring all liquid down before opening the tube.
4. Open the tube with sonicated 10% homogenate of scrapie material, mix the contents by careful pipetting up and down, and dispose of the tip.
5. With a new tip, transfer 10 μl of sonicated 10% scrapie homogenate into the next tube, and mix carefully but thoroughly by pipetting. This makes a 10^{-2} dilution of scrapie material.
6. Continue diluting tenfold, always changing pipette tips between dilutions.
7. After all dilutions of scrapie material are prepared, change gloves and open the tubes with the NBH substrate for the 1st PMCAb round. Keep PMCAb tubes with the substrate for the subsequent rounds at −80 °C.
8. To seed NBH with prepared dilutions of scrapie material, mix 10 μl of each dilution with 90 μl of the NBH substrate in the corresponding PMCA tubes. Start with the highest dilution and move toward the more infectious ones. Close PMCA tubes immediately after adding the seeds. Change tips between dilutions.
9. Space PMCAb tubes of the 1st round in a foam rack and place into a mictoplate horn of the sonicator (*see* **Notes 5** and **10**). Sonicate according to the protocol optimized for efficient amplification of the strain. For example, for 263 K scrapie material sonication for 20 s every 20 min within 24 h routinely allowed us to amplify 10^{12}-fold diluted brain material in only two to three rounds of PMCAb (*see* **Note** 7).
10. After one PMCAb round is completed, spin the tubes briefly and use to seed a subsequent round. Change gloves before opening the tubes with NBH substrate. In the case of tenfold dilutions between rounds (*see* **Note** 7), mix 10 μl from the completed reaction with 90 μl of the NBH substrate in the corresponding PMCA tubes for the following round. Start from the highest dilution and move toward the more infectious ones.
11. Repeat PMCAb rounds as long as needed. Analyze by immunoblotting (*see* Subheading 3.2, **step 12**). Usually, with every round of PMCAb, the amount of dilutions positive for PMCAb-active particles gradually increases. When after several subsequent PMCAb rounds no new positives are detected, the necessary number of rounds has been reached (*see* **Note 12**).
12. Calculate the titer of PMCAb-active particles or $PMCAb_{50}$ values. $PMCAb_{50}$ values are determined by regression analysis in Sigma Plot (other software can be employed, too) using

nonlinear least squares fitting of data to the sigmoidal equation:

$$F = (100^* \exp(-(A + B^* x)))/(1 + \exp(-(A + B^* x)))$$

where F is percent of positive PMCAb reactions, x is logarithm of the scrapie material dilution fold, A and B are two fitting parameters that define the position of a limiting dilution transition on the x axis and the slope of the transition, respectively. After A and B are determined, $PMCAb_{50}$ is calculated according to the equation:

$$PMCAb_{50} = A/B$$

Because the titer of scrapie material is calculated per gram of tissue, whereas the assay volumes used for PMCAb are typically 100 μl, the $PMCAb_{50}$ value has to be recalculated per gram of scrapie tissue (if PMCAb is conducted in 100 μl, the $PMCAb_{50}$ value should be multiplied by 10).

3.6 Estimation of Amplification Rate in PMCAb

The amplification rate differs for each prion strain or PrP^{Sc} subtype and can be used as an individual characteristic of a strain [23, 26, 27]. Evaluating the amplification rate can also be useful for optimization of the sonication protocol. The amplification rate is defined operationally as the highest dilution between serial PMCAb rounds at which amplification is still capable of compensating the effect of dilution [27]. To evaluate a strain-specific amplification rate, a set of serial PMCAb reactions are conducted, where the dilution factors between serial rounds could range from 3-fold to 10^6-fold (Figs. 2c and 3). However, in each serial PMCAb reaction, the dilution factor between rounds is fixed. The amplification rate is equal to the highest dilution factor at which PrP^{Sc} replication can be maintained steadily in serial PMCAb (Figs. 2c and 3).

1. Label PMCA tubes for all rounds of the experiment; place PTFE spheres in all the tubes (*see* Subheading 3.2, **steps 1–5**). To estimate amplification rate of one strain, four rounds are usually sufficient. Rounds 0 and 1 can be represented by only one tube each and contain 90 μl of NBH substrate. The number of tubes for rounds 2–4 and the volume of NBH in these tubes, however, depend upon the number of dilutions and on the dilution factors to be tested.
2. Seed 90 μl of NBH for rounds 0 and 1 with 10 μl of scrapie material (*see* Subheading 3.2, **step 6**), and subject to the desired sonication protocol.
3. Transfer a portion of the completed round to the tubes with fresh substrate for a subsequent round. For example, to test if amplification sustains a threefold dilution, have 60 μl of NBH in the tubes for rounds 2–4 and supplement them with 30 μl of

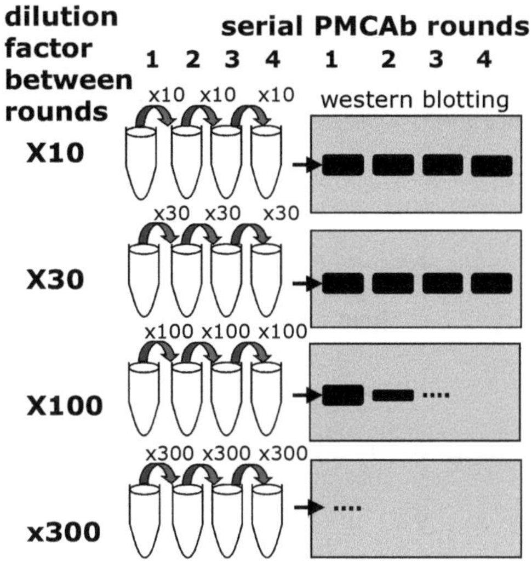

Fig. 3 Schematic representation of a PMCAb experiment for estimating a strain-specific amplification rate. Several serial PMCAb reactions with incrementing dilution factors between rounds are initiated using the same scrapie material. In each individual serial PMCAb reaction, the dilution factor between rounds is kept constant. The highest dilution factor at which PrP^{Sc} amplification fully compensates the effect of dilution determines the amplification rate

completed reaction from the previous round. In a typical assay, we increase the dilution factors with ~3-fold increments: ×3-fold, ×10-fold, ×30-fold, ×100-fold, etc. (Figs. 2c and 3). If the dilution factors call for a volume below 10 μl, prepare an intermediate dilution of the completed round, and use 10 μl of it to seed 90 μl of the substrate. For example, to check if the amplification can be sustained at a 1000-fold dilution between rounds, mix 10 μl of the completed round with 90 μl of conversion buffer, then use 10 μl of this dilution to mix with new 90 μl of conversion buffer, and then use 10 μl of the resulting dilution to seed 90 μl of NBH in a subsequent round. Repeat the same dilution process in each round.

4. After four rounds of amplification are completed, analyze all rounds together by immunodetection after PK digestion (*see* Subheading 3.2, **step 12**).

4 Notes

1. If brains are to be homogenized immediately after dissection, keep on ice in a 50 ml tube with PBS, pH 7.4 supplemented with 5 mM EDTA. Alternatively, brains can be flash frozen in liquid nitrogen and stored at −80 °C until needed.

Fig. 4 A floating rack for holding PMCAb tubes in the microplate horn is made from a 6 mm extra firm foam sheet available in craft stores

2. We observed that with time the quality of 10% normal brain homogenate prepared in conversion buffer decreases, even if stored at −80 °C without repeated freezing and thawing. To avoid this problem, we prefer to prepare a new batch of 10% normal brain homogenate every 3–6 months. Flash-frozen brains can be stored at −80 °C for at least 2 years.
3. To keep the tissue grinder cold, keep the glass tube and PTFE pestle on ice. Only when both parts of the tissue grinder are cooled will the pestle fit properly inside the glass tube and move up and down well during homogenization.
4. The temperature inside the horn can be maintained by coiled rubber tubing attached to a circulating water bath. Alternatively, the horn with the acoustic chamber can be placed in a 37 °C incubator.
5. We make floating racks by punching small holes in a piece of dense foam available in craft stores (Fig. 4). When inserting the tubes into the foam rack, maintain a 1 mm space between the lid and the surface of the foam to ensure no contact between the lid and water in the horn. The tubes should be positioned about 2 cm away from the center and from the edges of the horn and about 3 mm from the bottom. Make sure the rack floats in the center of the horn.
6. The surface of the horn weathers over time developing cracks and cavities. When positioning the tubes, avoid severely damaged areas. If PMCAb results show significant inconsistency, or overall efficiency of the sonication becomes low, replace the horn. Resurfacing of the horn by polishing can also prolong its life (Qsonica offers resurfacing services).
7. Different prion strains may require particular modifications to ensure maximum amplification efficiency. If a strain is difficult to amplify, lowering the starting dilution of seeds, lowering dilution between rounds, or shortening the sonication pulses with more cycles within one round of amplification may be useful [11, 23].

8. All reusable materials, e.g., dissection instruments and homogenizers, should be decontaminated thoroughly between uses. We decontaminate them by soaking them in sodium hydroxide (2 N NaOH) or undiluted household bleach for 1 h, then rinsing with water, and autoclaving at 134 °C for 1 h. If practical, designate separate sets of reusable materials for tissues infected with different strains and for noninfected tissues. Although the risk of contaminating PMCA samples through the water in the microplate horn is low, the horn should also be decontaminated before critical PMCA experiments by wiping with a generous amount of 2 N NaOH. Prolonged incubation with NaOH degrades the metal surface of the horn and should be avoided. Unless different strains need to be compared, avoid amplification of several strains with one microplate horn at the same time.
9. Twofold dilution of the PMCAb reaction before immunodetection permits better PK digestion and helps to avoid smearing of the bands in the gel. Optimal final concentration of SDS and PK, as well as the time and temperature of the PK digestion reaction, can be strain dependent.
10. We recommend sonicating no more than 30 tubes simultaneously, since overloading of the horn can change the efficiency of sonication.
11. If infectious material shows inconsistent amplification within one serial dilution, dilute scrapie material into 1% normal brain homogenate instead of conversion buffer to block possible sticking of PrP^{Sc} particles to the plastic tubes and tips. 1% normal brain homogenate is prepared by dilution of 10% precleared NBH into conversion buffer. It can be sonicated in 100 μl of aliquots to further increase homogeneity. 1% NBH can be prepared in advance and stored at −80 °C.
12. One should be aware that spontaneous misfolding of PrP^{C} into PK-resistant conformations can occasionally happen upon harsh and prolonged sonication. Sometimes spontaneously formed PK-resistant PrP states also propagate from round to round increasing the probability of false-positive signals. The sonication conditions should be optimized for each strain, so that the strain is amplified efficiently within the lowest possible number of rounds. Processing several non-seeded controls for the same number of rounds helps to control for false-positives due to spontaneous misfolding and cross-contamination.

Acknowledgments

We thank Pamela Wright for editing the chapter. This work was supported by the National Institute of Health grants R01 NS045585 and NS074998.

References

1. Castilla J, Saa P, Hetz C, Soto C (2005) In vitro generation of infectious scrapie prions. Cell 121:195–206
2. Saborio GP, Permanne B, Soto C (2001) Sensitive detection of pathological prion protein by cyclic amplification of protein misfolding. Nature 411:810–813
3. Deleault NR, Harris BT, Rees JR, Supattapone S (2007) Formation of native prions from minimal components in vitro. Proc Acad Natl Sci USA 104:9741–9746
4. Wang F, Wang X, Yuan CG, Ma J (2010) Generating a prion bacterially expressed recombinant prion protein. Science 327:1132–1135
5. Deleault NR, Piro JR, Walsh DJ et al (2012) Isolation of phosphatidylethanolamine as a solitary cofactor for prion formation in the absence of nucleic acids. Proc Acad Natl Sci USA 109:8546–8551
6. Deleault NR, Walsh DJ, Piro JR et al (2012) Cofactor molecules maintain infectious conformation and restrict strain properties in purified prions. Proc Acad Natl Sci USA 109: E1938–E1946
7. Nishina K, Deleault NR, Mahal S et al (2006) The stoichiometry of host PrPC glycoforms modulates the efficiency of PrPSc formation in vitro. Biochemistry 45:14129–14139
8. Makarava N, Savtchenko R, Baskakov IV (2013) Selective amplification of classical and atypical prions using modified protein misfolding cyclic amplification. J Biol Chem 288:33–41
9. Castilla J, Gonzalez-Romero D, Saa P et al (2008) Crossing the species barrier by PrPSc replication in vitro generates unique infectious prions. Cell 134:757–768
10. Green KM, Castilla J, Seward TS et al (2008) Accelerated high fidelity prion amplification within and across prion species barriers. PLoS Pathog 4:e1000139
11. Shikiya RA, Ayers JI, Schutt CR et al (2010) Coinfecting prion strains compete for a limiting cellular resource. J Virol 84:5706–5714
12. Meyerett C, Michel B, Pulford B et al (2008) In vitro strain adaptation of CWD prions by serial protein misfolding cyclic amplification. Virology 382:267–276
13. Gonzalez-Montalban N, Lee YJ, Makarava N et al (2013) Changes in prion replication environment cause prion strain mutation. FASEB J 27(9):3702–3710
14. Saa P, Castilla J, Soto C (2006) Ultra-efficient replication of infectious prions by automated protein misfolding cyclic amplification. J Biol Chem 281:35245–35252
15. Makarava N, Savtchenko R, Alexeeva I et al (2012) Fast and ultrasensitive method for quantitating prion infectivity titer. Nat Commun 3:741
16. Gonzalez-Romero D, Barria MA, Leon P et al (2008) Detection of infectious prions in urine. FEBS Lett 582:3161–3166
17. Murayama Y, Yoshioka M, Okada H et al (2007) Urinary excretion and blood level of prions in scrapie-infected hamsters. J Gen Virol 88:2890–2898
18. Saa P, Castilla J, Soto C (2006) Presymptomatic detection of prions in blood. Science 313:92–94
19. Haley NJ, Mathiason CK, Zabel MD et al (2009) Detection of sub-clinical CWD infection in conventional test-negative deer long after oral exposure to urine and feces from CWD+ deer. PLoS One 4:e7990
20. Cosseddu GM, Nonno R, Vaccari G et al (2011) Ultra-efficient PrP(Sc) amplification highlights potentialities and pitfalls of PMCA technology. PLoS Pathog 7:e1002370
21. Gonzalez-Montalban N, Makarava N, Ostapchenko VG et al (2011) Highly efficient protein misfolding cyclic amplification. PLoS Pathog 7:e1001277
22. Makarava N, Kovacs GG, Savtchenko R et al (2011) Genesis of mammalian prions: from non-infectious amyloid fibrils to a transmissible prion disease. PLoS Pathog 7:e1002419
23. Gonzalez-Montalban N, Baskakov IV (2012) Assessment of strain-specific PrPSc elongation rates revealed a transformation of PrPSc properties during protein misfolding cyclic amplification. PLoS One 7:0041210

24. Deleault NR, Kascsak R, Geoghegan JC, Supattapone S (2010) Species-dependent differences in cofactor utilization for formation of the protease-resistant prion protein in vitro. Biochemistry 49:3928–3934
25. Kovacs GG, Makarava N, Savtchenko R, Baskakov IV (2013) Atypical and classical forms of the disease-associated state of the prion protein exhibit distinct neuronal tropism, deposition patterns, and lesion profiles. Am J Pathol 183:1539–1547
26. Ayers JL, Schutt CR, Shikiya RA et al (2011) The strain-encoded relationship between PrP replication, stability and processing in neurons is predictive of the incubation period of disease. PLoS Pathog 7:e1001317
27. Gonzalez-Montalban N, Makarava N, Savtchenko R, Baskakov IV (2011) Relationship between conformational stability and amplification efficiency of prions. Biochemistry 50:7933–7940

Chapter 14

RT-QuIC Assays for Prion Disease Detection and Diagnostics

Christina D. Orrù, Bradley R. Groveman, Andrew G. Hughson, Matteo Manca, Lynne D. Raymond, Gregory J. Raymond, Katrina J. Campbell, Kelsie J. Anson, Allison Kraus, and Byron Caughey

Abstract

In coping with prion diseases, it is important to have tests that are practical enough for routine applications in medicine, agriculture, wildlife biology, and research, yet sensitive enough to detect minimal amounts of infectivity. *R*eal-*t*ime *qu*aking-*i*nduced *c*onversion (RT-QuIC) assays have evolved to the point where they fulfill these criteria in applications to various human and animal prion diseases. For example, RT-QuIC assays of cerebrospinal fluid and nasal brushings allow for highly sensitive (77–97%) and specific (99–100%) identification of human sCJD patients. Recent improvements have markedly enhanced sensitivity and reduced the assay time required for many samples to a matter of hours rather than days. By combining analyses of cerebrospinal fluid and nasal brushings, diagnostic sensitivities and specificities of nearly 100% can be achieved. RT-QuIC assays are based on prion-seeded amyloid fibril formation by recombinant prion protein (rPrP^{Sen}) in multiwell plates using a Thioflavin T fluorescence readout. Here we describe our current RT-QuIC methodologies as well as technical considerations in executing, troubleshooting, and adapting the assay to new strains of prions and sample types.

Key words RT-QuIC, TSE diagnosis, Prion, Recombinant PrP^{Sen}, Amyloid

1 Introduction

The ability to rapidly detect prions down to their smallest infectious doses has clear relevance to prion disease diagnosis, surveillance, and research. Whereas the gold standard method of detecting prions by animal bioassays takes months to years, a variety of ultrasensitive in vitro assays have shortened the time required to hours to weeks [1–25]. These assays are typically based on the inherent ability of infectious prions, composed primarily of abnormal disease-specific forms of prion protein (e.g., PrP^{Sc}), to induce, or seed, the conversion of native, largely α-helical prion protein (PrP^{C} or rPrP^{Sen}) into highly β-sheet forms that are similar, or identical, to PrP^{Sc}. With the scrapie cell assay (SCA), prion seeds

Victoria A. Lawson (ed.), *Prions: Methods and Protocols*, Methods in Molecular Biology, vol. 1658,
DOI 10.1007/978-1-4939-7244-9_14, © Springer Science+Business Media LLC 2017

propagate PrP^{Sc} in murine cell cultures over the course of 2–3 weeks [18, 21]. Although the SCA is so far restricted to the detection of a couple of murine-adapted scrapie prion strains, it appears to be a true measure of prion infectivity. In *p*rotein *m*isfolding *c*yclic *a*mplification (PMCA) assays, the prion-seeded conversion occurs in sonicated reactions with the PrP^{C} substrate being provided by brain homogenates [19, 20], cell lysates [26], or purified preparations from E. coli expressing recombinant PrP ($rPrP^{Sen}$) [12]. Conventional PMCA assays use Western blotting to detect reaction products that, like most types of PrP^{Sc}, are relatively protease resistant. However, other sensitive methods have been devised to detect PrP^{Sc} as well as the products of PMCA-based PrP^{Sc} amplifications [16]. In the amyloid seeding assay (ASA) [27] and *qu*aking-*i*nduced *c*onversion (QuIC) [7] assays, the prion-seeded reactions are shaken rather than sonicated, and the substrate is $rPrP^{Sen}$. Real-time QuIC (RT-QuIC) assays combine attributes of ASA and QuIC reactions, using shaking in a 96-well plate-based assay with a simple fluorescence readout that detects prion-seeded $rPrP^{Sen}$ amyloid fibrils (Fig. 1) [7, 9]. In contrast to the months-to-years timeframe required for animal bioassays, the latest RT-QuIC reactions now often require less than a day to obtain positive reactions [1] and are at least as sensitive as bioassays [2, 5, 7]. When coupled with an initial prion

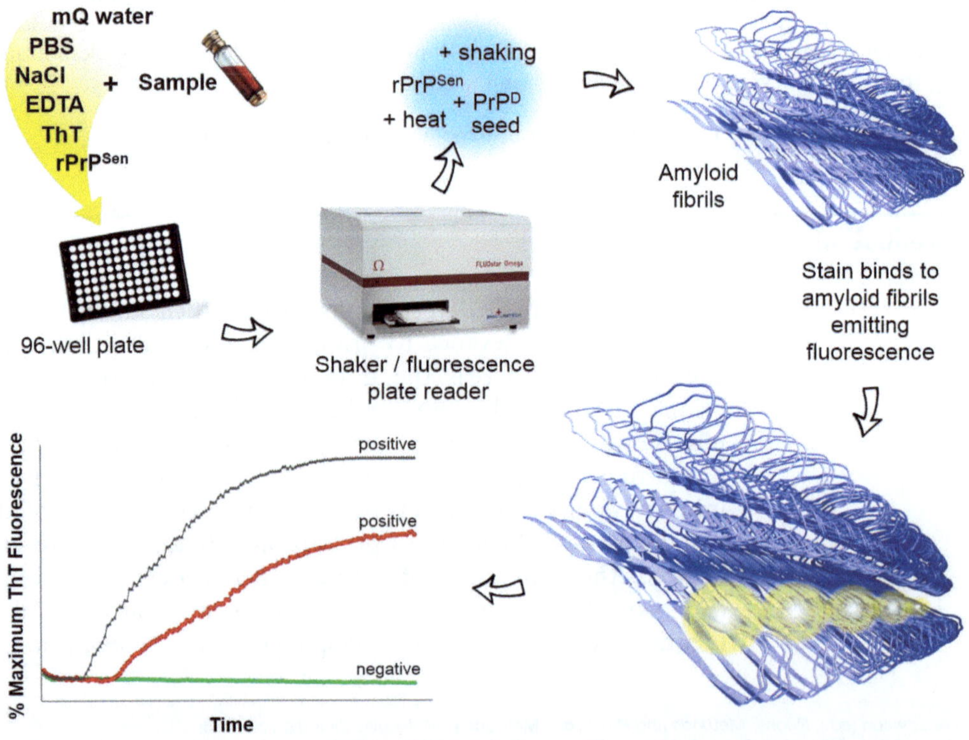

Fig. 1 Schematic representation of the real-time quaking-induced conversion (RT-QuIC) process

immune-capture step, RT-QuIC can be several orders of magnitude more sensitive than bioassay and capable of detecting prion seeding activity in 10^{-14}-fold dilutions of variant CJD or scrapie brain homogenates [5, 6].

RT-QuIC assays have been adapted and applied to virtually all known prion diseases of humans and animals [28], including detection of the following: (1) human Creutzfeldt-Jakob disease (CJD) in the blood [6], brain [2, 28, 29], cerebrospinal fluid (CSF) [3, 9, 13, 22–24, 29–33], or nasal brushings [1, 34, 35]; (2) sheep and goat scrapie in the brain [7, 28, 36]; (3) chronic wasting disease (CWD) in cervid brain [7], blood [37], saliva [38], urine, and feces [39]; and (4) rodent-adapted scrapie or transmissible mink encephalopathy in the brain [5, 40], CSF [7, 29], nasal fluids [40], and blood [5, 6]; (5) classical and atypical forms of bovine spongiform encephalopathy in cattle [28, 41, 42]. The applications to human CJD CSF specimens are the most extensively developed, with large-scale studies indicating diagnostic sensitivities of 77–91% and specificities of 99–100% [1, 3, 9, 13, 23, 24, 32, 34]. Exploratory RT-QuIC analyses of nasal brushings of the olfactory mucosa (OM) have indicated >97% sensitivity and 100% specificity, suggesting a relatively painless, noninvasive, and definitive means of diagnosing sporadic CJD in living patients [1, 34]. Moreover, quantitative analysis by end-point dilution RT-QuIC assays indicated that the concentrations of prion seeding activity in the nasal brushings were several orders of magnitude higher than those found in CSF specimens. Nonetheless, it is likely that RT-QuIC testing of CSF will continue to be a frontline diagnostic screen for CJD because of the common practice of performing lumbar punctures on patients with neurological symptoms [23, 24, 31]. Moreover, recently we have markedly improved the speed and sensitivity of RT-QuIC testing of human CSF and a variety of other potentially prion-infected specimens [22, 23, 25, 30, 31, 33]. In this chapter, we describe current RT-QuIC protocols as well as factors that can affect the implementation, improvement, and adaptation of RT-QuIC assays to various prion strains and specimens.

1.1 The Key Importance of rPrP^{Sen} Substrate and Sample Handling

We would emphasize first and foremost that the most crucial and demanding aspects of the overall RT-QuIC protocol are the preparation of the rPrP^{Sen} substrate and careful lab technique. A good rPrP^{Sen} substrate is efficiently and quickly converted to amyloid in the presence of prion-associated seed while being slow to undergo spontaneous, i.e., prion-independent, amyloidogenesis. The latter characteristic requires the substrate to be stable enough to resist nucleation during purification, storage, and prolonged RT-QuIC reactions. However, for efficient prion-seeded polymerization, the substrate cannot be too stable in its native monomeric α-helical conformation under RT-QuIC conditions. With a good substrate, RT-QuIC is relatively easy and quite reproducible once appropriate conditions have been identified for a given combination of prion strain and sample type. Practically speaking, the key issue is the difference in kinetics between prion-seeded versus spontaneous amyloid formation [7, 25, 27]. As is typical with seeded polymerizations of amyloid fibrils in general, the more dilute the prion seed concentration, the longer the lag phase prior to the exponential fibril growth phase [2, 7, 43]. Ultimately, the sensitivity of RT-QuIC assays is determined by the point at which the lag phase obtained from a particular (usually very low) concentration of prion seeds becomes indistinguishable from the lag phase of spontaneous amyloid formation in appropriate negative control reactions. This also means that great care should be taken to remove, or prevent the spontaneous nucleation, of nascent seeds in the $rPrP^{Sen}$ substrate prior to RT-QuIC assays because such seeds will give rise to false-positive reactions. We have estimated that in our most sensitive RT-QuIC assay, i.e., eQuIC [6], as few as 1–100 PrP^{CJD} or PrP^{Sc} seed particles (roughly 1×10^{-18} g of protease-resistant PrP) can be detected. This indicates not only the extraordinary amplification potential of the RT-QuIC-seeded polymerization reaction (a trillion-fold or more) but also the importance of keeping even miniscule amounts of seeds from forming spontaneously in the $rPrP^{Sen}$ substrate or being introduced inadvertently into the reaction from seed-positive samples. Having said that, we have made >120 independent $rPrP^{Sen}$ substrate preparations, almost all of which have worked excellently for many different researchers, ranging from novices to veterans. Even suboptimal preparations obtained using slightly different protocols were still adequate for applications that did not require the highest sensitivity. Our experience, and that of multiple other labs, is that once these issues are under control under well-vetted reaction conditions, false-positive reactions are extremely rare.

2 Materials

2.1 Transcardial Perfusion, Brain Collection, and Homogenization (See Note 1)

2.1.1 Transcardial Perfusion

1. Perfusion buffer: PBS with 5 mM EDTA.
2. 70% ethanol.
3. Hygienic surgical tools.

2.1.2 Brain Homogenization

1. Phosphate-buffered saline (pH 7.4): 137 mM NaCl, 2.7 mM KCl, 10 mM Na_2HPO_4 (sodium phosphate, dibasic), and 1.76 mM KH_2PO_4 (potassium phosphate monobasic).
2. Triton X-100 (Sigma T9284).

3. Protease inhibitors: Complete mini EDTA-free (Roche 04693159001).
4. Brain homogenization buffer (20 ml): 150 mM NaCl, 0.5% Triton X-100, and one tablet of protease inhibitors (Promega). Solubilize all components in PBS, pH 7.4. Alternatively, PBS alone can be used to make brain homogenates.
5. Homogenizing beads (Biospec products 11079110z).

2.2 rPrP^{Sen} Production and Purification

2.2.1 Bacterial Growth and Recombinant Prion Protein (rPrP^{Sen}) Expression

1. Bacteria: *E. coli* Rosetta cells (EMD Biosciences) transformed with pET41 vector (EMD Biosciences), e.g., containing Syrian golden hamster prion protein sequence (rPrP^{Sen} accession number K02234); (*see* **Note 2**).
2. Luria Broth (LB) culture agar plates with kanamycin and chloramphenicol.
3. LB-M Broth.
4. Kanamycin: 50 mg/ml in H_2O.
5. Mini-culture tubes (BD Falcon 352059).
6. Chloramphenicol: 34 mg/ml in 100% ethanol.
7. Overnight Express Autoinduction System 1 (Novagen 71300-4).
8. 2.5 l Tunair flasks with half-baffle for bacteria growth (Sigma Z710822).
9. 250 ml centrifuge tube with conical bottom (Corning 430776).

2.2.2 Preparation of Inclusion Bodies

1. Bacterial lysis buffer: Bug Buster Master Mix (Novagen 71456-4).
2. Homogenizer: OMNI International TH (Tissue Homogenizer) with plastic disposable tip (Omni international 32750H).
3. 50 ml conical tubes (BD Falcon 352098).
4. Centrifuge JA12 rotor for 50 ml tube centrifugation.

2.2.3 Protein Purification

1. 8 M guanidine hydrochloride (MP Biomedicals 820540): 38 g guanidine hydrochloride in 0.1 M $NaPO_4$ pH 8.0, bring volume up to 50 ml (*see* **Note 3**).
2. Nickel-NTA beads (QIAGEN 30450) (*see* **Note 4**).
3. Sodium phosphate monobasic (Sigma S8282): 0.2 M.
4. Sodium phosphate dibasic (Sigma S7907): 0.2 M.
5. Trizma base (Sigma T6791).
6. Imidazole (Sigma I2399).
7. Chromatography column (GE Healthcare XK16/40 for a 1X preparation, or XK 26/20 for a 3X preparation).
8. Denaturing buffer (pH 8.0, 300 ml): 100 mM sodium phosphate (142 ml 0.2 M dibasic and 8 ml 0.2 M monobasic), 10 mM Trizma base (0.36 g Trizma base), and 6 M guanidine

(171.9 g). Bring to approximately 295 ml volume, check and adjust pH (8.0), and bring to 300 ml final volume with Milli-Q H_2O. Filter buffer through prewashed 0.22 μm filter.

9. Refolding buffer (pH 8.0, 600 ml): 100 mM sodium phosphate (284.1 ml 0.2 M dibasic and 15.9 ml 0.2 M monobasic) and 10 mM Trizma base (0.73 g). Bring to 590 ml volume, check and adjust pH (8.0), and bring to 600 ml final volume with Milli-Q H_2O. Filter buffer through prewashed 0.22 μm filter.
10. Elution buffer (pH 5.8, 500 ml): 100 mM sodium phosphate (20 ml 0.2 M dibasic and 230 ml 0.2 M monobasic), 10 mM Trizma base (0.61 g), and 500 mM imidazole (17.02 g). Bring to ~480 ml volume, add 12 M HCl to adjust pH to 5.8 (*see* **Note 5**), and bring to 500 ml final volume with Milli-Q H_2O. Filter buffer through prewashed 0.22 μm filter.
11. Dialysis buffer (pH 5.8, 3.6 l): 10 mM sodium phosphate (14.3 ml 0.2 M dibasic and 166 ml 0.2 M monobasic). Bring volume to 3.6 L volume and check and adjust pH (5.8).
12. Filters for buffers: 0.22 μm bottle top filter (Nalgene 595-3320).
13. Filters to concentrate $rPrP^{Sen}$ on rare occasions when it might be needed: Amicon Ultra-15 MCWCO 10 kDa (Millipore UFC901008).
14. SnakeSkin Dialysis Tubing (7000 MWCO, Thermo Scientific 68700).

2.2.4 RT-QuIC

1. Sodium dodecyl sulfate (SDS, Sigma L6026).
2. N-2 cell culture media supplement 100× (Gibco 17502-048) (*see* **Note 6**).
3. Screw cap tubes (0.5 ml, Fisherbrand 02-681-334).
4. Screw cap tubes (1.5 ml Fisherbrand 02-681-339).
5. Plastic screw caps (Fisherbrand 02-681-361).
6. 10× phosphate-buffered saline (PBS, pH 7.4): 100 mM $NaH_2PO_4*H_2O$ (sodium phosphate, monobasic monohydrate), 100 mM $Na_2HPO_4*7H_2O$ (sodium phosphate, dibasic heptahydrate), and 130 mM NaCl.
7. 5× PBS (filter; *see* **Note 9**).
8. 0.1% SDS/PBS (50 ml): 0.5 g of SDS powder in 50 ml of milli-Q water to make 1% stock. Filter through prewashed 0.22 um filter (*see* **Note 9**), then dilute appropriately with filtered 5X PBS and water to achieve a final concentration of 0.1%.
9. 100 mM EDTA (*see* **Notes 7 & 9**).

10. 10 mM Thioflavin T: 0.032 g Thioflavin T (Sigma 3516) in 10 ml Milli-Q H_2O (*see* **Notes 8 & 9**). Dilute tenfold in Milli-Q water prior to adding into the RT-QuiC reaction mix.
11. 2 M sodium chloride: 5.85 g of NaCl (Sigma S7653) in 50 ml Milli-Q H_2O (*see* **Note 9**).
12. 60 ml sterile plastic syringes (BD 309653).
13. Syringe filters: 0.22 μm filter (Fisherbrand 09-719A) (*see* **Note 9**).
14. Nanosep centrifuge spin filters MWCO 100 kDa (Pall P/N OD1OOC35).
15. Reaction plate: 96-well optical bottom polymer base black w/o lid N/treated N/sterPS (Thermo Scientific Nunc 265301).
16. Plate sealer: Thermo Scientific Nunc 232702.
17. Shaker/fluorescence plate reader: FLUOstar OMEGA (BMG Labtech).

3 Methods

3.1 Transcardial Perfusion, Brain Tissue Collection, and Homogenization

3.1.1 Rodent Transcardial Perfusion (see Note 1) and Brain Tissue Collection

1. Deeply anesthetize the animal.
2. Position the animal on its back with continued anesthesia applied via nose cone.
3. Saturate the animal's chest fur with 70% ethanol.
4. Cut open the chest cavity to expose the heart.
5. Clip the hole in the right atrium to allow for blood outflow.
6. Pierce the left ventricle with needle and syringe and inject ~10 mls of cold perfusion buffer, under moderate pressure.
7. Turn the animal back over and saturate the head with 70% ethanol.
8. Excise the brain with hygienic surgical tools.
9. Flash-freeze the brain in liquid nitrogen and store at −80 °C.

3.1.2 Brain Homogenization

1. Partially thaw the whole brain.
2. Section out and weigh appropriate portion of the brain (*see* **Note 10**).
3. Load enough homogenizing beads to fill the conical bottom of 1.5 ml screw cap tube.
4. Add the brain and appropriate volume of brain homogenization buffer in with beads.
5. Homogenize the brain using a bead beater apparatus (Biospec) for 1 min (*see* **Note 11**) following manufacturer's instructions.
6. Spin the brain homogenate at 2000 × *g* for 2 min to partially clarify the sample from cellular debris.

7. Transfer supernatant to fresh tube.
8. Aliquot and store supernatant at −80 °C (*see* **Note 12**).

3.2 rPrP^{Sen} Expression and Purification

3.2.1 Bacterial Growth and Recombinant Prion Protein (rPrP^{Sen}) Expression

1. Add 1 ml kanamycin (stock at 50 mg/ml) and 1 ml chloramphenicol (stock at 34 mg/ml in 100% EtOH) to 1 L LB-M Broth. Transfer 4 ml to a small vessel to be used as a mini-culture to initiate bacterial growth.
2. Inoculate 4 ml mini-culture of LB-M Broth media with antibiotics either directly from the glycerol stock or from colonies on the culture plate (*see* **Notes 13** and **14**).
3. Grow mini-culture at 37 °C with ~225 rpm shaking in Innova 4300 incubator shaker for 5–6 h.
4. Add Overnight Express Autoinduction System 1 (EMD 71300-4) components following the manufacturer's instructions to the remainder of the original 1 L LB-M Broth media with antibiotic into a Tunair vessel.
5. Inoculate the 1 L Overnight Express culture by adding 4 ml of the turbid mini-culture.
6. Grow 1 L Overnight Express culture at ~225 rpm in Innova 4300 incubator shaker at 37 °C for ~20–24 h (*see* **Note 15**).
7. Split the resulting 1 L culture into 4 × 250 ml portions in conical centrifuge tubes and spin at 3250 × *g* for 20 min at room temperature (RT).
8. Decant supernatant and freeze pellets for storage at −20 °C.

3.2.2 Inclusion Bodies Isolation (See Note 16)

1. Thaw cell pellet by placing 250 ml tube at 37 °C for ~5 min (*see* **Note 17**).
2. Add 12 ml of Bug Buster Master Mix to 250 ml tube.
3. Using a disposable plastic spatula, scoop the majority of the pellet out of the tube and transfer it to a blue cap BD 50 ml tube, then swirl the remaining solution in the 250 ml tube to incorporate the residual pellet, and pour it into the same tube.
4. Homogenize cell pellet in BD tubes using tissue homogenizer with disposable plastic tips (*see* **Note 18**).
5. Incubate the BD 50 ml tube on rotating mixer for 20 min at RT.
6. Centrifuge at 7900 × *g* (7000 rpm; JA12 rotor) for 20 min at 4 °C.
7. Discard the supernatant.
8. Resuspend pellet in 10 ml of 1× Bug Buster Master Mix, reusing the same tip on the tissue homogenizer.
9. Incubate at RT on the rotator for 15 min.

10. Bring the volume to 40 ml with Bug Buster reagent diluted 1:10 (0.1×) in Milli-Q water and gently mix by inversion.
11. Centrifuge at 7900 × *g* for 15 min at 4 °C to pellet $rPrP^{Sen}$ in inclusion bodies.
12. Pour off and discard supernatant and resuspend pellet in 30 ml of the 0.1× Bug Buster reagent using the homogenizer.
13. Centrifuge at 7,900 × *g* (7000 rpm; JA12 rotor) for 15 min at 4 °C.
14. Pour off and discard supernatant and freeze inclusion body pellet at −20 °C for later use.

3.2.3 Protein Purification

1. Make two 3.6 L volumes of dialysis buffer and cool to 4 °C.
2. Make and filter all buffers prior to starting purification.
3. Add 14ml of 8 M GdnHCl to inclusion body pellet.
4. Homogenize pellet in 8 M GdnHCl solution using the homogenizer and disposable tip.
5. Mix by gentle inversion on rotator at room temperature for ~50 min.
6. Take a small volume of Nickel-NTA bead slurry and use the in house vacuum and a 100 ml Nalgene 0.22 μm bottle top filter to eliminate most of the bead storage buffer (*see* **Notes 19–21**).
7. Weigh out 18 g of beads and transfer them to a fresh 50 ml tube.
8. Bring volume up to 30 ml with denaturing buffer and equilibrate for ~50 min on a rotator at RT.
9. Spin solubilized inclusion body pellet at 7,000 × *g* for 5 min to remove insoluble debris, and transfer supernatant to the tube containing the Nickel-NTA equilibrated beads.
10. Incubate on rotator at RT for 40 min to allow protein binding to the beads.
11. Wash AKTA FPLC (GE Healthcare) with Milli-Q water to clean out both lines (A and B), and run denaturing buffer through both lines (A and B) to prime the system (the column is not yet attached).
12. Load protein-bound beads into column (e.g. XK 16/20) making sure to minimize air bubbles allowing excess fluid to flow out of the bottom of the column as it packs..
13. Attach column to AKTA FPLC (GE Healthcare) and run gradient refold at 0.75 ml/min for 240 min following the gradient refolding program:

    100% denaturing buffer (line A) and 0% refolding buffer (line B) to 100% refolding buffer and 0% denaturing buffer.

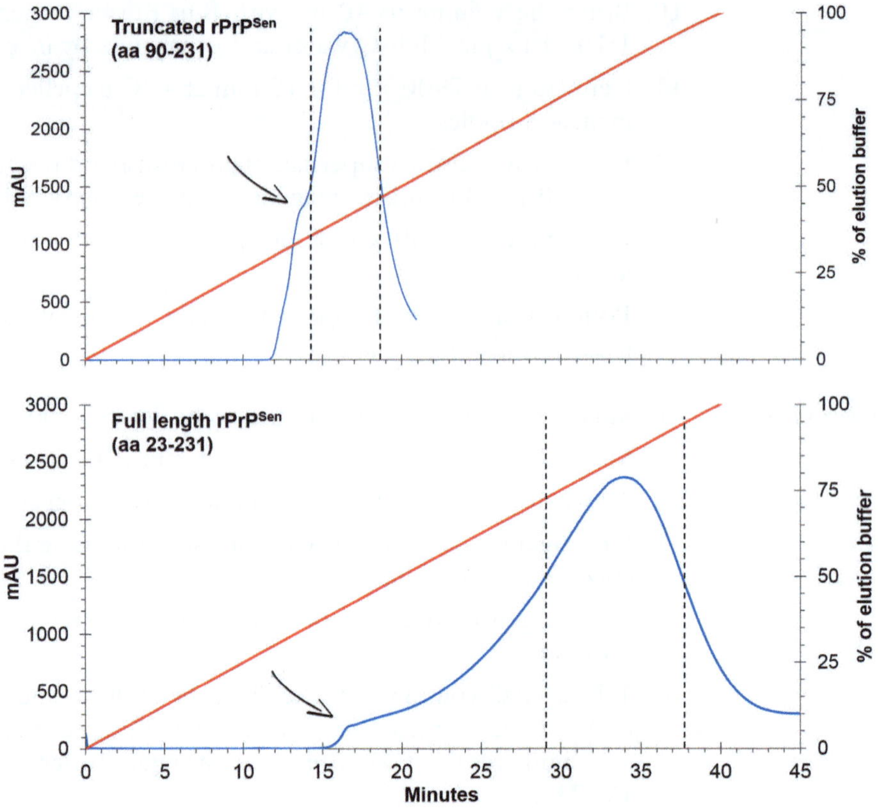

Fig. 2 Representative elution profile of truncated and full length rPrP^{Sen} purified by immobilized metal-ion affinity chromatography. *Blue lines* indicate the UV absorbance (mAU) at 280 nm plotted on the left axis. *Red lines* plotted on the right axis indicate the percentage of elution buffer applied to the column. *Dashed lines* indicate approximate portion of the elution peak that is collected. *Arrows* indicate leading shoulder in the elution profile that is not collected (*see* **Note 22**)

14. Following the gradient refolding, let 100% refolding buffer continue to flow through column for an additional 30 min at 0.75 ml/min.
15. Rinse line A initially with water and then with elution buffer (bypassing the column) to clean out the denaturing buffer.
16. To elute protein, run a gradient elution over 40 min from 100% refolding buffer (line B) to 100% elution buffer (now in line A) at a flow rate of 2 ml/min (*see* **Note 22**).
17. Monitor UV 280 signal and collect the center of the peak (Fig. 2); (*see* **Notes 23** and **24**). Pool the elution fractions that represent the center of the elution peak.
18. Prewash 0.22 μm syringe filter with dialysis buffer to remove filter contaminants such as glycerol. Filter pooled protein.
19. Gently pour pooled protein into SnakeSkin Dialysis Tubing and dialyze in 3.6 L of prechilled dialysis buffer over night at 4 °C.

20. The next day transfer to a fresh 3.6 L dialysis buffer volume for an additional 4 h of dialysis.
21. Following dialysis, filter protein with a prewashed 0.22 μm syringe filter.
22. Determine protein concentration by 280 nm absorbance on spectrophotometer using $rPrP^{Sen}$ extinction coefficient.
23. Concentrate if necessary using Amicon Ultra-15 tubes aiming for a concentration of the protein of about 0.5 mg/ml (*see* **Note 25**).
24. Following filtration, make 1 ml aliquots of the protein and immediately store at −80 °C (*see* **Note 26**).
25. Check protein on gel and/or with mass spec for contaminants or fragmentation.
26. When the aliquots are thawed, filter with a 100 kDa filter, as described in the assay protocol, to remove oligomeric $rPrP^{Sen}$.

3.3 RT-QuIC Protocol for brain tissue analysis (for CSF applications see [22, 30])

1. Serially dilute 10% brain homogenate (BH) from 263 K-infected and normal Syrian Golden hamsters into 0.1% SDS/1XN-2/PBS as follows:

	Final brain tissue dilution in the reaction
(a) Dilute 10% BH 1:10	1×10^{-2}
(b) Dilute previous 1:10	1×10^{-3}
(c) Dilute previous 1:10	1×10^{-4}
(d) Dilute previous 1:10	1×10^{-5}
(e) Dilute previous 1:10	1×10^{-6}
(f) Dilute previous 1:10	1×10^{-7}
(g) Dilute previous 1:10	1×10^{-8}
(h) Dilute previous 1:10	1×10^{-9}

2. Filter recombinant prion protein ($rPrP^{Sen}$) with 100 kDa Nanosep spin filters at 4000 × *g*, 4 °C for 5 min (*see* **Note 27**).
3. Prepare RT-QuIC reaction mix per well, as follows (CSF analysis requires the addition of SDS at a final concentration of 0.002% to the reaction mix, see [22, 30] for details):

Stocks	For one well (μl)	Final concentration
Milli-Q H_2O	X	–
5× PBS	20	1×
2 M NaCl	Y	–
100 mM EDTA	1	1 mM
1 mM ThT	1	0.01 mM
$rPrP^{Sen}$	Z	0.1 mg/ml
Total volume	98	

(*see* **Note 28**)

Example using hamster 90–231 $rPrP^{Sen}$ (0.55 mg/ml), 300 mM NaCl:

Stocks	For one well (μl)	For 24 wells (μl)	Final concentration
Milli-Q H_2O	49.5	1188	–
5× PBS	20	480	1×, 130 mM NaCl
2 M NaCl	8.5	204	170 mM
100 mM EDTA	1	24	1 mM
1 mM ThT	1	24	0.01 mM
$rPrP^{Sen}$	18	432	0.1 mg/ml
Total volume	98	2352	

(*see* **Note 29**)

4. Transfer 98 μl of RT-QuIC reaction mix into each plate well.
5. Seed RT-QuIC reactions with 2 μl of brain homogenate dilution directly into the well (*see* **Note 30**).
6. Seal the plate with plate sealer.
7. Incubate the plate in BMG FLUOstar plate reader at 42 °C or other temperature if desired, and measure ThT fluorescence every ~45 min with a shaking kinetic cycle (*see* **Note 31**).

4 Notes

1. Blood can inhibit amplification in RT-QuIC reactions. This phenomenon can be minimized by perfusion of the tissue prior to homogenization. Although perfusion is not always necessary for brain homogenate preparation, we recommend it particularly for samples suspected of having very low levels of prion seeding activity.

2. None of the $rPrP^{Sen}$ constructs we have purified so far have a polyhistidine-tag.
3. Guanidine-HCl has little effect on the pH of the solution and may damage the pH meter probe. Thus, we do not measure the pH of this solution once the GdnHCl has been added.
4. It is our practice to not clean and reuse Ni-NTA Superflow beads for additional $rPrP^{Sen}$ preparations. We tried the QIAGEN cleaning and recharging bead protocols but found that the subsequent purification batch had a poor yield.
5. Approximately 10 ml of concentrated HCl is required to adjust the pH to 5.8. We recommend leaving enough room in the solution to add this volume and then bring the solution to the final volume.
6. N-2 supplement is added to the sample diluent buffer to even out seed distribution and prevent protein loss in tubes at extreme dilutions. N-2 addition likely reduces the absorption of low amounts of protein to the walls of the tubes and may help prevent the nonspecific aggregation of sample proteins.
7. To solubilize EDTA, a basic pH is required.
8. The Thioflavin (ThT) stock can be kept at 4 °C up to 4 weeks.
9. Occasionally all of the solutions used for the RT-QuIC should be filtered using a 0.22 μm filter. In experiments where particulate material was intentionally introduced in the RT-QuIC reaction, an increased rate of spontaneous conversion of the substrate was observed. As a result, it is recommended that reasonable efforts be made to eliminate potential sources of particulate matter such as dust. Prewashing sterilization filters, using Eppendorf dual-filter pipette tips, and only opening tube bags in the hood with gloves on are some of the measures that we take in the lab to reduce the rate of spontaneous false-positive RT-QuIC reactions.
10. We typically generate 10% (w/v) brain homogenates. For example, 900 μl of brain homogenate buffer is added to 0.1 g of brain material.
11. The Bead Beater (Biospec) method is convenient with volumes of 1 ml or less. Similar results have been achieved using dounce or tissue homogenizer techniques; the method of choice is often dictated by volume. All three of these techniques are sufficient to homogenize most biological tissues and efficiently lyse the cells. In all cases, we standardly employ a 2 min, 2000 × *g* spin of the homogenate to clear cellular debris prior to freezing aliquots.
12. Our practice is to store the 10% brain homogenates in small aliquots (i.e., 5–12 μl) to avoid multiple freeze/thaws of the

same aliquot. In our experience, multiple freeze/thaws can lead to a significant loss of seeding activity.

13. During this incubation a 4 ml mini-culture will grow through a lag phase and into growth phase, resulting in a turbid culture.
14. Alternatively LB agar culture plates with kanamycin and chloramphenicol can be streaked with the bacterial glycerol stock and incubated overnight upside down in a 37 °C incubator to grow colonies and then be used to inoculate a 4 ml mini-culture.
15. The bacteria culture will have a short lag phase followed by the growth phase. During the growth phase, the bacteria will be forced to switch carbon sources, stimulating the expression of $rPrP^{Sen}$. Near the end of the recommended growth time, the culture will approach the end of the growth phase with a 600 nm OD of at least 3.0 (using LB broth as a blank). Ideally, the bacteria are grown to maximum density, just prior to the start of stationary phase.
16. The bacteria cell pellet of a single 250 ml conical centrifuge tube, derived from 250 ml of the original 1 l Overnight Express culture, typically yields 15–20 mgs of $rPrP^{Sen}$. We have scaled up protein purifications to 3× batches, using 3 pellets, which typically yield 50–70 mgs of $rPrP^{Sen}$. Here we describe the inclusion body isolation protocol for each cell pellet from a 250 ml conical centrifuge tube containing a quarter of the original 1 L culture.
17. Cell pellets are approximately 3–4 g per 250 ml tube.
18. With each homogenization, first break up all the big chunks at low speed, and then continue to blend at a slightly higher speed (around 50% power) until chunks are no longer visible. This step takes no more than 2 min per tube. The homogenizing tip should be submerged at all times, and the use of full speed should be avoided to prevent foam formation.
19. In our hands the *QIAGEN Ni-NTA Superflow* beads have proven to produce a greater yield of purified prion recombinant protein.
20. Do not dry out the beads. Pull the vacuum just until cracking/white color first becomes apparent. Other methods to obtain 18 g of semi-dry Ni-NTA resin and then rehydrate with Denaturing Buffer may be employed.
21. This is for a ~3–4 g cell pellet. If the cell pellets are smaller, combine them to get approximately 4 g total.
22. The elution of the protein with an imidazole gradient rather than an isocratic step gradient is done to exclude a leading shoulder peak in the protein elution profile (see Fig. 2). The precise identity of the leading shoulder peak is unknown. No

major protein contaminant has yet been identified or can be seen on a protein stain gel. In our experience, the center of the protein elution profile is a superior substrate for the RT-QuIC assay. Protein UV 280 nm elution peak should start about 1/3 of the way through the gradient.

23. It is important to move quickly after protein elution from the column and to keep the protein cold from this time forward. Collect the eluted protein in a chilled tube held in an ice bucket. The protein is directly eluted into cold dialysis buffer (approximately 5 ml of dialysis buffer for 40 ml of eluted protein) to reduce precipitation by completing the pH correction to 5.8. This also reduces the concentration of the protein and the likelihood of precipitation. The operator is to move directly from gently pooling the eluted protein to filtering and gently transferring to the dialysis tubing. Even with these efforts, there will likely be some visible precipitation of the protein. Filtering again immediately after recovering the protein solution from the dialysis tubing the next day will remove most of the protein aggregates that could contribute to accelerating protein precipitation. These filtration steps are necessary to remove nascent protein aggregates that would promote further aggregation and precipitation of protein. In our experience, less than 10% of the $rPrP^{Sen}$ is lost during these steps.

24. There could be some noticeable cloudiness when the eluted protein is placed in dialysis tubing and dialyzed against dialysis buffer (10 mM Na phosphate, pH 5.8). This could be interpreted as loss of protein by precipitation. However, we do not experience a major loss of protein at this stage, and we have a reason to think that the cloudiness is likely caused by imidazole, not protein, precipitation.

25. This is uncommon and only done when there is poor protein yield. Our typical yield from a 4 g cell pellet is ~15–30 mg. The protein concentration of our final purification product is typically around 0.4–0.6 mg/ml following dialysis. We do not store the $rPrP^{Sen}$ at concentrations higher than 0.8 mg/ml since we have observed higher rates of protein precipitation at such high concentrations.

26. Our practice is to freeze 0.5–1 ml aliquots of ~0.5 mg/ml $rPrP^{Sen}$. While freezing and eventually thawing $rPrP^{Sen}$ may result in some level of protein precipitation (usually less than 10% loss), this practice provides for a consistent result from a batch over an extended period of time (months and even years). Alternatively, storing a batch of protein at 4 °C for weeks or months may result in loss of $rPrP^{Sen}$ due to precipitation or degradation. This protein concentration is also convenient in that a full 96-well plate of reaction volumes requires

nearly 1 mg of $rPrP^{Sen}$. Having aliquots at ~0.5 mg/ml allows for some flexibility to use one aliquot for a partial plate or two aliquots for a full plate.

27. Filtering $rPrP^{Sen}$ with a 100 kDa spin filter just prior to adding it to the RT-QuIC reaction mixture serves to limit the initial substrate to monomers and possible small oligomers (dimers and trimers). The use of this step serves to normalize the starting substrate state prior to individual experiments, ensuring more consistent assay results over time and between substrate batches. It is possible to use the filter up to two times by increasing the time of centrifugation on the second spin. Keep filtered protein on ice.

28. Each reaction well has a final volume of 100 μl (98 μl of cocktail + 2 μl of sample). Z is the volume of $rPrP^{Sen}$ that is added to each well volume to have a final concentration of 0.1 mg/ml. This volume varies depending on the concentration of the $rPrP^{Sen}$ batch. The volume of NaCl (Y) will also vary depending on the $rPrP^{Sen}$ used. Therefore, the volume of water per reaction (X) will have to be calculated based on the volume of $rPrP^{Sen}$ and NaCl needed for that specific reaction.

29. Vortex the first five components of the reaction mix for 5 s prior to adding the $rPrP^{Sen}$. Gently invert reaction mix after adding $rPrP^{Sen}$, or, alternatively, mix by pipetting up and down, avoiding the generation of bubbles.

30. Vortex each brain homogenate dilution ~10 s prior to seeding the reaction.

31. The shaking program should be set for 1 min shaking at 700 rpm double orbital and 1 min resting. Fluorescence measurement settings: excitation, 450 nm; emission, 480 nm; bottom read; and number of flashes, 20. Manual gain may vary depending on the final concentration of SDS in the well, usually set at 1600-2100 if using 0.1% SDS/PBS as dilution buffer for the samples. Integration time: 20 μs. It is important to note that the shaking energy input is dependent on the displacement distance of the shaking mechanism as well as the revolutions per minute. While it can be assumed that identical shaking incubators will impart similar shaking energies at matching rpms, the comparison of different shaking incubators will likely require empirical testing. Orbital and double orbital shaking have both been shown to work in different laboratories.

5 Concluding Remarks

Several critical parameters need to be balanced in the RT-QuIC to achieve a high level of sensitivity while maintaining 100% specificity. In optimizations for specific analytes, we tend to focus primarily on varying the following parameters: (1) recombinant PrP substrate, (2) NaCl concentration, (3) SDS concentration, and (4) sample preparation, i.e., isolating prions from confounding tissue components if necessary. Increases in temperature (to 55–60 °C) and mechanical agitation often shorten the lag phases for both prion-seeded and unseeded (spontaneous) fibrillization of the substrate proportionally such that a clear distinction can be maintained between them. However, in attempting to shorten reaction times, care must be taken to avoid conditions that expedite spontaneous, relative to prion-seeded, fibrillization of $rPrP^{Sen}$. In any case, depending upon the goals of a particular RT-QuIC assay application, either sensitivity or specificity can be emphasized in the selection of criteria used to discriminate positive and negative specimens.

Acknowledgments

This work was supported by the Intramural Research Program of the National Institute of Allergy and Infectious Diseases.

References

1. Orru CD, Bongianni M, Tonoli G et al (2014) A test for Creutzfeldt-Jakob disease using nasal brushings. New Engl J Med 371(6):519–529
2. Peden AH, McGuire LI, Appleford NE et al (2012) Sensitive and specific detection of sporadic Creutzfeldt-Jakob disease brain prion protein using real-time quaking induced conversion. J Gen Virol 93:438–449
3. McGuire LI, Peden AH, Orru CD et al (2012) RT-QuIC analysis of cerebrospinal fluid in sporadic Creutzfeldt-Jakob disease. Ann Neurol 72(2):278–285
4. Orru CD, Wilham JM, Vascellari S et al (2012) New generation QuIC assays for prion seeding activity. Prion 6(2):147–152
5. Vascellari S, Orru CD, Hughson AG et al (2012) Prion seeding activities of mouse scrapie strains with divergent PrPSc protease sensitivities and amyloid plaque content using RT-QuIC and eQuIC. PLoS One 7(11):e48969
6. Orru CD, Wilham JM, Raymond LD et al (2011) Prion disease blood test using immunoprecipitation and improved quaking-induced conversion. MBio 2(3): e00078–e00011
7. Wilham JM, Orrú CD, Bessen RA et al (2010) Rapid end-point quantitation of prion seeding activity with sensitivity comparable to bioassays. PLoS Pathog 6(12):e1001217
8. Sano K, Satoh K, Atarashi R et al (2013) Early detection of abnormal prion protein in genetic human prion diseases now possible using real-time QUIC assay. PLoS One 8(1):e54915
9. Atarashi R, Satoh K, Sano K et al (2011) Ultrasensitive human prion detection in cerebrospinal fluid by real-time quaking-induced conversion. Nat Med 17(2):175–178
10. Atarashi R, Sano K, Satoh K, Nishida N (2011) Real-time quaking-induced conversion: a highly sensitive assay for prion detection. Prion 5(3):150–153
11. Atarashi R, Wilham JM, Christensen L et al (2008) Simplified ultrasensitive prion detection by recombinant PrP conversion with shaking. Nat Methods 5(3):211–212

12. Atarashi R, Moore RA, Sim VL et al (2007) Ultrasensitive detection of scrapie prion protein using seeded conversion of recombinant prion protein. Nat Methods 4(8):645–650
13. Cramm M, Schmitz M, Karch A et al (2014) Characteristic CSF prion seeding efficiency in humans with prion diseases. Mol Neurobiol 51 (1):396–405
14. Rubenstein R, Chang B, Gray P et al (2011) Prion disease detection, PMCA kinetics, and IgG in urine from sheep naturally/experimentally infected with scrapie and deer with preclinical/clinical chronic wasting disease. J Virol 85(17):9031–9038
15. Rubenstein R, Chang B, Gray P et al (2010) A novel method for preclinical detection of PrPSc in blood. J Gen Virol 91(Pt 7):1883–1892
16. Chang B, Gray P, Piltch M et al (2009) Surround optical fiber immunoassay (SOFIA): an ultra-sensitive assay for prion protein detection. J Virol Methods 159(1):15–22
17. Edgeworth JA, Farmer M, Sicilia A et al (2011) Detection of prion infection in variant Creutzfeldt-Jakob disease: a blood-based assay. Lancet 377:487–493
18. Edgeworth JA, Jackson GS, Clarke AR et al (2009) Highly sensitive, quantitative cell-based assay for prions adsorbed to solid surfaces. Proc Natl Acad Sci U S A 106 (9):3479–3483
19. Saa P, Castilla J, Soto C (2006) Ultra-efficient replication of infectious prions by automated protein misfolding cyclic amplification. J Biol Chem 281(46):35245–35252
20. Saborio GP, Permanne B, Soto C (2001) Sensitive detection of pathological prion protein by cyclic amplification of protein misfolding. Nature 411(6839):810–813
21. Klohn PC, Stoltze L, Flechsig E et al (2003) A quantitative, highly sensitive cell-based infectivity assay for mouse scrapie prions. Proc Natl Acad Sci U S A 100(20):11666–11671
22. Orrú CD, Groveman BR, Hughson AG, Zanusso G, Coulthart MB, Caughey B (2015) Rapid and Sensitive RT-QuIC Detection of Human Creutzfeldt-Jakob Disease Using Cerebrospinal Fluid. mBio 6(1): e02451–e02414
23. Zanusso G, Monaco S, Pocchiari M, Caughey B (2016) Advanced tests for early and accurate diagnosis of Creutzfeldt-Jakob disease. Nat Rev Neurol 12(6):325–333
24. Schmitz M, Cramm M, Llorens F, Muller-Cramm D, Collins S, Atarashi R, Satoh K, Orru CD, Groveman BR, Zafar S, Schulz-Schaeffer WJ, Caughey B, Zerr I (2016) The real-time quaking-induced conversion assay for detection of human prion disease and study of other protein misfolding diseases. Nat Protoc 11(11):2233–2242
25. Orrú C, Hughson A, Groveman B, Campbell K, Anson K, Manca M, Kraus A, Caughey B (2016) Factors that improve RT-QuIC detection of prion seeding activity. Viruses 8(5):140
26. Jones M, Peden AH, Yull H et al (2009) Human platelets as a substrate source for the in vitro amplification of the abnormal prion protein (PrP) associated with variant Creutzfeldt-Jakob disease. Transfusion 49 (2):376–384
27. Colby DW, Zhang Q, Wang S et al (2007) Prion detection by an amyloid seeding assay. Proc Natl Acad Sci U S A 104 (52):20914–20919
28. Orrú CD, Groveman BR, Raymond LD, Hughson AG, Nonno R, Zou W, Ghetti B, Gambetti P, Caughey B (2015) Bank vole prion protein as an apparently universal substrate for RT-QuIC-based detection and discrimination of prion strains. PLoS Pathog 11 (6):e1004983
29. Orru CD, Hughson AG, Race B et al (2012) Time course of prion seeding activity in cerebrospinal fluid of scrapie-infected hamsters after intratongue and intracerebral inoculations. J Clin Microbiol 50:1464–1466
30. Groveman BR, Orru CD, Hughson AG, Bongianni M, Fiorini M, Imperiale D, Ladogana A, Pocchiari M, Zanusso G, Caughey B (2017) Extended and direct evaluation of RT-QuIC assays for Creutzfeldt-Jakob disease diagnosis. Ann Clin Trans Neurol 4(2):139–144
31. Bongianni M, Orrù C, Groveman BR, Sacchetto L, Fiorini M, Tonoli G, Triva G, Capaldi S, Testi S, Ferrari S, Cagnin A, Ladogana A, Poleggi A, Colaizzo E, Tiple D, Vaianella L, Castriciano S, Marchioni D, Hughson AG, Imperiale D, Cattaruzza T, Fabrizi GM, Pocchiari M, Monaco S, Caughey B, Zanusso G (2017) Diagnosis of human prion disease using real-time quaking-induced conversion testing of olfactory mucosa and cerebrospinal fluid samples. JAMA Neurol 74(2):155
32. Lattanzio F, Abu-Rumeileh S, Franceschini A, Kai H, Amore G, Poggiolini I, Rossi M, Baiardi S, McGuire L, Ladogana A, Pocchiari M, Green A, Capellari S, Parchi P (2017) Prion-specific and surrogate CSF biomarkers in Creutzfeldt-Jakob disease: diagnostic accuracy in relation to molecular subtypes and analysis of neuropathological correlates of ptau and Abeta42 levels. Acta Neuropathol 133 (4):559–578
33. Foutz A, Appleby BS, Hamlin C, Liu X, Yang S, Cohen Y, Chen W, Blevins J, Fausett C,

Wang H, Gambetti P, Zhang S, Hughson A, Tatsuoka C, Schonberger LB, Cohen ML, Caughey B, Safar JG (2017) Diagnostic and prognostic value of human prion detection in cerebrospinal fluid. Ann Neurol 81(1):79–92

34. Zanusso G, Bongianni M, Caughey B (2014) A test for Creutzfeldt-Jakob disease using nasal brushings. N Engl J Med 371(19):1842–1843
35. Redaelli V, Bistaffa E, Zanusso G, Salzano G, Sacchetto L, Rossi M, Luca CMGD, Di Bari M, Portaleone SM, Agrimi U, Legname G, Roiter I, Forloni G, Tagliavini F, Moda F (2017) Detection of prion seeding activity in the olfactory mucosa of patients with Fatal Familial Insomnia. Sci Rep 7:46269
36. Dassanayake RP, Orru CD, Hughson AG, Caughey B, Graca T, Zhuang D, Madsen-Bouterse SA, Knowles DP, Schneider DA (2016) Sensitive and specific detection of classical scrapie prions in the brains of goats by real-time quaking-induced conversion. J Gen Virol 97 (3):803–812
37. Elder AM, Henderson DM, Nalls AV et al (2013) In vitro detection of prionemia in TSE-infected cervids and hamsters. PLoS One 8(11):e80203
38. Henderson DM, Manca M, Haley NJ et al (2013) Rapid antemortem detection of CWD prions in deer saliva. PLoS One 8(9):e74377
39. John TR, Schatzl HM, Gilch S (2013) Early detection of chronic wasting disease prions in urine of pre-symptomatic deer by real-time quaking-induced conversion assay. Prion 7 (3):253–258
40. Bessen RA, Shearin H, Martinka S et al (2010) Prion shedding from olfactory neurons into nasal secretions. PLoS Path 6:e1000837
41. Masujin K, Orru CD, Miyazawa K, Groveman BR, Raymond LD, Hughson AG, Caughey B (2016) Detection of atypical H-type bovine spongiform encephalopathy and discrimination of bovine prion strains by real-time quaking-induced conversion. J Clin Microbiol 54 (3):676–686
42. Orrú CD, Favole A, Corona C, Mazza M, Manca M, Groveman BR, Hughson AG, Acutis PL, Caramelli M, Zanusso G, Casalone C, Caughey B (2015) Detection and discrimination of classical and atypical L-type bovine spongiform encephalopathy by real-time quaking-induced conversion. J Clin Microbiol 53 (4):1115–1120
43. Shi S, Mitteregger-Kretzschmar G, Giese A, Kretzschmar HA (2013) Establishing quantitative real-time quaking-induced conversion (qRT-QuIC) for highly sensitive detection and quantification of PrPSc in prion-infected tissues. Acta Neuropathol Commun 1(1):44

Chapter 15

A Quick Method to Evaluate the Effect of the Amino Acid Sequence in the Misfolding Proneness of the Prion Protein

Natalia Fernández-Borges, Hasier Eraña, Saioa R. Elezgarai, Chafik Harrathi, Vanesa Venegas, and Joaquín Castilla

Abstract

Prion diseases or transmissible spongiform encephalopathies (TSEs) are a group of neurodegenerative diseases where the misfolding of the prion protein (PrP) is a crucial event. Based on studies in TSE-affected humans and the generation of transgenic mouse models overexpressing different mutated versions of the PrP, we conclude that both wild-type and mutated PrPs exhibit differential propensity to misfold in vivo. Here, we describe a new method in vitro to assess and quantify the PrP misfolding phenomenon in order to better understand the molecular mechanisms involved in this process.

Key words In vitro prion propagation, Spontaneous recombinant prion, PMCA, recPMCA, Protein misfolding

1 Introduction

Prion diseases or transmissible spongiform encephalopathies (TSEs) are a group of rare and fatal neurodegenerative disorders that affect both humans and animals. They are characterized by the accumulation of a pathogenic infectious protein called PrP^{Sc} that induces abnormal folding of the normal cellular protein PrP^{C}.

Scrapie, affecting sheep and goats, bovine spongiform encephalopathy (BSE) in cows, and chronic wasting disease (CWD) in deer are good examples of how prions are spread in nature. Humans can also be affected by various forms of TSEs: (1) sporadic, i.e., Creutzfeldt-Jakob disease (sCJD) [1]; (2) iatrogenic, i.e., ***kuru*** [2, 3] or vCJD [4]; and (3) inherited prion diseases, i.e., familial fatal insomnia (FFI) [5]. Around the 85% of human prion diseases are grouped as sporadic, affecting worldwide approximately one person per million per year and without any currently recognized cause. There are two major possibilities considered regarding the

Victoria A. Lawson (ed.), *Prions: Methods and Protocols*, Methods in Molecular Biology, vol. 1658,
DOI 10.1007/978-1-4939-7244-9_15, © Springer Science+Business Media LLC 2017

onset of the disease: spontaneous *PRNP* gene somatic mutations or stochastic PrP protein structural changes or misfolding [6].

Over 40 different mutations involved in inherited prion diseases have been described in human PrP, comprising the 15% of the human affecting prion diseases [7]. The most prevalent diseases enclosed in this group are genetic CJD (gCJD), fatal familial insomnia (FFI), and Gerstmann-Sträussler-Scheinker (GSS), each one represented by distinct mutations that might trigger a spontaneous conversion from PrP^{C} to PrP^{Sc} [8].

There are dozens of studies that, through the generation of transgenic mouse models overexpressing different mutated versions of the PrP, establish a correlation between a human *PRNP* mutation and PrP misfolding. These studies have been very valuable to understand how prions work but are strongly limited because of cost and time consuming.

In this context, we should assume that both wild-type and mutated versions of PrP have a certain tendency to misfold in vivo. A decade ago, the PrP^{C} to PrP^{Sc} conversion process could be replicated in vitro using the protein misfolding cyclic amplification (PMCA) methodology. The PMCA requiring brain homogenates as source of PrP^{C} mimicked the most important features of prions, i.e., infectivity [9], species barrier [10], and strain phenomenon [11]. This technology has been of paramount importance to better understand the misfolding (PrP^{C} to PrP^{Sc}) process. However, the need of brain homogenates has limited other potential and more accurate studies. In recent years, a new technology that requires recombinant PrP instead of brain homogenate as source of protein is allowing further studies focused on assessing and quantifying the misfolding phenomenon.

In this article we will detail the use of an adapted methodology, recPMCA, to evaluate the misfolding proneness of the prion protein containing different amino acid substitutions. In a similar way that the in vitro studies using brain PMCA predicted the behavior of certain prions in vitro [12, 13], this new technology will contribute to understand the molecular mechanisms involved in the conversion process: transmission barrier, strain phenomenon, and spontaneous generation of infectivity.

2 Materials

2.1 Chemicals

1. Ethylenediaminetetraacetic acid (EDTA).
2. Sodium chloride (NaCl).
3. Sodium hydroxide (NaOH).
4. Guanidine hydrochloride (GdnHCl).
5. *N*-Lauroylsarcosine sodium salt (Sarkosyl).

6. Triton X-100.
7. PBS 1× (w/o calcium, magnesium).
8. Complete protease inhibitor cocktail (Roche).
9. Distilled water.
10. Proteinase K (Roche).

2.2 Solutions

1. Perfusion buffer: 5 mM EDTA, PBS 1×.
2. Conversion buffer: 150 mM NaCl, 1% Triton X-100, complete protease inhibitor cocktail (Roche), PBS 1×.
3. PrP^C digestion buffer: 10% sarkosyl, 85 μg/ml PK, PBS 1× (*see* **Note 1**).
4. NuPAGE LDS sample buffer (4×) (Invitrogen).
5. Tris/Glycine/SDS running buffer (Bio-Rad).
6. SuperSignal West Pico (Thermo Scientific).
7. Specific primary and secondary antibodies.
8. Bleach at 20,000 ppm (commercial).
9. 1 N NaOH.

2.3 Biologicals

1. 4–10-week-old PrP *knockout* (PrP-KO) mice (129/OLAxC-BAxC57BL6) (*see* **Note 2**).

2.4 Equipment

1. 1.5 ml centrifuge tubes.
2. 0.2 ml PCR tubes.
3. 15 ml centrifuge tubes.
4. 20 G needle.
5. Slide-A-Lyzer Dialysis Cassette, 10K MWCO (Thermo Scientific Pierce).
6. PVDF Trans-Blot Turbo transfer packs (Bio-Rad).
7. 4–15% Criterion TGX gel (Bio-Rad).
8. Surgical materials (scissors and forceps).
9. 30 ml glass tissue grinder Potter-Elvehjem (Thermo Scientific).
10. Single and multichannel pipettes.
11. Plastic paraffin film.
12. Permanent markers.
13. Temperature probe.
14. Digital laboratory timer.
15. Criterion cell (Bio-Rad).
16. PowerPac Universal power supply (Bio-Rad).
17. Trans-Blot Turbo system (Bio-Rad).

18. Sonicator and microplate horn system S-700MPX-010 (Qsonica, LLC) (*see* **Note 3**).
19. Sonicator tube holder and cover (Qsonica, LLC).
20. Thermomixer comfort (Eppendorf).
21. Thermoblock dry heater (Fisher Scientific).
22. Magnetic stirrer.
23. Refrigerated centrifuge (ST 16R, Thermo Sorvall).
24. UV-Vis spectrophotometer NanoDrop 2000 (Thermo Scientific).
25. CO_2 mice euthanasia chamber.
26. −80 °C freezer.
27. −20 °C freezer.
28. Ice.
29. Dry ice.

3 Methods

The methodology involved in the spontaneous generation of PK-resistant (misfolded) recombinant prion protein (rec-PrP^{res}) by recombinant PMCA (recPMCA) requires two steps: (1) a first step comprising the preparation of the rec-PrP-based PMCA substrate and (2) a second step concerning the unseeded in vitro amplification/propagation.

The whole preparation of the rec-PrP-based PMCA will be described through two different procedures (Subheadings 3.1 and 3.2): (1) preparation of rec-PrP and (2) preparation of the helper brain homogenate. Both procedures should be performed in parallel in order to avoid withdrawal periods where the protein and/or helper molecules could be degraded. Thus, it is worthy to stress the importance of maintaining all the compounds at 4 °C before the substrate can be stored at −80 °C.

Regarding the second step involving in vitro PMCA, similarly to what happens in a conventional PMCA, numerous variables will play an important role in the process (power and duration of sonication, temperature, etc.) increasing the importance of thorough observation of the system and the development and implementation of traceability of the process [13, 14].

The use of a completely prion-free area and prion-free instruments is highly recommended for conducting the whole protocol. This methodology is extremely efficient amplifying pre-seeded samples making it difficult to discriminate these acting as contaminants from the bona fide de novo unseeded samples [15]. We also recommend cleaning regularly the bench and the equipment using commercial bleach (at 20,000 ppm) and/or 1 N NaOH with the purpose to prevent cross-contamination of the samples [16, 17].

3.1 Preparation of rec-PrP as Source of Protein for recPMCA

Taking advantage of the presence of histidines in the octarepeat region of the PrP, we recommend the use of standard protocols for the expression and histidine-affinity-based column purification of the rec-PrP (*see* **Note 4**). Importantly, the purified rec-PrP should be stored at approximately 4 mg/ml in 6 M GdnHCl in aliquots (100 μl or similar) at −20 °C or lower.

1. Add 100 μl of 6 M GdnHCl rec-PrP (4 mg/ml) previously thawed at 4 °C to 400 μl of PBS 1×, and gently invert the tube several times at 4 °C (*see* **Note 5**).
2. Load the 500 μl sample into a 0.1–0.5 Slide-A-Lyzer Dialysis Cassette, 10K MWCO using a 20 G needle.
3. Follow 1 h dialysis in a precooled 2 L PBS container at 4 °C and under continuous agitation in a magnetic stirrer.
4. Remove the sample from the dialysis cassette to a precooled 1.5 ml Eppendorf tube using a 20 G needle. Depending on the type of rec-PrP, a white precipitation should be expected since the initial mixing increasing until the end of the dialysis.
5. Discard the naturally aggregated (likely amorphous aggregation) rec-PrP by 19,000 × *g* centrifugation for 15 min at 4 °C. Supernatant is kept at 4 °C.
6. Protein concentration in the supernatant is measured by spectrophotometry (NanoDrop 2000). Around 2 mg/ml of rec-PrP is optimal for the spontaneous recombinant PMCA methodology (*see* **Note 6**).

In order to avoid spontaneous amorphous aggregation after the centrifugation (**step 5**), we recommend to maintain the sample at 4 °C and use it as soon as possible to prepare the recPMCA substrate (Subheading 3.3). If the supernatant containing the dialyzed protein will be used after 24 h, we recommend to freeze it at −80 °C and repeat the **step 5** before continuing with the process described in Subheading 3.3.

3.2 Preparation of Helper Brain Homogenate

Although rec-PrP might spontaneously misfold in the absence of brain homogenate complementing the substrate (*see* **Note 7**), the reliability of the recPMCA results is improved by the addition of mouse PrP-KO brain homogenate (*see* **Note 2**).

1. Euthanize a PrP-KO mouse using CO_2 (avoid other procedures), and perfuse (intracardiac) the whole animal using 30 ml of PBS + 5 mM EDTA (perfusion buffer). The aim of this process is to remove completely the blood avoiding the presence of cations and certain blood components that can seriously affect the PMCA process [18]. A correct perfusion will keep the brain completely white before its extraction (*see* **Note 8**). We recommend performing the whole perfusion process in a prion-free animal facility in order to avoid cross-contaminations.

2. Freeze the extracted brain immediately using dry ice or liquid nitrogen, and store it at −80 °C.
3. Homogenize the brain at 10% (w/v) using conversion buffer and pestling at 4 °C with a 30 ml glass tissue grinder Potter-Elvehjem. A standard brain will yield around 4.5–5 ml of brain homogenate.
4. Aliquot the brain homogenate in 1.5 ml tubes, and centrifuge the sample at 19,000 × *g* for 15 min at 4 °C in order to discard large pieces of tissues and undissolved cellular membranes. Discard the pellet and use the supernatant to prepare the recPMCA substrate.

3.3 Preparation of recPMCA Substrate

1. Mix by pipetting 500 μl of dialyzed rec-PrP (see Subheading 3.1) with 4500 μl of PrP-KO mouse brain homogenate supernatant (see Subheading 3.2) in a 15 ml centrifuge tube. In order to avoid recurrent freezing-thawing, aliquot samples after proper homogenization in convenient experimental volumes, preferably in PCR tubes containing 45–50 μl of substrate. Use directly in recPMCA or freeze immediately at −80 °C before use.
2. Previous to the experimental recPMCA procedure, analyze the presence of rec-PrP using 1 μl of substrate (around 0.2 μg of rec-PrP) by Western blotting following the protocol detailed in Subheading 3.5 (**steps 4** and **5**).

3.4 Serial Rounds of recPMCA Using Unseeded Samples

This is the most critical procedure to generate bona fide misfolded rec-PrP (PK-resistant rec-PrP; rec-PrP^{res}) spontaneously. The protocol requires serial rounds of recPMCA in order to refresh the rec-PrP contained in the substrate (Fig. 1). Depending on the propensity of the protein to misfold spontaneously, the first rec-PrP^{res}-positive tube will appear in any round and will continue appearing in the consecutive rounds (*see* **Note 9** and Fig. 2).

1. Thaw the substrate aliquots at 4 °C or use them directly after their preparation as described in Subheading 3.3 avoiding recurrent freezing-thawing.
2. In order to obtain reliable and statistically reproducible results, use 6–8 tube replicates per sample. We recommend to group the PCR tube strips/caps two by two in order to seal them conveniently with plastic paraffin film. This will avoid the caps opening during the sonication process.
3. Place the tubes over the sonicator PCR tube-adapted rack. We recommend to have the sonicator switched on and distilled water on the horn plate constantly circulating at 39 °C. The amplitude of the sonicator should be adjusted to 40–70%. This parameter will control the power of the ultrasonic wave that should work in the range of 180–220 W. The ultrasound

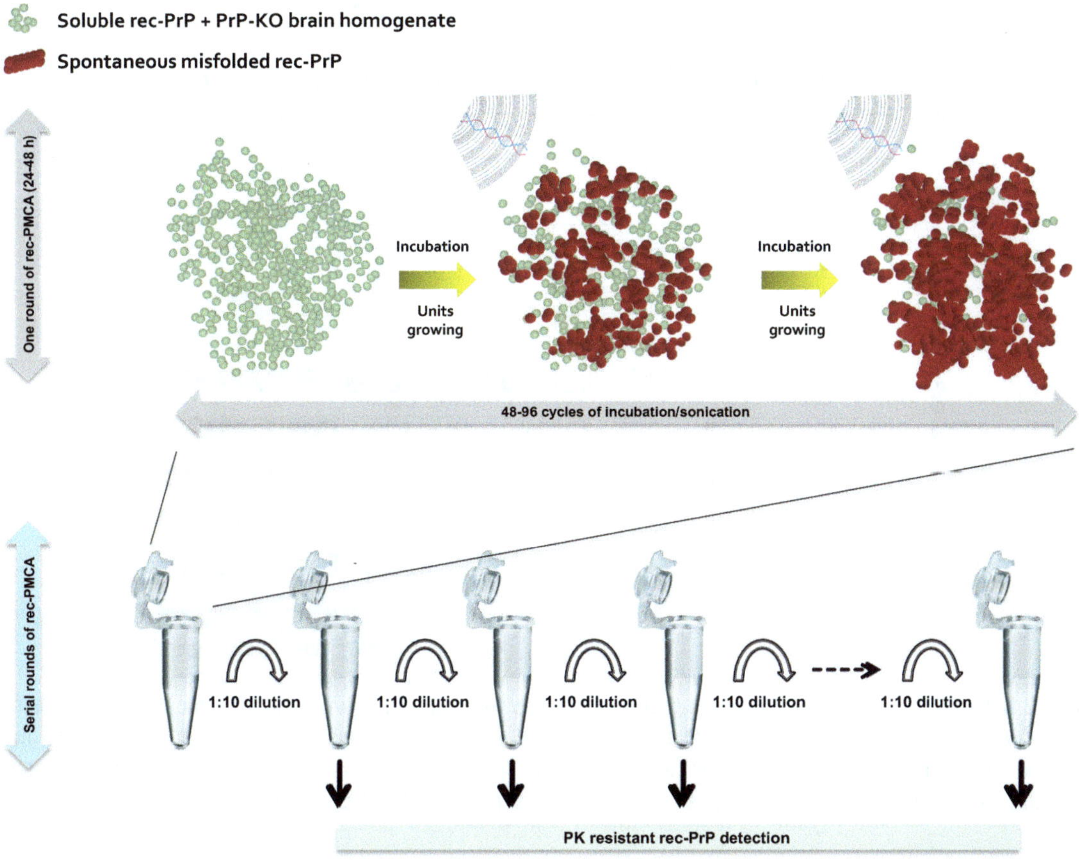

Fig. 1 Scheme of serial rounds of recPMCA. Unseeded recPMCA substrate based on rec-PrP and PrP-KO brain homogenate is subjected to a round of recPMCA (cycles of sonication-incubation steps). After 24–48 h occurring the first round, a 1:10 dilution is performed to refresh the rec-PrP. This process is repeated serially and the samples are analyzed for the presence of PK-resistant rec-PrP^{res}

strength of the sonication should be adapted previously for each sonicator trying to obtain a homogeneous ultrasound (*see* **Note 10**).

4. Run a PMCA program using the following parameters: 20 s of sonication and 30 min of incubation per cycle at 39 °C. Keep the samples at least for 24 h (48 cycles) per PMCA round (*see* **Note 10**).
5. Stop the PMCA round taking out the treated tubes, and keep them at 4 °C. Remove carefully the plastic paraffin film and spin down before opening the caps. We recommend cleaning the external areas of the tubes using 20,000 ppm of commercial bleach and washing them before opening.
6. Perform the second round of recPMCA by adding (mix with intense pipetting) 5 μl from each treated tube to new thawed tubes containing 45–50 μl of fresh substrate. Run the PMCA

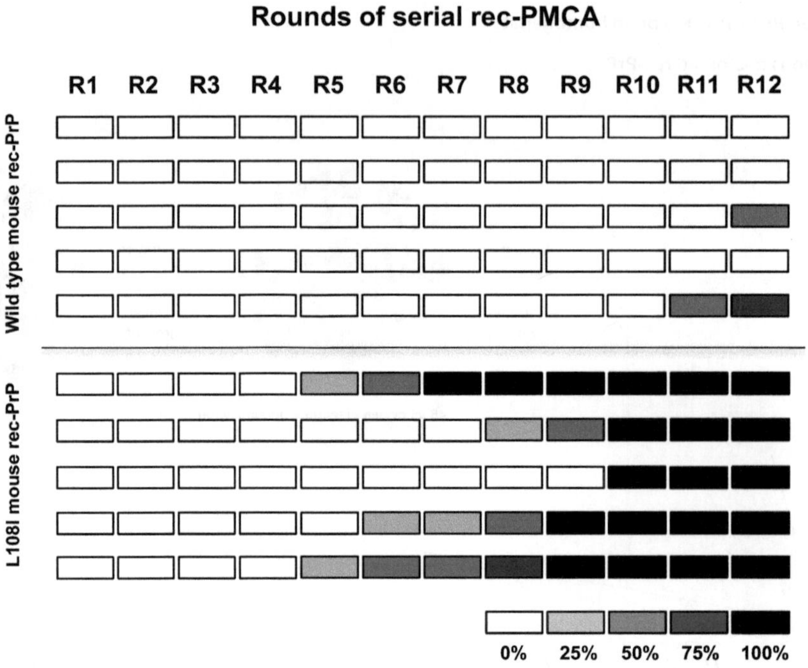

Fig. 2 Comparative analysis of misfolding proneness between wild-type (wt) and L108I mouse rec-PrP using serial recPMCA. Two different substrates based on wt and L108I mouse rec-PrP were subjected to five independent experiments of 12 rounds of serial recPMCA. Each experiment contained four tube replicates. Legend indicates the percentage of positive tubes that turned rec-PrP^{res} positive after their PK digestion

program again as described in the **step 4**. This serial process should be repeated according to the number of rounds that the experiment requires (*see* **Note 11**). Analyze each tube from each PMCA round (see Subheading 3.5) for the presence of -PK-resistant PrP (rec-PrP^{res}). Take out from the serial recPMCA experiment those positive tubes since they could be a source of cross-contaminations.

It is critical to keep stable the horn plate water bath temperature and sonication power during the whole serial recPMCA experiment in order to obtain reliable results.

3.5 PK-Resistant rec-PrP Biochemical Detection

1. Mix 10 μl of recPMCA-treated sample with 10 μl of PrP^{C} digestion buffer, and perform the digestion for 1 h at 42 °C with shaking (450 rpm).
2. Stop the digestion by adding 1:3 (v/v) NuPAGE LDS sample buffer 4x, and heat the samples for 10 min at 100 °C.
3. Perform the protein electrophoresis on Criterion 4–15% TGX gel (Bio-Rad), and transfer the proteins to a PVDF membrane through the Bio-Rad Trans-Blot Turbo system using the manufacturer conditions.

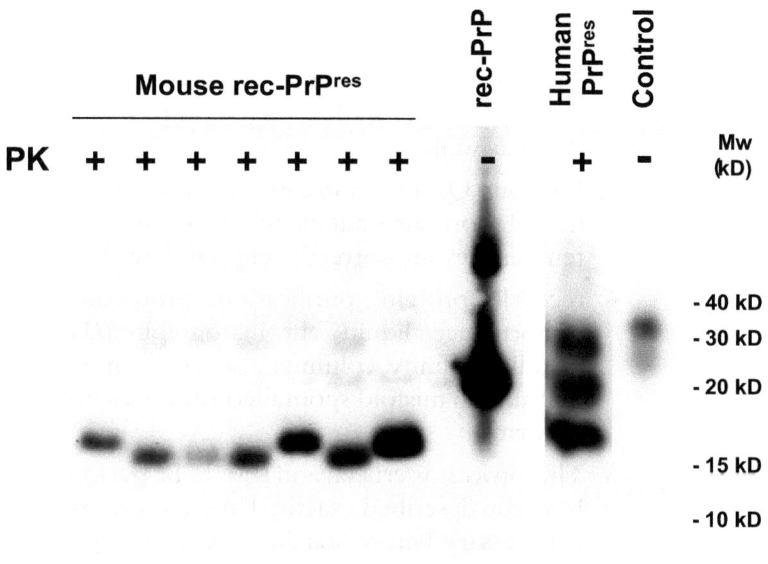

Fig. 3 Biochemical analysis of different recombinant mouse prions. Unseeded recombinant mouse PrP^{C} substrates from different serial recPMCA experiments were digested with 85 μl/ml of PK for 1 h at 42 °C. Different electrophoretic patterns are observed. *Human PrP^{res}* Human brain from a sCJD-affected patient was used as control after its digestion with 85 μl/ml of PK for 1 h at 42 °C. *rec-PrP* PK-undigested mouse rec-PrP. *Control* PK-undigested mouse brain homogenate. Human PrP^{res} and Control were run in a different gel

4. Develop the membrane with the SuperSignal West Pico after immunoblotting using specific antibodies anti-PrP.

The presence of a PK-resistant 15–17 kDa band will be indicative of the spontaneous generation of a misfolded recombinant protein (*see* **Note 12** and Fig. 3). Other smaller bands, like rec-PrP^{res} fragments, can also appear although the bona fide (infectious) misfolded protein should weigh 15–17 kDa and resist at least 50–100 μg/ml of PK in a similar way than mammalian prions.

4 Notes

1. Seeded rec-PCMA has shown (unpublished data) a distinct PK sensitivity for each species/strain rec-PrP^{res}. Thus, PK concentration (see Subheading 3.5) for digestion should be adapted according to the species of the rec-PrP being studied.

   Although we recommend the use of 10% sarkosyl detergent for the PrP^{C} digestion buffer, other detergents like 2% Tween, 2% NP40 work properly as well.
2. Although we recommend the use of PrP-KO mice (129/OLAxCBAxC57BL6) [19], in principle, other PrP-KO mice from any origin might be used as source of brains. Alternative

to PrP-KO brains, other brain homogenate sources containing an in vitro inconvertible PrP^C might be used. Thus, perfused chicken (*Gallus gallus domesticus*) brains result an adequate source of helper molecules to complement the recPMCA substrate as well.

3. Previous Qsonica sonicator versions of the S-4000MPX-010 model work also adequately when the sonication power and temperature are correctly set up as described in Subheading 3.4.
4. rec-PrP protein purification protocols based on high-performance liquid chromatography (HPLC) coupled to histidine-affinity columns also yield an adequate folded rec-PrP able to misfold spontaneously into a bona fide PK-resistant isoform.
5. This process is crucial and should be performed in the way that has been described exactly. Thus, a previous 1:5 dilution in PBS is necessary before starting the dialysis process. This mixture should be performed adding 1 volume of rec-PrP over 4 volumes of PBS but not in the opposite way. We believe that the first quick folding step (a quick change in the chaotropic conditions from 6 M GdnHCl to 1.2 M GdnHCl) is crucial to create misfolded molecules that will start the nucleation and seeding process along the serial recPMCA.
6. The PrP concentration is a critical factor for prion propagation in vitro [20]. Thus, lower- or higher-recombinant PrP concentrations in the recPMCA substrate might yield distinct results in the number of PMCA rounds where the first spontaneous rec-PrP^{res} appears. We recommend the use of 2 mg/ml of rec-PrP as the optimal concentration and to keep the same concentration when comparative studies are performed between different rec-PrPs.
7. The absence of brain homogenate in the substrate significantly decreases the ability to generate recombinant PrP^{res} spontaneously. This fact has been evaluated by a lower number of positive replicates and a later passage in the appearance of the first positive tube. We strongly recommend the use of PrP *knockout* mouse brain homogenate or alternative brains (see **Note 2**) to prepare the recPMCA substrate.
8. If the perfusion quality has not been optimal or has been impossible to conduct, we recommend to immerse the extracted brain into a recipient containing perfusion buffer and cut it into small pieces with appropriate surgical material. This process should be performed to eliminate blood vessels and blood traces as much as possible.
9. Since the nucleation process by which the misfolded proteins assemble to grow before acting as seed is a stochastic process, we should expect the following scenarios after performing

repetitive serial recPMCA experiments: (1) the first rec-PrP^{res}-positive tube does not appear always in the same round of recPMCA even if we repeat the experiment using the same substrate (other aliquots) and similar conditions (the PMCA parameters cannot be maintained identical during the whole serial recPMCA process), and (2) not all the tubes (duplicates) in the same experiment turn positive at the same time. Although this situation could reflect that we are handling an irreproducible procedure, the proper interpretation would be that we are facing a genuine stochastic process. Thus, the use of the wild-type rec-PrP as control in all the repetitive experiments is of paramount importance. In this case, we will observe differential tendencies of the different mutated rec-PrP to misfold spontaneously (Fig. 2).

10. Alternative PMCA parameters might work properly especially, as it was mentioned in **Note 3**, if a different sonicator is used. Thus, temperature in the water bath lower than 39 °C and longer sonication periods than 20 s can be used. However, it is important to be aware that the results could vary significantly if the same sample is treated comparatively using different parameters. For this reason, all the changes should be assessed using wild-type rec-PrP controls comparing them with the rec-PrP-mutated versions in terms of their proneness to misfold. The parameters should be maintained along the whole serial recPMCA experiment. The trickiest parameter is the ultrasonic power, which should be calibrated manually trying to keep it with a strong and homogeneous tone during the whole experimental process.

11. There is not information a priori about the number of rounds that a particular rec-PrP will require to misfold spontaneously. Our unpublished results indicate that most of the rec-PrP but not all are able to misfold in less than 30 rounds of recPMCA. There are rec-PrP species that require much less rounds of recPMCA than others. For that, it is crucial to use the proper control for each recPMCA experiment. We recommend performing a maximum of 30 rounds of recPMCA in those cases that a new rec-PrP is tested.

12. In a similar way that the mammalian prion strains can appear in distinct electrophoretic migrations observed by Western blot (WB), recombinant prions might also show a differential pattern by WB. Since rec-PrP is an unglycosylated protein, we should not expect so many differences comparing to what we observe in PrPs containing all the glycosylated variants. According to our unpublished results, some (may be all) rec-PrP^{res} from the same species can be misfolded showing different electrophoretic patterns between 15 and 17 kDa, presumably suggesting the existence of different recombinant prion strains (Fig. 3).

Acknowledgments

This work was financially supported by a national grant from Spain (AGL2012-37988-C04-01), a CTP grant (CTP2013-P05), and an InteRreg grant (EFA218/11). The authors thank the support from IKERBasque foundation and vivarium and maintenance from CIC bioGUNE.

The authors declare no competing financial interests.

References

1. Cali I, Castellani R, Yuan J et al (2006) Classification of sporadic Creutzfeldt-Jakob disease revisited. Brain 129:2266–2277
2. Brandner S, Whitfield J, Boone K et al (2008) Central and peripheral pathology of kuru: pathological analysis of a recent case and comparison with other forms of human prion disease. Philos Trans R Soc Lond Ser B Biol Sci 363:3755–3763
3. Liberski PP, Brown P (2009) Kuru: its ramifications after fifty years. Exp Gerontol 44:63–69
4. Smith PG, Cousens SN, d' Huillard Aignaux JN et al (2004) The epidemiology of variant Creutzfeldt-Jakob disease. Curr Top Microbiol Immunol 284:161–191
5. Lugaresi E, Medori R, Montagna P et al (1986) Fatal familial insomnia and dysautonomia with selective degeneration of thalamic nuclei. N Engl J Med 315:997–1003
6. Knight RS, Will RG (2004) Prion diseases. J Neurol Neurosurg Psychiatry 75(Suppl 1):36–42
7. Colby DW, Prusiner SB (2001) Prions. Cold Spring Harb Perspect Biol 3:a006833
8. Poggiolini I, Saverioni D, Parchi P (2013) Prion protein misfolding, strains, and neurotoxicity: an update from studies on Mammalian prions. Int J Cell Biol 2013:910314
9. Castilla J, Saá P, Hetz C, Soto C (2005) In vitro generation of infectious scrapie prions. Cell 121:195–206
10. Castilla J, Gonzalez-Romero D, Saá P et al (2008) Crossing the species barrier by PrP(Sc) replication in vitro generates unique infectious prions. Cell 134:757–768
11. Castilla J, Morales R, Saá P et al (2008) Cell-free propagation of prion strains. EMBO j 27:2557–2566
12. Chianini F, Fernández-Borges N, Vidal E, Gibbard L et al (2012) Rabbits are not resistant to prion infection. Proc Natl Acad Sci U S A 109:5080–5085
13. Saá P, Castilla J, Soto C (2005) Cyclic amplification of protein misfolding and aggregation. Methods Mol Biol 299:53–65
14. Castilla J, Saá P, Morales R et al (2006) Protein misfolding cyclic amplification for diagnosis and prion propagation studies. Methods Enzymol 412:3–21
15. Generation of a new mouse model showing a fast, spontaneous and highly transmissible prion disease (2014) Oral communication in Neuroprion meeting Trieste
16. Käsermann F, Kempf C (2003) Sodium hydroxide renders the prion protein PrPSc sensitive to proteinase K. J Gen Virol 84:3173–3176
17. Taylor DM (2000) Inactivation of transmissible degenerative encephalopathy agents: a review. Vet J 159:10–17
18. Saá P, Castilla J, Soto C (2006) Ultra-efficient replication of infectious prions by automated protein misfolding cyclic amplification. J Biol Chem 281:35245–35252
19. Fernández-Borges N, Castilla J (2010) PMCA. A decade of in vitro prion replication. Curr Chem Biol 4:200–207
20. Manson JC, Clarke AR, Hooper ML et al (1994) 129/Ola mice carrying a null mutation in PrP that abolishes mRNA production are developmentally normal. Mol Neurobiol 8:121–127

Part IV

In Vivo Investigation of Prion Disease

Chapter 16

Insights into Mechanisms of Transmission and Pathogenesis from Transgenic Mouse Models of Prion Diseases

Julie A. Moreno and Glenn C. Telling

Abstract

Prions represent a new paradigm of protein-mediated information transfer. In the case of mammals, prions are the cause of fatal, transmissible neurodegenerative diseases, sometimes referred to as transmissible spongiform encephalopathies (TSEs), which frequently occur as epidemics. An increasing body of evidence indicates that the canonical mechanism of conformational corruption of cellular prion protein (PrP^C) by the pathogenic isoform (PrP^{Sc}) that is the basis of prion formation in TSEs is common to a spectrum of proteins associated with various additional human neurodegenerative disorders, including the more common Alzheimer's and Parkinson's diseases. The peerless infectious properties of TSE prions, and the unparalleled tools for their study, therefore enable elucidation of mechanisms of template-mediated conformational propagation that are generally applicable to these related disease states. Many unresolved issues remain including the exact molecular nature of the prion, the detailed cellular and molecular mechanisms of prion propagation, and the means by which prion diseases can be both genetic and infectious. In addition, we know little about the mechanism by which neurons degenerate during prion diseases. Tied to this, the physiological role of the normal form of the prion protein remains unclear and it is uncertain whether or not loss of this function contributes to prion pathogenesis. The factors governing the transmission of prions between species remain unclear, in particular the means by which prion strains and PrP primary structure interact to affect interspecies prion transmission. Despite all these unknowns, advances in our understanding of prions have occurred because of their transmissibility to experimental animals, and the development of transgenic (Tg) mouse models has done much to further our understanding about various aspects of prion biology. In this review, we will focus on advances in our understanding of prion biology that occurred in the past 8 years since our last review of this topic.

Key words Prions, Transmissible spongiform encephalopathies (TSEs), Scrapie, Chronic wasting disease (CWD), Creutzfeldt–Jakob disease (CJD), Bovine spongiform encephalopathy (BSE), Transgenic mice

1 Introduction

Prions challenge fundamental concepts of inheritance and infection. Prion-mediated phenotypes, in organisms such as yeast and aplasia, as well as a variety of mammalian neurodegenerative diseases, result

Victoria A. Lawson (ed.), *Prions: Methods and Protocols*, Methods in Molecular Biology, vol. 1658,
DOI 10.1007/978-1-4939-7244-9_16, © Springer Science+Business Media LLC 2017

from the ability of the prion conformation to interact with and induce further structural conversion of counterpart endogenous proteins in their normal states. The prototype prion diseases are the transmissible spongiform encephalopathies (TSEs) of animals and humans, which include scrapie in sheep, bovine spongiform encephalopathy (BSE), chronic wasting disease (CWD) of deer and elk, transmissible mink encephalopathy (TME), and various human disorders, the most common being Creutzfeldt–Jakob disease (CJD). They share a number of common features, the most consistent being that disease is generally naturally and/or experimentally transmissible; the extraordinary physical properties of the infectious agent that account for its extreme resistance to conventional disinfection procedures; and the neuropathologic changes that accompany disease in the central nervous system (CNS). These features typically consist of neuronal vacuolation and degeneration, a reactive proliferation of astroglia, and, in some conditions such as the human prion diseases kuru, Gerstmann–Straussler–Scheinker (GSS) syndrome, and variant CJD (vCJD), the deposition of amyloid plaques.

Central to all prion states are the protean conformational properties of particular host-encoded proteins. In the case of TSEs, this is the prion protein (PrP). The normal form, referred to as PrP^C, is a sialoglycoprotein of molecular weight ~ 33 kDa that is attached to the surface of neurons and other cell types by means of a glycophoshphatidyl inositol (GPI) anchor. Experimentally, PrP^C is sensitive to protease treatment, can be released from cell surfaces by treatment with phosphoinositide phospholipase C (PIPLC), and is soluble in detergents as a monomer. During the disease, a conformational variant of PrP^C, referred to as PrP^{Sc}, accumulates in infected brains and, in most cases, tissues of the lymphoreticular system. PrP^{Sc} is partially resistant to protease treatment, is resistant to PIPLC treatment, is insoluble in detergents, and is prone to aggregate. In most examples of prion disease, experimental protease cleavage of the amino-terminal 66, or so, amino acids of PrP^{Sc} gives rise to a protease-resistant core, originally referred to as PrP27–30. Considerable evidence now supports the once unorthodox hypothesis that prions lack nucleic acid, and are composed largely, if not entirely, of PrP^{Sc}, and that replication involves corruption of the benign cellular form of the prion protein by its abnormally conformed infective counterpart, which results in the exponential accumulation of prions and inevitable demise of the infected host as a result of profound CNS neurodegeneration [1].

2 Prion Transmission Barriers

A crucial aspect of prion disorders is their transmissibility. Inoculation of prion disease brain material into individuals of the same species will typically reproduce disease with a remarkably synchronous time to disease onset. Although transmission efficacy between species is, by comparison, generally significantly less than when prions propagate within their natural hosts, cross-species infection is nevertheless known to have featured in the etiologies of several epidemics, in particular BSE, vCJD, and most likely prions disease in mink, referred to as transmissible mink encephalopathy (TME).

2.1 PrP Primary Structure, Prion Strains, and the Species Barrier

Although the elements controlling interspecies prion transmission are not completely understood, seminal studies in Tg mice [2, 3] and cell-free systems [4] suggested that minimal amino acid divergences may have a major impact on the transmission efficiency and that barriers to transmission between species resulted from primary structural incompatibilities between PrP^{Sc} constituting the prion and substrate PrP^{C} expressed in the newly infected host.

Experimentally, the barrier to prion transmission between species may be absolute, in which case no transmission is recorded or, more commonly, primary interspecies transmissions are characterized by variable rates of infection and protracted, inconstant intervals to disease onset. Mechanistically, this is thought to reflect a two-step process in which initial stochastic conversion of host PrP^{C} by structurally mismatched PrP^{Sc} is followed by effective PrP^{C} conversion by the resulting structurally compatible PrP^{Sc}, leading to efficient replication of adapted prions, neurodegeneration, and inevitable death of the host. Consistent with this notion, serial passage of such adapted prions to additional animals expressing the same PrP^{C} results in a relatively short, synchronous time to disease in all inoculated recipients.

Transgenic approaches have also been used to address the notion that PrP primary structures expressed in particular species are inherently resistant to infectious conversion. Vidal and coworkers generated Tg mice overexpressing rabbit PrP (TgRab) under the control of the mouse prion protein gene promoter [5]. Their rationale for this approach was that rabbits have generally been considered to be resistant to prion infection. This group showed, however, that PMCA-derived rabbit PrP^{Sc} caused a prion disease in rabbits [6]. In subsequent studies, the susceptibility of TgRab to different prions was assessed. Susceptible prions included BSE, sheep BSE, L-type BSE, and in vitro generated rabbit prions, as well as ME7 and RML. SSBP-1, CWD and atypical scrapie did not cause disease.

The influence of intraspecies PrP polymorphisms on prion disease susceptibility in mice [7], sheep [8], and humans [9–11] supported the concept that the primary structure of PrP was a crucially important determinant of prion species barriers. Nonetheless, while the foregoing mechanism is in accordance with the outcomes of the majority of prion interspecies transmission barriers, other circumstances indicate that our knowledge of the parameters controlling interspecies prion transmission remains incomplete. For example, the criteria governing PrP primary structural control over transmission appear to be moot in the case of the bank voles, which are susceptible to prions from a number of mammals with divergent PrP primary structures [12].

Mammalian prion strains are an equally important component affecting prion transmission. Seminal studies indicated that distinct heritable prion strain phenotypes, including the incubation time to disease, and the profile of CNS lesions, are enciphered within the conformation of PrP^{Sc} [13, 14]. This provided investigators with the ability to use the biochemical properties of PrP^{Sc} as a means of tracking, and to some extent identifying strains by assessing of PrP^{Sc} conformation, and/or the extent of PrP^{Sc} glycosylation [15–18]. Since a change in PrP^{Sc} conformation accompanies the emergence of newly adapted prions following passage across species barriers [16], the concept of a strain barrier was introduced to account for the influence of distinct conformations of PrP^{Sc} molecules [19]. The unexpectedly wide host range of BSE prions, exemplified by transmission to humans as variant CJD (vCJD), as well as the restricted transmission of vCJD prions in human PrP Tg mice [20, 21] compared to bovine PrP Tg mice [22], exemplify the capacity of particular prion strains to overcome the influence of PrP primary structure.

While mutational events in agent-associated nucleic acid were originally cited as the cause of strain instability [23], more recently, changes in the conformation of PrP^{Sc} have been shown to be associated with the acquisition of new strain properties [16]. To account for the phenomena of prion transmission barriers, strain instability, heterogeneity, and adaptation in the context of PrP^{Sc} conformation, the conformational selection model [24, 25] was proposed to reconcile how PrP primary structure and prion strain conformations interact to control transmission barriers to prion propagation. This model postulates that only a subset of PrP^{Sc} conformations is compatible with each individual PrP primary structure [24, 26]. By extension, this leads to the notion that prions exist as an array of quasispecies conformations, a model first applied to populations of a virus within its host [27]. There is a mounting body of experimental evidence for this concept [28–30]. Quasispecies acquire fitness when populations of prion conformers are subjected to selective pressure, for example, during propagation in a host expressing a different PrP primary structure following interspecies transmission. From this perspective, PrP^{C} primary structure influences the portfolio of thermodynamically preferred PrP^{Sc} conformations that are kinetically selected during propagation, such that only a subset of such PrP^{Sc} conformers are optimized for fitness in a particular host [31, 32].

The issue of strain effects on the efficacy of cross-species transmission was elegantly addressed in transgenic experiments by

Béringue and colleagues who compared the ability of brain and lymphoid tissues from ovine and human PrP Tg mice to replicate foreign, inefficiently transmitted prions [33]. They observed that lymphoid tissue was consistently more permissive than the brain for CWD and BSE prions. Furthermore, when the transmission barrier was overcome through strain shifting in the brain, a distinct agent propagated in the spleen, which retained the ability to infect the original host.

An additional factor affecting the efficiency of interspecies transmissions was revealed by recent experiments using Tg mice expressing PrP^{C} lacking the GPI moiety that tethers the protein to the cell surface. Previous in vitro studies showed that an artificially mutant version of PrP that lacks the terminal sequence for addition of the GPI lipid anchor could acquire resistance to protease digestion and therefore resembled PrP^{Sc} [4, 34]. To determine whether this mutant PrP can support prion propagation and PrP^{Sc} production in vivo, Chesebro and colleagues produced Tg mice expressing "anchorless" PrP [35]. As expected, anchorless PrP was not expressed on the surface of cells derived from Tg mice; instead it was secreted. When inoculated with prions, the mice accumulated mutant PrP in the form of abundant amyloid plaques throughout the brain as early as 70 days after inoculation. But despite their accumulation of amyloid-forming protease-resistant PrP (in many cases at levels higher than PrP^{Sc} found in the brains of clinically sick wild-type mice), these Tg mice failed to develop neurologic signs of prion disease up to 600 days following prion inoculation—long after inoculated control mice succumbed to disease. To test the effect of lack of GPI anchoring on a species barrier model, anchorless 22L mouse prions derived from $TgGPI^{-}$ mice were more infectious than wild-type 22L prions in tg66 mice expressing HuPrP at levels 8- to 16-fold above normal, but not in tgRM transgenic mice, which expressed human PrP at two- to fourfold above normal.

2.2 Variable Effects of Endogenous Mouse PrP Expression in Transgenic Mouse Models

The availability of *Prnp*$^{0/0}$ knockout mice, and the characterization of increasing numbers of Tg mice expressing different PrP alleles, revealed interesting protective functions of wild-type PrP on various PrP-related pathologies. In some cases, the disease can be partially or fully suppressed by co-expression of wild-type PrP. This effect was first observed during the characterization of Tg mice expressing human (Hu) PrP, referred to as Tg(HuPrP) mice, which only became susceptible to CJD prions when endogenous wild-type mouse PrP was eliminated by crossing the HuPrP transgene array to *Prnp*$^{0/0}$ knockout mice [36, 37]. Because of this dominant-negative effect, subsequent Tg mouse models expressing foreign PrP coding sequences have generally been either produced in or ultimately crossed with *Prnp*$^{0/0}$ knockout mice. Coincidentally, these observations also formed the partial basis for the model of prion replication involving an auxiliary factor, referred to as protein X, which was proposed to bind to PrP^C and facilitate conversion to PrP^{Sc} [21, 37, 38]. While subsequent experimental evidence questioned the requirement for protein X [39], this model remains a guiding principle that has informed structural analyses of PrP^C [40, 41].

Interactions between wild-type and mutant prion proteins have also been shown to modulate neurodegeneration in Tg mouse models expressing mutated PrP, the seminal observations being made in Tg mouse models of the inherited human prion disease called Gerstmann–Straussler–Scheinker (GSS) syndrome, caused by mutation of codon 102 which results in substitution of proline for leucine [42]. In contrast, neurodegeneration induced by expression of a nine-octapeptide insertion associated with familial Creutzfeldt–Jakob disease (CJD), designated Tg(PG14), is unaffected by wild-type mouse PrP expression [43]. These studies are also consistent with a model in which mutated and wild-type PrPs compete for a hypothetical binding partner that, in this case, serves to transduce neurotoxic signals. Interestingly, while Tg mice expressing PrP tagged at its amino terminus with green fluorescent protein (GFP) supported compromised prion replication, prion propagation was facilitated by co-expression of wild-type PrP, suggesting that wild-type PrP rescued an altered amino-terminal function in the tagged PrP [44]. In contrast, the effect of tagging PrP at the C-terminus with GFP was that Tg mice expressing this construct were incapable of sustaining prion infection and the PrP-GFP chimera acted as a dominant-negative inhibitor of wild-type PrP conversion to PrP^{Sc} [45]. In a similar fashion, co-expression of wild-type mouse PrP was shown to rescue the neurodegenerative phenotype of mice expressing amino-terminal deletions [46–48].

2.3 Transgenic Mouse Models of Mammalian Prion Diseases

Seminal experiments by Scott and coworkers abrogated the resistance of mice to hamster prions by transgenic expression of hamster PrP^C in mice [2, 49] and paved the way for the development of a variety of facile Tg mouse models that authentically recapitulate known transmission barriers [50]. The lack of species barrier during homotypic transmission, i.e., when the host expresses a PrP gene identical to that of the infecting species, led to the development of many different facile mouse models used in the study of the biology of mammalian prions, including sheep, bovine, human, cervid, and mink prions, by transgenic expression in mice of PrP coding sequences from these various species. These Tg resources largely circumvented the imprecise approach of studying the biology of various mammalian prion diseases by transmission to generally non-susceptible experimental animals, or, where possible, the natural host, but has also greatly contributed to our understanding of various mechanistic aspects of these extraordinary pathogens [51]. Generally, an inverse correlation exists between the length of prion incubation time in these Tg mouse models and transgene expression level. Transgenes are usually expressed on a *Prnp* knock-out background (*Prnp*$^{0/0}$) [52] in order to avoid partial or full suppression of disease caused by co-expression of wild-type PrP. A variety of different transgenic mice expressing chimeric versions of PrP in which specific regions of mouse PrP primary structure were replaced by the corresponding elements from human, sheep, and bovine PrP have also been created [36, 37, 53–55]. Inclusion of certain mouse PrP primary structural elements in such constructs, in particular MHu2MPrP, also countermanded the inhibitory effect of mouse PrP co-expression [36, 37].

In an alternative approach, expression of foreign PrP genes in mice has been accomplished by gene replacement methods. This approach ensures that the PrP coding sequence is controlled by the same regulatory elements as wild-type mouse PrP, in which case gene expression is expected to recapitulate authentic PrP^{C} expression. In both cases, PrP sequence identity between the transgenic host and donor usually lead to a higher transmission rate as compared to wild-type mice. When considering such transmission models, it is important to consider that outcomes might be limited by the life span of the recipient host, which might be exceeded by the incubation period, although pathological examination, PrP^{Sc} detection, and secondary passage may identify infected animals. These problems can be partially overcome by transgenic mice over-expressing PrP^{C} which might be desirable to fully assess the extent of a species barrier, since it results in highly reduced incubation times. For example, while human PrP knock-in mice did not register disease when challenged with prions from cattle affected with BSE [56], infection did occur in Tg mice expressing higher levels of human PrP [20].

2.3.1 Human Prion Diseases

The initial transmission of human prions to experimental primates has a rich history beginning with William Hadlow's recognition of the similarity between kuru and scrapie and his prediction that patient brain extracts would cause disease in inoculated nonhuman primates after a prolonged incubation period [57]. Seven years later, Gajdusek, Gibbs, and Alpers demonstrated the transmissibility of kuru to chimpanzees after incubation periods ranging from 18 to 21 months [58] and subsequently showed that CJD was similarly infectious [59]. For many years, the transmission of human prion diseases was investigated using this approach [60], but the expense of housing these animals for long time periods, as well as ethical concerns surrounding their use, severely limited these studies. Inoculations of laboratory rodents produced variable results [61–65]. Generally, only ~10% of intracerebrally inoculated mice developed CNS dysfunction with incubation times of >500 days (d) [36], limiting the ability of this approach to characterize specific agent strains.

While seminal transgenic investigations by Prusiner and colleagues studied the transmission properties of scrapie prions experimentally adapted to hamsters or mice [3, 49], Tg mice expressing human PrP^C, referred to as Tg(HuPrP) mice, were the first to abrogate a species barriers to naturally occurring prions, in this case prions causing human diseases such as human disorders such as sporadic and iatrogenic CJD (sCJD and iCJD) [36, 37]. As previously described for the animal prion diseases, human prion disease susceptibility is strongly influenced by polymorphic variation of *PRNP*. In particular, homozygosity at *PRNP* codon 129, which encodes methionine (M) or valine (V), predisposes to the development of sporadic and acquired CJD [10, 11]. Surprisingly, two lines of Tg(HuPrP) mice expressing HuPrP with V at residue 129 (HuPrP-V129), referred to as Tg(HuPrP)152 and Tg(HuPrP) 110, inoculated with CJD prions failed to develop CNS dysfunction more frequently than non-transgenic controls [36]. Subsequently, mice expressing a chimeric human/mouse PrP transgene, designated MHu2M, were constructed, because earlier studies had shown that a chimeric hamster/mouse PrP gene supported transmission of either mouse or hamster prions [3, 66]. These Tg (MHu2M)5378 mice were found to be highly susceptible to human prions suggesting that Tg(HuPrP) mice have considerable difficulty converting $HuPrP^C$ into PrP^{Sc} [36]. However, Tg (HuPrP)152 mice, and another line designated Tg(HuPrP)440, which expresses HuPrP with M at 129, when crossed with $Prnp^{0/0}$ [52], were rendered susceptible to human prions [37]. These observations demonstrated that Tg(HuPrP) mice were resistant to human prions because mouse PrP^C inhibited the conversion of $HuPrP^C$ into PrP^{Sc}. In contrast, Tg(MHu2M)5378 mice crossed onto the null background were only slightly more susceptible to human prions compared to Tg(MHu2M)5378 mice that expressed both chimeric and $MoPrP^C$. Furthermore, Tg(MHu2M) mice inoculated with either Hu or chimeric MHu2M prions exhibited similar incubation times.

The availability of such susceptible Tg mice made possible the rapid and relatively inexpensive transmission of human prion diseases for the first time. Based on these findings, several additional similar Tg mouse models expressing HuPrP have been produced, with identical results [20, 67, 68]. Gene-targeting approaches have also been employed to produce mice expressing human PrP [69]. This approach has the obvious advantage of expressing "normal" levels of transgene-encoded PrP under the control of the *Prnp* transcriptional elements and, for example, to conveniently model the effects of heterozygosity at codon 129. Transmission of a limited number of sporadic CJD cases in these mice has provided evidence for four distinct prion strains. Evidence of strain variation in sCJD has also come from laboratory studies in bank voles [12].

Early transmission studies to Tg(MHu2M)5378 mice provided evidence that different human prion strains, namely, in patients with fatal familial insomnia (FFI), caused by mutation at codon 178 (D178N), and familial CJD, caused by mutation at codon 200 (E200K), are enciphered by different conformational states of PrP^{Sc} [13], a concept first elaborated following transmission of the hamster-adapted strains of TME called hyper (HY) and drowsy (DY) [14]. While the conformational enciphering hypothesis is supported by considerable additional experimental evidence, defining human prion strain prevalence has been hampered by difficulties in arriving at an internationally accepted classification system for human prion strains [70, 71] and by the observation that multiple PrP^{Sc} subtypes coexist in the same brain [29, 72]. Coincidentally, their transmissions also supported the concept that genetically programmed prion diseases are also transmissible. Other examples of inherited human prion diseases that have been transmitted under similar circumstances include transmission of an inherited form of CJD caused by mutation of codon 210 which changes valine to isoleucine at this residue [73].

Given the importance of infectious transmission in prion diseases, the availability of Tg mice with susceptibility to human prions has increased the number of known sporadic prion diseases. Transmission to Tg(MHu2M) mice of an unusual case of human prion diseases that presented with insomnia but no *PRNP* abnormalities led to the discovery of a novel prion disease referred to as sporadic fatal insomnia [74]. More recently transmission of a new prion disorder, referred to as variably protease-sensitive prionopathy (VPSPr), a seemingly sporadic disease that is distinct from CJD but shares features of GSS, confirmed that VPSPr might be the long-sought sporadic form of GSS [75]. While inefficient transmission of VPSPr was recorded on first passage to Tg mice overexpressing HuPrP, infectivity was not serially transmissible. Thus, while VPSPr is an authentic prion disease, Tg(HuPrP) mice do not appear able to sustain replication beyond the first passage. Similar findings using gene-targeted mice expressing HuPrP confirmed the interpretation that VPSPr has limited potential for human-to-human transmission [76].

A significant goal of Tg development was to produce mice in which the incubation time of prions was as rapid as possible using additional chimeric mouse/human PrP transgenes [21]. Korth and coworkers refined the MHu2M PrP approach and optimized human prion transmission by replacing key human PrP residues with the equivalent residues from mouse. The resulting chimera, referred to as Tg(MHu2M,M165V,E167Q) mice, resulted in shortening the incubation time to approximately 110 days for prions from sCJD patients and divergence into two strain types [21]. Even shorter incubation times and CJD strain evolution were also observed in another line, termed Tg1014 in which a single additional residue (M111V) was reverted to mouse [77].

Development of Tg mice with susceptibility to human prions was timely, as it occurred in the context of significant human exposure to BSE prions, at least in the UK, and the consequential occurrence of vCJD in young adults and teens [78]. Tg(HuPrP) created by Prusiner and colleagues and similar mice subsequently generated by other groups [20] were used to characterize vCJD prions and to model human susceptibility to BSE [17, 18, 22, 79]. As previously described for the animal prion diseases, human prion disease susceptibility is strongly influenced by polymorphic variation of *PRNP*. In particular, homozygosity at *PRNP* codon 129, which encodes M or V, predisposes to the development of sporadic and acquired CJD [10, 11]. Transmission studies of human CJD cases to transgenic mice confirm the influence of this polymorphism. While mice expressing V129 are susceptible to all PrP^{Sc} types and PrP 129 genotypes [17, 18, 21, 37, 79], mice expressing the HuPrP-129M allele are susceptible to prions from M129 homozygous patients, transmissions from patients mismatched at this codon, or heterozygotes are generally more inefficient [20, 21, 37].

Strikingly, all neuropathologically confirmed vCJD cases studied so far have been homozygous for M at codon 129 [80]. Transmission studies of human CJD cases to Tg mice confirm the influence of this polymorphism. While mice expressing V129 are susceptible to all PrP^{Sc} types and PrP 129 genotypes [17, 18, 21, 37, 79], mice expressing the HuPrP-129M allele are susceptible to prions from M129 homozygous patients, transmissions from patients mismatched at this codon, or heterozygotes are generally more inefficient [20, 21, 37]. Although initial BSE transmissions to Tg(HuPrP)152 mice were uniformly negative, suggesting a substantial species barrier in humans to BSE prions [79], subsequent BSE transmission to Tg mice expressing M at human PrP codon 129 revealed inefficient transmission, characterized by low attack rates and long incubation times. Moreover, a strain shift was occasionally observed in these transmissions, producing a sCJD-like phenotype in a proportion of inoculated Tg mice [20, 56]. In contrast, gene-targeted transgenic mice expressing human PrP were not susceptible to BSE prions [56]. However, these mice were susceptible to sheep-adapted BSE prions suggesting increased susceptibility of humans to BSE prions following passage through sheep [81], an effect that is unrelated to increased titer of BSE prions in sheep brain [82]. The effect of codon 129 heterozygosity on human prion susceptibility has been further examined in Tg mice co-expressing transgene arrays expressing HuPrP-M129 and HuPrP-V129 [83].

The emergence of vCJD in the late twentieth century renewed interest in kuru, another acquired human prion disease among the Fore peoples of the inner highlands of Papua New Guinea.

Kuru exemplifies the epidemic nature of prion diseases. By the mid-twentieth century, it was the leading cause of death in this region, killing over 3000 people in the exposed population of 30,000. While its etiology was initially puzzling, studies show that kuru patient brain extracts produced a progressive neurodegenerative condition in inoculated chimpanzees after a prolonged incubation period [58], supporting the notion that kuru was an infectious disorder. Despite neuropathological similarities, scrapie was clearly not a candidate for the origin of this infectious disorder. Moreover, kuru predated the BSE/vCJD epidemic by several decades. In fact kuru was a devastating consequence of ritualistic endocannibalism, where brains and other body parts of tribal elders were consumed as an act of mourning.

Remarkably, a novel *PRNP* variant was found uniquely among unaffected individuals in the kuru-exposed population, in which V at residue 127 was replaced by glycine (G) [84]. This polymorphism was hypothesized to be an acquired prion disease resistance factor, selected in response to the kuru epidemic. Asante and colleagues modeled this kuru resistance polymorphism, referred to as HuPrP V^{127}, in Tg mice [85]. Tg mice having the genotype associated with disease resistance, specifically heterozygosity (G/V) at position 127, and M/M homozygosity at 129, referred to as $G^{127}M^{129}/V^{127}M^{129}$, were completely protected against kuru prions. In contrast all kuru-inoculated mice expressing G/G at 127 and either M/M or V/V at 129 ($G^{127}M^{129}/G^{127}M^{129}$ and $G^{127}V^{129}/G^{127}V^{129}$) developed disease after ~200 days. Complete recapitulation of genetic susceptibility to kuru therefore validated this transgenic modeling approach. Given the previously recognized similarities between kuru and sCJD, it came as no surprise that $G^{127}M^{129}/V^{127}M^{129}$ were also protected from CJD prions. Interestingly however, these same mice were incompletely protected against vCJD prions. Remarkably, mice homozygous for G^{127} M^{129} were completely protected against all forms of human prion diseases, including vCJD, and failed to manifest either clinical or subclinical signs of prion disease. As such, they behaved like mice which fail to express PrP as a result of disruption of the PrP gene and yet still express HuPrP. The availability of Tg mice expressing different levels of HuPrP M^{129} and V^{129} showed that the mechanism underlying the profound inhibitory effects of HuPrP V^{127} occur independently from M129V, which is thought to influence PrP interactions during the replicative process. Moreover, HuPrP V^{127} acts as a dominant-negative inhibitor of prion conversion.

2.3.2 Transgenic Models of Inherited Human Prion Diseases

Approximately 10–20% of human prion diseases exhibit an autosomal dominant mode of inheritance resulting from missense or insertion mutations in the coding sequence of the human PRNP. Several mutations are genetically linked to loci controlling familial CJD, GSS syndrome, and FFI. GSS syndrome, which is characterized clinically by ataxia and dementia and neuropathologically by the deposition of PrP amyloid, most commonly results from mutation at codon 102 of PRNP resulting in the substitution of leucine (L) for proline (P) [86]. GSS linked to this mutation is transmissible to nonhuman primates [60], wild-type mice [63], Tg mice expressing a chimeric mouse–human PrP gene expressing the GSS mutation [37], and *Prnp* gene-targeted mice referred to as 101LL [87].

Initial studies in Tg(GSS) mice which attempted to understand how an inherited disease could also be infectious suggested that prions in the brains of spontaneously sick Tg(GSS) mice could be transmitted to Tg mice expressing lower levels of mutant protein, referred to as Tg196 mice [42, 88]. The disease was also induced in Tg196 mice by a mutant synthetic peptide comprising MoPrP residues 89–103 refolded into a beta-sheet conformation [89], and this disease was subsequently propagated to additional Tg196 mice [90]. While these studies lent support to the prion hypothesis, since they suggested that pathogenic PrP gene mutations resulted in the spontaneous formation of PrP^{Sc} and de novo production of prions [91], this explanation was controversial for several reasons. Although protease-sensitive forms of PrP^{Sc} have been identified using biochemical and immunological methods [15, 92], the lack of protease-resistant PrP^{Sc} in the brains of spontaneously sick or recipient mice eliminated a property that, to some, was synonymous with prion infectivity. Moreover, *Prnp* gene-targeted 101LL mice expressing MoPrP-P101L failed to spontaneously develop neurodegenerative disease [87]. Finally, disease transmission from spontaneously sick mice to wild-type mice did not occur, and spontaneous disease was eventually registered in aged Tg196 mice [89, 90] complicating the interpretation of the original transmission experiments.

To address the apparent dissociation of prion infectivity and PrP^{Sc} in this well-established Tg model, subsequent studies attempted to use means other than differential resistance to proteinase K treatment to detect disease-associated forms of PrP in spontaneously sick Tg mice expressing MoPrP-P101L. Using the prototype PrP^{Sc}-specific monoclonal antibody (Mab) reagent referred to as 15B3 [93], Nazor and coworkers showed that disease in Tg mice overexpressing MoPrP-P101L results from the spontaneous conversion of mutant PrP^{C} to protease-sensitive $MoPrP^{Sc}$-P101L, defined by its reactivity with 15B3, which accumulates as aggregates in the brains of sick Tg mice [94]. To understand the influence of mutant PrP expression levels on the transmissibility of spontaneously generated pathogenic MoPrP-P101L, they produced mice in which transgene copy numbers and levels of MoPrP-P101L expression were carefully defined. While inoculation of disease-associated MoPrP-P101L accelerated disease in Tg mice expressing MoPrP-P101L from multiple transgenes, disease transmission neither occurred to wild-type nor Tg mice expressing

MoPrP-P101L from two transgene copies that did not develop disease spontaneously in their natural life span.

Since disease transmission from spontaneously sick Tg(GSS) mice depended on recipient mice expressing MoPrP-P101L at levels greater than that produced by two transgene copies and since such levels of overexpression ultimately resulted in spontaneous disease in older uninoculated recipients, these results suggest that the phenomenon of disease transmission from spontaneously sick Tg(GSS) mice might be more appropriately viewed as disease acceleration whereby inoculation of disease-associated MoPrP-P101L promotes the aggregation of precursors of pathological MoPrP-P101L that result from transgene overexpression. Such a scheme is consistent with a nucleated polymerization mechanism of prion replication originally postulated from cell-free conversion systems [4] and subsequently demonstrated to be the basis of prion propagation in lower eukaryotes [95]. According to this model, PrP^C is in equilibrium with PrP^{Sc}, or its precursor, and the equilibrium normally favors PrP^C. Also, PrP^{Sc} is stable only in its aggregated form that can "seed" polymerization of additional PrP^C, thus converting it into additional PrP^{Sc}. Further, more definitive work combining studies in Tg(GSS) and *Prnp* gene-targeted 101LL mice indicated that clinical or profound neuropathological changes were absent in gene-targeted mice inoculated with brain extracts of spontaneously sick Tg(GSS) mice, indicating that de novo formation of abnormally aggregated PrP in the host does not always result in a transmissible prion disease [96]. In a related issue, the role of PrP overexpression in the production and transmission of synthetic mammalian prions (SMP) [97] originating from *E. coli*-derived recombinant MoPrP remains to be determined. While the transmission properties and protease-resistance of MoPrP (89–230) SMPs are clearly different from disease-associated MoPrP-P101L, it may be significant that the Tg mice in which these SMPs were initially derived expressed MoPrP (89–231) at levels 16 times higher than normal.

Interestingly, unlike mice expressing the GSS P102L mutation in the context of the mouse PrP primary structure, Tg mice expressing human PrP with the P102L mutation failed to develop disease spontaneously with increasing age. However, these mice were susceptible to infection from patients with the homotypic pathogenic mutation, as well as CJD, producing distinct prion strains with transmission properties distinct from sporadic and acquired human prion disease [98]. In contrast to reports in the gene-targeted mouse model, GSS-102L prions produced in this study were incapable of transmitting disease to wild-type mice [87].

Transgenic expression of other disease-associated mutations in the context of mouse or human PrP has been met with varying success. While Tg mice expressing mouse PrP containing the most common E200K fCJD mutation (E199K in mouse PrP) did not develop disease spontaneously [42], a Tg mouse expressing chimeric MHu2M PrP [37] containing the E199K mutation PrP developed neurological signs at 5–6 months of age and deteriorated to death several months thereafter [99]. Inoculation of brain extracts from diseased Tg(MHu2M-E199K) mice induced a distinct fatal prion disease in wild-type mice. Mice expressing a mouse PrP version of a nine octapeptide insertion associated with familial CJD, designated Tg(PG14), exhibited a slowly progressive neurological disorder characterized by apoptotic loss of cerebellar granule cells, gliosis but no spongiosis [100]. Whether the brains of sick Tg (PG14) mice, like sick Tg(GSS) mice, contain 15B3-immunoprecipitable PrP has not yet been reported; however, in both models, mutated PrP adopts different pathologic conformations either spontaneously or following inoculation with authentic prions [94, 101]. Like Tg(GSS) mice, brain homogenates from spontaneously sick Tg(PG14) mice failed to transmit disease to Tg mice that express low levels of mutated PrP that do not become sick spontaneously. Whether differences in the state of aggregation of PG14spon compared to MoPrP-P101L will affect its ability to accelerate disease progression in overexpressor Tg(PG14) mice remains to be determined.

Chiesa's group also described a Tg mouse model of inherited CJD expressing the mouse homolog of the D178N in combination with the V129 polymorphism. These Tg(CJD) mice had EEG and sleep abnormalities, memory impairment, motor dysfunction, and striking morphological alterations of the neuronal endoplasmic reticulum (ER) associated with ER retention of mutant PrP [102]. This study was followed by reports of the properties of Tg mice expressing the mouse PrP homolog of the same D178N mutation in cis with M129, which is associated with FFI [103]. Spontaneous disease in these so-called Tg(FFI) mice was different from Tg(CJD) mice. Tg(FFI) synthesize misfolded mutant PrP in their brains and, like FFI, illness is associated with sleep disruption. However, unlike this form of fCJD and FFI, bioassay and protein misfolding cyclic amplification (PMCA) showed that prions were not produced in Tg(FFI) and Tg(CJD) brains, suggesting to the authors that the disease-encoding properties of mutant PrP do not depend on its ability to propagate its misfolded conformation. Tg (FFI) mice complement a gene-targeted model of this disease that provided a different outcome. Knock-in mice carrying the FFI mutation [104] developed biochemical, physiological, behavioral, and neuropathological abnormalities similar to FFI, and this spontaneous disease could be transmitted, and serially passaged, to mice expressing physiological amounts of PrP without the mutation. Likewise, the same investigators produced knock-in mouse models of CJD caused by the N178/V129 variants. These mice differed phenotypically from FFI knock-in mice, producing a spontaneous, transmissible CJD-like disease [105]. The reasons for the discrepant properties of Tg and knock-in models with respect to spontaneous prion formation are unclear.

Transgenic mice expressing a mutation at codon 117 associated with a telencephalic form of GSS [106] also spontaneously developed neurodegenerative disease and accumulated an aberrant, neurotoxic form of PrP termed CtmPrP, which appears to be distinct from conventional protease-resistant PrP^{Sc} [107]. Mastrianni and colleagues also constructed Tg mice that express PrP carrying the mouse homolog of this GSS mutation. These Tg(A116V) mice express approximately six times the endogenous levels of PrP and recapitulate many clinicopathologic features of GSS(A117V) that are distinct from CJD [108]. More recently, Asante and coworkers produced Tg mice expressing the A117V mutation in the context of HuPrP. Unlike previous attempts at transmission, prions from human patients with GSS A117V transmitted to these Tg mice, producing appropriate neuropathology and accumulation of PrP^{Sc} [109].

2.3.3 Bovine Prion Disease

Several lines of Tg mice expressing bovine PrP, referred to as Tg (BoPrP) mice, have been independently produced [55, 110, 111]. All Tg(BoPrP) mice are susceptible to BSE prions, and their availability during the peak years of the BSE and vCJD epidemics were invaluable for ascertaining the pathogenesis of BSE and the properties of these and other bovine prion diseases.

While BSE initially appeared to be a homogeneous disease, the large-scale testing of livestock nervous tissues for the presence of PrP^{Sc} led to the recognition, in Europe, Japan, and the USA, of two additional bovine PrP^{Sc} variants termed H- and L-types [112]. The molecular signature of bovine PrP^{Sc} from animals with the bovine amyloidotic spongiform encephalopathy (BASE) variant corresponds to L-type and appears similar to a distinct subtype of sporadic CJD [113]. L-type has a tendency to form amyloid plaques in cattle brain and has a distribution of brain pathology distinct from BSE [113]. These "atypical BSE" cases have been detected in aged asymptomatic cattle during systematic testing at slaughterhouse. The etiology of these atypical forms remains unexplained but could involve either (1) a change in the biological properties of the BSE agent; (2) infection of cattle with prions from another source, such as scrapie or CWD; or (3) previously unrecognized sporadic forms of prion disease in cattle. A case of BSE in the USA with an H-type PrP^{Sc} signature in an approximately 10-year-old cow from Alabama was also associated with mutation of glutamate (E) to lysine (K) at 211, referred to as E211K [114]. Of particular significance, the identical substitution at the equivalent codon 200 in human *PRNP* is linked to the most frequent form of familial CJD with clusters described in Chileans, Moravian Slovaks, Libyan Jews, Britons, and Japanese.

The development of various Tg mouse models expressing bovine PrP was invaluable for characterizing and titrating BSE infectivity in a variety of tissues [55, 111, 115] and provided compelling evidence for a relationship between vCJD and BSE [22]. While these Tg mouse models were characterized by rapid incubation times and 100% attack rates, mice expressing bovine PrP generated by gene replacement of the mouse PrP coding sequence had long incubation times (>500 days) and incomplete attack rates [56]. The experimental transmission of H- and L-type cases to bovine PrP Tg mice unambiguously demonstrated their infectious nature and revealed strain properties distinct from BSE [116–119]. While BSE and BASE transmitted readily to Tgbov XV mice, they produced different clinical, neuropathological, and molecular disease phenotypes [119]. Interestingly, the same study indicated that BASE prions were able to convert into BSE prions upon serial transmission in inbred mice. The relationship of this finding to the apparently protean nature of BSE prions in aforementioned transmission studies [20, 21, 120] remains to be determined.

Strikingly, serial passage of the L-type strain to wild-type mice and mice expressing the VRQ allele of ovine PrP induced a disease phenotype indistinguishable from that of BSE [117, 119], suggesting a possible etiological relationship between atypical and classical BSE. The relevance of these findings to studies in Tg mice, which consistently reveal the existence of more than one molecular type of PrP^{Sc} [20, 21, 120] and suggest that more than one BSE prion strain might infect humans [20], remains to be determined.

Challenge of two lines of Tg mice expressing human PrP with M at codon 129 with L-type isolates produced a molecular phenotype distinct from classical BSE [67, 68]. In one case, L-type transmitted with no transmission barrier [68], and in both cases the L-type PrP^{Sc} biochemical signature was conserved upon transmission. In contrast, the transmission efficiency of classical BSE and H-type isolates to transgenic mice expressing human PrP is relatively low [20, 68, 121]. Increased pathogenicity of sheep-passaged BSE occurred in Tg mice expressing porcine PrP [122] or human PrP [82], raising the possibility that BSE may gain virulence by passage in another species.

2.3.4 Ovine Prion Disease

Transgenic mice expressing ovine PrP are susceptible to prions from scrapie-affected sheep [123–128]. A clear link to codon 136 genotype and susceptibility/resistance to different sheep scrapie isolates has been described in multiple previous studies. Generally, increased susceptibility to scrapie is associated with expression of sheep PrP with valine (V) at residue 136, referred to as OvPrP-V136 compared to alanine (A) at residue 136, referred to as OvPrP-A136, with A/A136 being the most resistant, and V/V136 the most susceptible genotypes. In the case of SSBP/1, incubation periods are ~170 days in V/V136 sheep, while transmission to

A/A136 sheep is relatively inefficient, with no disease recorded after >1000 days [129]. The most widely characterized models include Tg mice expressing OvPrP-V136, including tg338 [127], and Tgov59 mice [130] or Tgov4 [125] mice expressing OvPrP-A136. In the case of tg338 mice, the transgene was comprised of a bacterial artificial chromosome insert of 125 kb of sheep DNA, while in the case of Tgov59 and Tgov4 mice, the neuron-specific enolase promoter was used to drive OvPrP expression. These lines are maintained on different heterogeneous genetic backgrounds, and CNS expression levels in tg338 mice are ~ eight- to tenfold higher than wild type, while Tgov59 and Tgov4 lines each over-express OvPrP-ARQ at levels ~ two- to fourfold higher than those found in sheep brain. Spontaneous neurological dysfunction has been reported in Tg lines overexpressing OvPrP [123, 127].

Subsequently, two additional Tg models, referred to as Tg (OvPrP-A136)3533$^{+/-}$ and Tg(OvPrP-V136)4166$^{+/-}$ mice, were produced and characterized that express either OvPrP-V136 or OvPrP-A136 approximately equivalent to PrP levels normally expressed in the CNS of wild-type mice [131]. Importantly, the influence of residue 136 on the transmission of SSBP/1 and CH1641 prions in Tg(OvPrP-A136)3533$^{+/-}$ and Tg(OvPrP-V136)4166$^{+/-}$ mice is in accordance with the properties of these isolates in sheep of various genotypes [132]. While SSBP/1 eventually transmits to Tg(OvPrP-A136)3533$^{+/-}$ mice with incubation times exceeding 400 days, the general effects of the A/V136 dimorphism on SSBP/1 transmission observed in sheep are recapitulated in Tg(OvPrP-A136)3533$^{+/-}$ and Tg(OvPrP-V136) 4166$^{+/-}$ mice. Similarly, CH1641, which propagates efficiently in A/A136 sheep [129], preferentially propagates in Tg(OvPrP-A136)3533$^{+/-}$ mice. In other studies, CH1641 transmitted to TgOvPrP4 mice with an ~250 days mean incubation time [133].

Tg(OvPrP-A136)3533$^{+/-}$ and Tg(OvPrP-V136)4166$^{+/-}$ lines were produced using the cosSHa.Tet cosmid vector which drives expression from the PrP gene promoter [66], and therefore it was expected that expression of OvPrP^{C}-A136 and OvPrP^{C}-V136 occurs in identical neuronal populations. Accordingly, homozygous Tg mice were mated to produce mice, referred to as Tg (OvPrP-A/V) mice, that co-express both alleles [131]. Previous studies reported on Tg mice expressing OvPrP with V at 136, referred to as Tg(OvPrP)14882$^{+/-}$ mice, that were also produced in a *Prnp*$^{0/0}$/FVB background using the cosSHa.Tet cosmid vector [128]. However, in that study, comparable Tg mice expressing OvPrP-A136 were not reported. While SSBP/1 incubation times are prolonged in A/V136 compared to V/V 136 sheep [129], incubation times were shorter in Tg(OvPrP-A/V) than in Tg (OvPrP-V136)4166$^{+/-}$ mice.

A novel mAb PRC5, the epitope of which comprises residue 136 [134], was used to monitor conversion of OvPrP^{C}-A136 in compound heterozygous. Surprisingly, in contrast to its relatively slow conversion when OvPrP^{C}-A136 is expressed in isolation, co-expression with OvPrP^{C}-V136 in Tg(OvPrP-A/V136) mice facilitated rapid conversion of OvPrP^{C}-A136 to OvPrP^{Sc}-A136. The conformation and diffuse CNS distribution of the resulting OvPrP^{Sc}-A136(U) were equivalent to that of OvPrP^{Sc}-V136(U) and not OvPrP^{Sc}-A136(S). These results demonstrate that under conditions of allele co-expression a dominant conformer may alter the conversion potential of an otherwise resistant PrP polymorphic variant to an unfavorable prion strain [131].

Ovine Tg mice have been shown to be a useful tool for discriminating scrapie strains [135] and in particular for differentiating BSE in sheep from natural scrapie isolates [126, 130, 136]. Tg mouse models are also at the forefront of characterizing a relatively newly emerging "atypical" scrapie strain [137] related to so-called Nor98 cases first identified in Norwegian sheep. Atypical scrapie was confirmed to be a prion disease following transmission to Tg338 mice expressing OvPrP-V136, and revealed the uniformity of features between atypical scrapie cases [137], confirming limited studies in the natural host [138]. In one study, Tg mice that overexpress human prion protein were found to be susceptible to BSE prions, but not classical or atypical scrapie prions [139]. In contrast, Andrioletti and coworkers show that a panel of classical sheep scrapie prions transmit to several Tg mice expressing human PrP mouse models with an efficiency comparable to that of cattle BSE indicating that classical scrapie prions have zoonotic potential [140].

2.3.5 Cervid Prion Disease

Some 15 years ago, CWD was perhaps the least understood of all the prion diseases of animals and humans. Known to be highly contagious, its origins and mode of transmission were unclear, and it was not known whether multiple CWD strains exist or whether CWD prions pose a risk to other animals or humans. The main objectives at that time were to develop rapid and sensitive bioassays for CWD prions and to experimentally address the risks that CWD prions pose to humans and other species. The development of Tg mice that were the first reliable bioassay for rapid and sensitive detection of CWD prions was a significant advance. With these resources in hand, investigators have obtained important information about CWD pathogenesis and the molecular mechanisms of prion propagation, species barriers, and strains. Using CWD-susceptible Tg mice, it became possible to bioassay CWD prions in tissues, body fluids, and secretions of deer and elk, which provided insights into the mode of transmission of this highly contagious disease. The study of inter-mammalian species barriers in Tg mice allows investigators to model the risks posed to humans and livestock from exposure to CWD prions, and this information helps facilitate management decisions designed to minimize interspecies prion transmission. During this time, the development of more facile, sensitive approaches to amplify prions in vitro, such as PMCA and RT-QuIC, has revolutionized our ability to detect prions, even at extremely low titers. In concert, cell culture models have provided an alternate means of CWD titration that has largely superseded bioassay in Tg mice and provided insights into strategies for developing compounds that inhibit CWD propagation. The development of compounds such as IND24 provides significant optimism for treating this currently incurable disease. Finally, studies of CWD using these newly developed tools have provided unexpected mechanistic insights into PrP^{C}-to-PrP^{Sc} conversion, particularly the role of the β2–α2 loop/α-helix 3 epitope and the proposal that prion strains exist as a continuum of conformational quasispecies.

The expense of housing cervids under prion-free conditions for long periods and the highly communicable nature of CWD present significant challenges for using deer as experimental hosts [141]. As noted above, transmission to other species has yielded mixed results. The resistance of mice [142] and the inefficient transmission of CWD to ferrets [143] are examples of species barriers to CWD prions, albeit of varying extent. The prototype Tg(CerPrP) mice developed signs of neurological dysfunction ~230 days following intracerebral inoculation with four CWD isolates [142]. The brains of sick Tg mice recapitulated the cardinal neuropathological features of CWD. As part of a larger study of CWD pathogenesis, Tg(CerPrP) mice were used as a sensitive means to show that skeletal muscles of CWD-infected deer harbor infectious prions, demonstrating that humans consuming or handling meat from CWD-infected animals are at risk to prion exposure [144]. Similar analyses of skeletal muscle BSE-affected cattle in a larger study of BSE pathogenesis using Tgbov XV mice did not reveal high levels of BSE infectivity [115]. Since the seminal reports of accelerated CWD transmission from deer and elk to Tg(CerPrP) mice [142], several other groups have reported similar results using comparable Tg mouse models [145–148]. CWD has also been transmitted, albeit with less efficiency, to Tg mice expressing mouse [149] or Syrian hamster PrP [150].

The generation of CWD-susceptible Tg mice, in concert with the development of PMCA-based approaches for amplifying CWD infectivity using PrP^{C} expressed in the CNS of those mice [151, 152], has also provided crucial information about the biology of CWD and cervid prions. Not only was amplification in vitro shown to maintain CWD prion strain properties, but it also provided a means of generating novel cervid prion strains [151–154].

Transgenic and in vitro amplification approaches have also facilitated our understanding of the mechanism of CWD transmission among deer and elk [141, 155–157]. Transmission studies in Tg (CerPrP)1536$^{+/-}$ and similar Tg mice demonstrated that CWD prions were present in urine and feces and saliva [147, 155] and these findings are substantiated by in vitro amplification techniques [155, 158, 159]. Tg approaches have been essential for assessing the potential risk of human exposure to CWD prions [144, 160, 161]. The availability of CWD-susceptible transgenic mouse models, for the first time, also provided a means of quantifying CWD infectivity by end-point titration [160].

As demonstrated in other species in which prion diseases occur naturally, susceptibility to CWD is highly dependent on polymorphic variation in deer and elk *PRNP*. Polymorphisms at codons 95 [glutamine (Q) or histidine (H)] [162], 96 [glycine (G) or serine (S)] [162, 163], and 116 [alanine (A) or glycine (G)] [164] in white-tailed deer have been reported. While all major genotypes were found in deer with CWD, the Q95, G96, A116 allele (QGA) was more frequently found in CWD-affected deer than the QSA allele [162, 165], suggesting a protective effect of the counterpart polymorphisms. The elk *PRNP* coding sequence is polymorphic at codon 132 encoding either methionine (M) or leucine (L) [166, 167]. This position is equivalent to human *PRNP* codon 129. Studies of free-ranging and captive elk with CWD [168], as well as oral transmission experiments [169, 170], indicate that the L132 allele protects against CWD.

Transgenic mouse modeling provided a means of assessing the role of these cervid PrP gene polymorphisms on CWD pathogenesis. In recent work combining studies in Tg mice, the natural host, cell-free prion amplification, and molecular modeling approaches, we analyzed the effects of deer polymorphic amino acid variations on CWD propagation and susceptibility to prions from different species [50]. Reflecting the general authenticity of the Tg modeling approach, the properties of CWD prions were faithfully maintained in deer following their passage through Tg mice expressing cognate PrP. Moreover, the protective influences of naturally occurring PrP polymorphisms on CWD susceptibility were accurately reproduced in Tg mice or during cell-free amplification. The resistance of Tg mice expressing deer PrP S96 to CWD, referred to as Tg(DeerPrP-S96) 7511 mice, is consistent with previously generated tg60 mice expressing serine at residue 96 [145]. In the studies of Angers and colleagues, whereas substitutions at residues 95 and 96 affected CWD propagation, their protective effects were overridden during replication of sheep prions in Tg mice and, in the case of residue 96, deer.

To more fully address the influence of the elk 132 polymorphism, transmissibility of CWD prions was assessed in Tg mice expressing cervid PrP^{C} with L or M at residue 132 [171].

While Tg mice expressing CerPrP-L132 afforded partial resistance to CWD, SSBP/1 sheep scrapie prions transmitted efficiently to Tg mice expressing CerPrP-L132, suggesting that the elk 132 polymorphism also controls prion susceptibility at the level of prion strain selection. The contrasting ability of CWD and SSBP/1 prions to overcome the inhibitory effects of the CerPrPL132 allele is reminiscent of studies describing the effects of the human codon 129 methionine (M)/valine (V) polymorphism on vCJD/BSE prion propagation in Tg mice expressing human PrP, which concluded that human PrP V129 severely restricts propagation of the BSE prion strain [120]. It therefore appears that amino acid substitutions in the unstructured region of PrP affect PrP^{C}-to-PrP^{Sc} conversion in a strain-specific manner.

The susceptibility of Tg(DeerPrP-S96)7511 mice, albeit with incomplete attack rates and long incubation times, are at odds with previous work showing complete resistance of tg60 mice [145, 172]. This apparent discrepancy is most likely related to the low transgene expression in tg60 mice, reported to be 70% the levels found in deer. CWD occurs naturally in deer homozygous for the PrP-S96 allele [173], which is clearly inconsistent with a completely protective effect of this substitution, suggesting that Tg (DeerPrP-S96)7511 mice represent an accurate Tg model in which to assess the effects of the S96 substitution.

In accordance with a role for this region in strain selection, in subsequent studies, Tg mice expressing wild-type deer PrP (tg33) or tg60 were challenged with CWD prions from experimentally infected deer with varying polymorphisms at residues 95 and 96 [174]. Passage of deer CWD prions into tg33 mice expressing wild-type deer PrP resulted in 100% attack rates, with CWD prions from deer expressing H95 or S96 having significantly longer incubation periods. Remarkably, otherwise resistant tg60 mice [145, 172] developed disease only when inoculated with prions from deer expressing H95/Q95 and H95/S96 PrP genotypes. Serial passage in tg60 mice resulted in propagation of a novel CWD strain, referred to as H95(+), while transmission to tg33 mice produced two disease phenotypes consistent with propagation of two strains.

High-resolution structural studies showed the loop region linking the second beta-sheet (β2) with the alpha2-helix (α2) of cervid PrP to be extremely well defined compared to most other species, raising the possibility that this structural characteristic correlates with the unusually facile contagious transmission of CWD [175]. Tg mice expressing mouse PrP in which the β2–α2 loop was replaced by the corresponding region from cervid species spontaneously developed prion disease [176]. Additional studies consistently point to the importance of the β2–α2 loop in regulating transmission barriers, including that of humans to CWD [154, 177–179]. Subsequent work suggested a more complex mechanism in which the β2–α2 loop participates with the distal region of α-helix 3 to form a solvent-accessible contiguous epitope [41]. These and later studies [40] ascribed greater importance to the plasticity of this discontinuous epitope. For example, substitution of D found in horse, which contains a similarly structured loop region, by S at residue 170 of mouse (elk PrP numbering) loop, increased not only the structural order of the loop but also the long-range interaction with Y228 in α-helix 3. Underscoring the importance of long-range β2–α2 loop/α-helix 3 interactions, similar structural connections occur between this residue, which is alanine (A) in tammar wallaby PrP, and residue 169, which is isoleucine (I) in this species (19). Stabilizing long-range interactions between the β2–α2 loop and α-helix 3 also occur in rabbit PrP, a species generally regarded as resistant to prion infection. X-ray crystallographic analyses showed the rabbit β2–α2 loop to be clearly ordered and indicated that hydrophobic interactions between the side chains of V169 and, in this case Y221 (elk PrP numbering) of α-helix 3, contributed to the stability of the β2–α2 loop/α helix 3 epitope [180].

In mule deer, polymorphism at codon 225 encoding serine (S) or phenylalanine (F) influences CWD susceptibility, the 225F allele being relatively protective. The occurrence of CWD was found to be 30-fold higher in deer homozygous for serine at position 225 (225SS) than in heterozygous (225SF) animals; the frequency of 225SF and 225FF genotypes in CWD-negative deer was 9.3%, but only 0.3% in CWD-positive deer [181]. Recent studies comparing CWD susceptibility in mule deer of the two residue 225 genotypes (225SS, 225FF) showed that 225FF mule deer had differences in clinical disease presentation, as well as more-subtle, atypical traits [182]. Immediately adjacent to the protective mule deer PrP polymorphism at 225, residue 226 encodes the singular primary structural difference between Rocky Mountain elk and deer PrP. Elk PrP contains glutamate (E) and deer PrP Q at this position.

Recent findings show that residues 225 and 226 play a critical role in PrP^{C}-to-PrP^{Sc} conversion and strain propagation but that their effects are distinct from those produced by the H95Q, G96S, and M132L polymorphisms [50]. Structural analyses confirm that residues 225 and 226 are located in the distal region of α-helix 3 that participates with the β2–α2 loop to form a solvent-accessible contiguous epitope [41]. Consistent with a role for this epitope in PrP conversion, these polymorphisms severely impact replication of both SSBP/1 and, to variable degrees, CWD. In the case of Tg mice expressing deerPrP-F225, referred to as Tg(DeerPrP-F225), SSBP/1 incubation times were prolonged threefold, whereas inoculation with CWD produced incomplete attack rates or prolonged and variable incubation times in small numbers of mice. In those Tg (DeerPrP-F225) mice that did succumb to CWD, PrP^{Sc} distribution patterns were altered compared with Tg(DeerPrP) mice.

To address the effects of substitution of E for Q at residue 226, we assessed whether Tg mice expressing wild-type elk or deer PrP differed in their responses to CWD. These studies showed that differences at residue 226 also affected CWD replication, but to a lesser degree than the residue 225 polymorphism, with disease onset prolonged by 20–46% in CWD-inoculated Tg(DeerPrP) compared with Tg(ElkPrP) mice, and PrP^{Sc} distribution and neuropathology varying in each case [160]. In contrast to Tg (DeerPrP) mice which are susceptible to SSBP/1 [171], Tg (ElkPrP) were completely resistant [160], although the resistance of elk PrP^{C} to propagation of SSBP/1 was overcome following adaptation in deer or Tg(DeerPrP) mice. Passage in Tg mice expressing E226 or Q226 profoundly affected the ability of SSBP/1 to reinfect Tg mice expressing sheep PrP^{C}. These studies paralleled aspects of our previously published studies indicating that amino acid differences at residue 226 controlled the manifestation of CWD quasispecies or closely related strains [29]. These findings therefore collectively point to an important role for residues 225 and 226 in PrP^{C}-to-PrP^{Sc} conversion and the manifestation of prion strain properties and substantiate the view that long-range interactions between the β2–α2 loop and α-helix 3 provide protection against prion infection and suggest a likely mechanism to account for the protective effects of the F225 polymorphism. Molecular dynamics analyses [50] showed that the S225F and E226Q substitutions in deer alter the orientations of D170 in the β2–α2 loop and Y228 in α-helix 3. This structural change allows hydrogen bonding between the side chains of these residues, which results in reduced plasticity of the β2–α2 loop/α-helix 3 epitope compared with deer or elk PrP structures. This suggests that the increased stability of this tertiary structural epitope precludes PrP^{C}-to-PrP^{Sc} conversion of deerPrP-F225.

Although our seminal studies in Tg mice [142] and subsequent work [146] raised the possibility of CWD strain variation, the limited number of isolates and the lack of detailed strain analyses in those studies meant that this hypothesis remained speculative. Subsequent studies supported the feasibility of using Tg(CerPrP) $1536^{+/-}$ mice for characterizing naturally occurring CWD strains, and novel cervid prions generated by PMCA [152]. To address whether different CWD strains occur in various geographic locations or in different cervid species, bioassays in transgenic mice were used to analyze CWD in a large collection of captive and wild mule deer, white-tailed deer, and elk from various geographic locations in North America [29]. These findings provided substantial evidence for two prevalent CWD prion strains, referred to as CWD1 and CWD2, with different clinical and neuropathological properties. Remarkably, primary transmissions of CWD prions from elk produced either CWD1 or CWD2 profiles, while transmission of deer inocula favored the production of mixed intra-study incubation times and CWD1 and CWD2 neuropathologies. These findings indicate that elk may be infected with *either* CWD1 *or* CWD2, while deer brains tend to harbor CWD1/CWD2 strain mixtures.

The different primary structures of deer and elk PrP at residue 226 provides a framework for understanding these differences in strain profiles of deer and elk. Because of the role played by residue 226, the description of a lysine polymorphism at this position in deer [183] and its possible role on strain stability may be significant. It is unknown whether CWD1 and CWD2 interfere or act synergistically or whether their co-existence contributes to the unparalleled efficiency of CWD transmission. Interestingly, transmission results reported in previous studies suggested that cervid brain inocula might be composed of strain mixtures [147].

Additional studies support the existence of multiple CWD strains. CWD has also been transmitted, albeit with varying efficiency, to transgenic mice expressing mouse PrP [147, 149]. In the former study, a single mule deer isolate produced disease in all inoculated Tga20 mice, which express mouse PrP at high levels. On successive passages, incubation times dropped to ~160 days. In the second study, one elk isolate from a total of eight deer and elk CWD isolates induced disease in 75% of inoculated Tg4053 mice, which also overexpress mouse PrP. The distribution of lesions in both studies appeared to resemble the CWD1 pattern. Low efficiency CWD prion transmission was also recorded in hamsters and transgenic mice expressing Syrian hamster PrP [150]. In that study, during serial passage of mule deer CWD, fast and slow incubation time strains with different patterns of brain pathology and PrP^{Sc} deposition were also isolated. In yet other studies, serial passages of CWD from white-tailed deer into transgenic mice expressing hamster PrP, and then Syrian golden hamsters, produced a strain, referred to as "wasting" (WST), characterized by a prominent preclinical wasting disease, similar to cachexia, which the authors propose is due to a prion-induced endocrinopathy [184]. These same investigators identified a second strain, defined as "cheeky" (CKY), derived from infection of Tg mice that express hamster PrP [185]. The CKY strain had a shorter incubation period than WST, but after transmission to hamsters, the incubation period of CKY became <150 days longer than WST. In this case, PK digestion revealed strain-specific PrP^{Sc} signatures that were maintained in both hosts, but the solubility and conformational stability of PrP^{Sc} differed for the CWD strains in a host-dependent manner. In addition to supporting the view that there are multiple CWD strains, these findings suggest the importance of host-specific pathways, independent of PrP, that participate in the selection and propagation of distinct strains.

Studies in cell culture also support the existence of CWD strains. Quinacrine altered the transmission properties of CWD prions, as well as the biochemical characteristics of the constitutive PrP^{Sc} [30]. Despite accumulating significantly higher prion titers, quinacrine-treated (Q-CWD) prions from $Elk21^{+}$ cells produced prolonged incubation times in Tg(DeerPrP) and Tg(ElkPrP) mice compared to CWD from untreated $Elk21^{+}$ cells. Since kinetics of disease onset in prion-infected animals is inversely related to the titer of a given prion strain, this unusual outcome is consistent with quinacrine affecting the intrinsic properties of the CWD prion. In accordance with this notion, while the deposition patterns of PrP^{Sc} in the brains of diseased Tg(DeerPrP) and Tg(ElkPrP) mice receiving prions from $Elk21^{+}$ are concordant with our previously published descriptions following transmissions of naturally occurring or PMCA-generated CWD prions [152], these distinctive patterns were not recapitulated in either line of Tg mice receiving Q-CWD prions. Prion incubation times and neuronal targeting are the biological criteria by which prion strains are defined. Previous studies showed that strain properties are enciphered within the conformation of PrP^{Sc} [13, 152] and that cervid prion incubation times positively correlate with PrP^{Sc} conformational stability [152]. It is therefore significant that the longer incubation times of Q-CWD prions and altered patterns of PrP^{Sc} deposition in both Tg models were associated with an increase in the relative stability of PrP^{Sc} conformers constituting Q-CWD prions.

The properties of Q-CWD prions provided convincing evidence for conformational mutation that were suggested, but could not be observed, in previous studies of mouse-adapted scrapie prions in cell cultures treated with swainsonine, an inhibitor of Golgi α-mannosidase II [28]. Since the swainsonine-induced properties of cell-derived mouse prions reverted to their original characteristics when propagated in vivo, and there were no detectable differences in the conformations of PrP^{Sc} generated under conditions of swainsonine treatment compared to untreated cells, claims of strain mutation were indirectly based on the differing properties of prions from swainsonine-treated and swainsonine-untreated cells using a cell panel assay. In contrast, CWD and Q-CWD prions are comprised of distinct PrP^{Sc} conformers that produce discernable phenotypes in two lines of Tg mice.

In accordance with Western blotting-based analyses showing higher conformational stabilities of PrP^{Sc} produced in the brains of CWD-infected Tg(DeerPrP) mice compared with Tg(ElkPrP) mice [29], C-CSA confirmed that the conformation of deer PrP^{Sc} in RK13 cells expressing deer PrP (RKD^{+}) was more stable than elk PrP^{Sc} produced during infection of $Elk21^{+}$. Because $Elk21^{+}$ and RKD^{+} propagated CWD prions from the same source, differences in conformational stability result from the effects of residue 226, the sole primary structural difference between elk and deer PrP.

The identification and characterization of distinct CWD strains, and the influence of PrP primary structure on their stabilities, are of importance when considering the potential for interspecies transmission. The appearance of vCJD following human exposure to BSE [18, 186] places the human species barrier to other animal prion diseases, particularly CWD, at the forefront of public health concerns. Since North American hunters harvest thousands of deer and elk each year, and it is not currently mandatory to have these animals tested, the demonstration of CWD prions in skeletal muscle and fat of deer [144, 161] means it is likely that humans consume CWD prions. The substantial market for elk antler velvet in traditional Asian medicine also warrants concern [160].

Estimates of the zoonotic potential of CWD are currently mixed. Surveillance currently shows no evidence of transmission to humans [187, 188]. While initial cell-free conversion studies suggested that the ability of CWD prions to transform human PrP^{C} into protease-resistant PrP was low [163], subsequent results showed that cervid PrP^{Sc} induced the conversion of human PrP^{C} by protein misfolding cyclic amplification (PMCA), following CWD prion strain stabilization by successive passages in vitro or in vivo [189]. These results have implications for the human species barrier to CWD and underscore the role of strain adaptation on interspecies transmission barriers. Additional studies using transgenic mice expressing human PrP^{C} showed that CWD failed to induce disease following intracerebral CWD infection [147, 148, 190]. However, CWD transmission was reported to nonhuman primates through the intracerebral inoculation of squirrel monkeys (*Saimiri sciureus*) [191, 192]. Moreover, using RT-QuIC to model the transmission barrier of CWD to human rPrP, the results of the work of Davenport and coworkers suggest that, at the level of protein–protein interactions, CWD adapts to a new species more readily than does BSE and that the barrier preventing transmission of CWD to humans may be less robust than estimated [193].

References

1. Prusiner SB (1998) Prions (Les Prix Nobel Lecture). In: Frängsmyr T (ed) Les Prix Nobel. Almqvist & Wiksell International, Stockholm, Sweden, pp 268–323
2. Prusiner SB, Scott M, Foster D et al (1990) Transgenetic studies implicate interactions between homologous PrP isoforms in scrapie prion replication. Cell 63:673–686
3. Scott M, Groth D, Foster D et al (1993) Propagation of prions with artificial properties in transgenic mice expressing chimeric PrP genes. Cell 73:979–988
4. Kocisko DA, Come JH, Priola SA et al (1994) Cell-free formation of protease-resistant prion protein. Nature 370:471–474
5. Vidal E, Fernandez-Borges N, Pintado B et al (2015) Transgenic mouse bioassay: evidence that rabbits are susceptible to a variety of prion isolates. PLoS Pathog 11:e1004977
6. Chianini F, Fernandez-Borges N, Vidal E et al (2012) Rabbits are not resistant to prion infection. Proc Natl Acad Sci U S A 109:5080–5085
7. Westaway D, Goodman PA, Mirenda CA et al (1987) Distinct prion proteins in short and long scrapie incubation period mice. Cell 51:651–662
8. Hunter N, Goldmann W, Marshall E et al (2000) Sheep and goats: natural and

experimental TSEs and factors influencing incidence of disease. Arch Virol Suppl 16:181–188
9. Baker HF, Poulter M, Crow TJ et al (1991) Amino acid polymorphism in human prion protein and age at death in inherited prion disease. Lancet 337:1286
10. Collinge J, Palmer MS, Dryden AJ (1991) Genetic predisposition to iatrogenic Creutzfeldt-Jakob disease. Lancet 337:1441–1442
11. Palmer MS, Dryden AJ, Hughes JT et al (1991) Homozygous prion protein genotype predisposes to sporadic Creutzfeldt-Jakob disease. Nature 352:340–342
12. Nonno R, Di Bari MA, Cardone F et al (2006) Efficient transmission and characterization of Creutzfeldt-Jakob disease strains in bank voles. PLoS Pathog 2:e12
13. Telling GC, Parchi P, Dearmond SJ et al (1996) Evidence for the conformation of the pathologic isoform of the prion protein enciphering and propagating prion diversity. Science 274:2079–2082
14. Bessen RA, Marsh RF (1994) Distinct PrP properties suggest the molecular basis of strain variation in transmissible mink encephalopathy. J Virol 68:7859–7868
15. Safar J, Wille H, Itri V et al (1998) Eight prion strains have PrP^{Sc} molecules with different conformations. Nat Med 4:1157–1165
16. Peretz D, Williamson RA, Legname G et al (2002) A change in the conformation of prions accompanies the emergence of a new prion strain. Neuron 34:921–932
17. Collinge J, Sidle KCL, Meads J et al (1996) Molecular analysis of prion strain variation and the aetiology of "new variant" CJD. Nature 383:685–690
18. Hill AF, Desbruslais M, Joiner S et al (1997) The same prion strain causes vCJD and BSE. Nature 389:448–450
19. Scott MR, Peretz D, Nguyen HO et al (2005) Transmission barriers for bovine, ovine, and human prions in transgenic mice. J Virol 79:5259–5271
20. Asante EA, Linehan JM, Desbruslais M et al (2002) BSE prions propagate as either variant CJD-like or sporadic CJD-like prion strains in transgenic mice expressing human prion protein. EMBO J 21:6358–6366
21. Korth C, Kaneko K, Groth D et al (2003) Abbreviated incubation times for human prions in mice expressing a chimeric mouse-human prion protein transgene. Proc Natl Acad Sci U S A 100:4784–4789
22. Scott MR, Will R, Ironside J et al (1999) Compelling transgenetic evidence for transmission of bovine spongiform encephalopathy prions to humans. Proc Natl Acad Sci U S A 96(26):15137–15142
23. Bruce ME, Dickinson AG (1987) Biological evidence that the scrapie agent has an independent genome. J Gen Virol 68:79–89
24. Collinge J, Clarke AR (2007) A general model of prion strains and their pathogenicity. Science 318:930–936
25. Haldiman T, Kim C, Cohen Y et al (2013) Co-existence of distinct prion types enables conformational evolution of human PrPSc by competitive selection. J Biol Chem 288:29846–29861
26. Collinge J (1999) Variant Creutzfeldt-Jakob disease. Lancet 354:317–323
27. Eigen M (1996) On the nature of virus quasispecies. Trends Microbiol 4:216–218
28. Li J, Browning S, Mahal SP et al (2010) Darwinian evolution of prions in cell culture. Science 327:869–872
29. Angers RC, Kang HE, Napier D et al (2010) Prion strain mutation determined by prion protein conformational compatibility and primary structure. Science 328:1154–1158
30. Bian J, Kang HE, Telling GC (2014) Quinacrine promotes replication and conformational mutation of chronic wasting disease prions. Proc Natl Acad Sci U S A 111:6028–6033
31. Makarava N, Baskakov IV (2008) The same primary structure of the prion protein yields two distinct self-propagating states. J Biol Chem 283:15988–15996
32. Rezaei H, Choiset Y, Eghiaian F et al (2002) Amyloidogenic unfolding intermediates differentiate sheep prion protein variants. J Mol Biol 322:799–814
33. Beringue V, Herzog L, Jaumain E et al (2012) Facilitated cross-species transmission of prions in extraneural tissue. Science 335:472–475
34. Rogers M, Yehiely F, Scott M et al (1993) Conversion of truncated and elongated prion proteins into the scrapie isoform in cultured cells. Proc Natl Acad Sci U S A 90:3182–3186
35. Chesebro B, Trifilo M, Race R et al (2005) Anchorless prion protein results in infectious amyloid disease without clinical scrapie. Science 308:1435–1439
36. Telling GC, Scott M, Hsiao KK et al (1994) Transmission of Creutzfeldt-Jakob disease from humans to transgenic mice expressing chimeric human-mouse prion protein. Proc Natl Acad Sci U S A 91:9936–9940

37. Telling GC, Scott M, Mastrianni J et al (1995) Prion propagation in mice expressing human and chimeric PrP transgenes implicates the interaction of cellular PrP with another protein. Cell 83:79–90
38. Perrier V, Kaneko K, Safar J et al (2002) Dominant-negative inhibition of prion replication in transgenic mice. Proc Natl Acad Sci U S A 99:13079–13084
39. Geoghegan JC, Miller MB, Kwak AH et al (2009) Trans-dominant inhibition of prion propagation in vitro is not mediated by an accessory cofactor. PLoS Pathog 5:e1000535
40. Christen B, Hornemann S, Damberger FF et al (2009) Prion protein NMR structure from tammar wallaby (Macropus eugenii) shows that the beta2-alpha2 loop is modulated by long-range sequence effects. J Mol Biol 389:833–845
41. Perez DR, Damberger FF, Wuthrich K (2010) Horse prion protein NMR structure and comparisons with related variants of the mouse prion protein. J Mol Biol 400:121–128
42. Telling GC, Haga T, Torchia M et al (1996) Interactions between wild-type and mutant prion proteins modulate neurodegeneration in transgenic mice. Genes Dev 10:1736–1750
43. Chiesa R, Drisaldi B, Quaglio E et al (2000) Accumulation of protease-resistant prion protein (PrP) and apoptosis of cerebellar granule cells in transgenic mice expressing a PrP insertional mutation. Proc Natl Acad Sci U S A 97:5574–5579
44. Bian J, Nazor KE, Angers R et al (2006) GFP-tagged PrP supports compromised prion replication in transgenic mice. Biochem Biophys Res Commun 340:894–900
45. Barmada SJ, Harris DA (2005) Visualization of prion infection in transgenic mice expressing green fluorescent protein-tagged prion protein. J Neurosci 25:5824–5832
46. Shmerling D, Hegyi I, Fischer M et al (1998) Expression of amino-terminally truncated PrP in the mouse leading to ataxia and specific cerebellar lesions. Cell 93:203–214
47. Baumann F, Tolnay M, Brabeck C et al (2007) Lethal recessive myelin toxicity of prion protein lacking its central domain. EMBO J 26:538–547
48. Li A, Christensen HM, Stewart LR et al (2007) Neonatal lethality in transgenic mice expressing prion protein with a deletion of residues 105–125. EMBO J 26:548–558
49. Scott M, Foster D, Mirenda C et al (1989) Transgenic mice expressing hamster prion protein produce species-specific scrapie infectivity and amyloid plaques. Cell 59:847–857
50. Angers R, Christiansen J, Nalls AV et al (2014) Structural effects of PrP polymorphisms on intra- and interspecies prion transmission. Proc Natl Acad Sci U S A 111:11169–11174
51. Telling GC (2010) Nucleic acid-free mutation of prion strains. Prion 4:252–255
52. Büeler H, Fischer M, Lang Y et al (1992) Normal development and behaviour of mice lacking the neuronal cell-surface PrP protein. Nature 356:577–582
53. Gombojav A, Shimauchi I, Horiuchi M et al (2003) Susceptibility of transgenic mice expressing chimeric sheep, bovine and human PrP genes to sheep scrapie. J Vet Med Sci 65:341–347
54. Kupfer L, Eiden M, Buschmann A et al (2007) Amino acid sequence and prion strain specific effects on the in vitro and in vivo convertibility of ovine/murine and bovine/murine prion protein chimeras. Biochim Biophys Acta 1772:704–713
55. Scott MR, Safar J, Telling G et al (1997) Identification of a prion protein epitope modulating transmission of bovine spongiform encephalopathy prions to transgenic mice. Proc Natl Acad Sci U S A 94:14279–14284
56. Bishop MT, Hart P, Aitchison L et al (2006) Predicting susceptibility and incubation time of human-to-human transmission of vCJD. Lancet Neurol 5:393–398
57. Hadlow WJ (1959) Scrapie and kuru. Lancet 2:289–290
58. Gajdusek DC, Gibbs CJ Jr, Alpers M (1966) Experimental transmission of a kuru-like syndrome to chimpanzees. Nature 209:794–796
59. Gibbs CJ Jr, Gajdusek DC, Asher DM et al (1968) Creutzfeldt-Jakob disease (spongiform encephalopathy): transmission to the chimpanzee. Science 161:388–389
60. Brown P, Gibbs CJ Jr, Rodgers-Johnson P et al (1994) Human spongiform encephalopathy: the National Institutes of Health series of 300 cases of experimentally transmitted disease. Ann Neurol 35:513–529
61. Gibbs CJ Jr, Gajdusek DC, Amyx H (1979) Strain variation in the viruses of Creutzfeldt-Jakob disease and kuru. In: Prusiner SB, Hadlow WJ (eds) Slow transmissible diseases of the nervous system, vol 2. Academic, New York, pp 87–110
62. Tateishi J, Sato Y, Ohta M (1983) Creutzfeldt-Jakob disease in humans and laboratory animals. In: Zimmerman HM (ed)

Progress in neuropathology, vol 5. Raven Press, New York, pp 195–221
63. Tateishi J, Kitamoto T (1995) Inherited prion diseases and transmission to rodents. Brain Pathol 5:53–59
64. Manuelidis E, Kim J, Angelo J et al (1976) Serial propagation of Creutzfeldt-Jakob disease in guinea pigs. Proc Natl Acad Sci U S A 73:223–227
65. Manuelidis E, Gorgacz EJ, Manuelidis L (1978) Interspecies transmission of Creutzfeldt-Jakob disease to Syrian hamsters with reference to clinical syndromes and strains of agent. Proc Natl Acad Sci U S A 75:3422–3436
66. Scott MR, Köhler R, Foster D et al (1992) Chimeric prion protein expression in cultured cells and transgenic mice. Protein Sci 1:986–997
67. Kong Q, Zheng M, Casalone C et al (2008) Evaluation of the human transmission risk of an atypical bovine spongiform encephalopathy prion strain. J Virol 82:3697–3701
68. Beringue V, Herzog L, Reine F et al (2008) Transmission of atypical bovine prions to mice transgenic for human prion protein. Emerg Infect Dis 14:1898–1901
69. Bishop MT, Will RG, Manson JC (2010) Defining sporadic Creutzfeldt-Jakob disease strains and their transmission properties. Proc Natl Acad Sci U S A 107:12005–12010
70. Hill AF, Joiner S, Wadsworth JD et al (2003) Molecular classification of sporadic Creutzfeldt-Jakob disease. Brain 126:1333–1346
71. Parchi P, Castellani R, Capellari S et al (1996) Molecular basis of phenotypic variability in sporadic Creutzfeldt-Jakob disease. Ann Neurol 39:767–778
72. Polymenidou M, Stoeck K, Glatzel M et al (2005) Coexistence of multiple PrPSc types in individuals with Creutzfeldt-Jakob disease. Lancet Neurol 4:805–814
73. Mastrianni JA, Capellari S, Telling GC et al (2001) Inherited prion disease caused by the V210I mutation: transmission to transgenic mice. Neurology 57:2198–2205
74. Mastrianni J, Nixon F, Layzer R et al (1997) Fatal sporadic insomnia: fatal familial insomnia phenotype without a mutation of the prion protein gene. Neurology 48(Suppl): A296
75. Notari S, Xiao X, Espinosa JC et al (2014) Transmission characteristics of variably protease-sensitive prionopathy. Emerg Infect Dis 20:2006–2014
76. Diack AB, Ritchie DL, Peden AH et al (2014) Variably protease-sensitive prionopathy, a unique prion variant with inefficient transmission properties. Emerg Infect Dis 20:1969–1979
77. Giles K, Glidden DV, Patel S et al (2010) Human prion strain selection in transgenic mice. Ann Neurol 68:151–161
78. Will RG, Ironside JW, Zeidler M et al (1996) A new variant of Creutzfeldt-Jakob disease in the UK. Lancet 347:921–925
79. Collinge J, Palmer MS, Sidle KC et al (1995) Unaltered susceptibility to BSE in transgenic mice expressing human prion protein. Nature 378:779–783
80. Mead S, Poulter M, Uphill J et al (2009) Genetic risk factors for variant Creutzfeldt-Jakob disease: a genome-wide association study. Lancet Neurol 8:57–66
81. Plinston C, Hart P, Chong A et al (2011) Increased susceptibility of human-PrP transgenic mice to bovine spongiform encephalopathy infection following passage in sheep. J Virol 85:1174–1181
82. Plinston C, Hart P, Hunter N et al (2014) Increased susceptibility of transgenic mice expressing human PrP to experimental sheep bovine spongiform encephalopathy is not due to increased agent titre in sheep brain tissue. J Gen Virol 95:1855–1859
83. Asante EA, Linehan JM, Gowland I et al (2006) Dissociation of pathological and molecular phenotype of variant Creutzfeldt-Jakob disease in transgenic human prion protein 129 heterozygous mice. Proc Natl Acad Sci U S A 103:10759–10764
84. Mead S, Whitfield J, Poulter M et al (2009) A novel protective prion protein variant that colocalizes with kuru exposure. N Engl J Med 361:2056–2065
85. Asante EA, Smidak M, Grimshaw A et al (2015) A naturally occurring variant of the human prion protein completely prevents prion disease. Nature 522:478–481
86. Hsiao K, Baker HF, Crow TJ et al (1989) Linkage of a prion protein missense variant to Gerstmann-Sträussler syndrome. Nature 338:342–345
87. Manson JC, Jamieson E, Baybutt H et al (1999) A single amino acid alteration (101L) introduced into murine PrP dramatically alters incubation time of transmissible

spongiform encephalopathy. EMBO J 18:6855–6864
88. Hsiao KK, Groth D, Scott M et al (1994) Serial transmission in rodents of neurodegeneration from transgenic mice expressing mutant prion protein. Proc Natl Acad Sci U S A 91:9126–9130
89. Kaneko K, Ball HL, Wille H, Zhang H, Groth D, Torchia M, Tremblay P, Safar J, Prusiner SB, Dearmond SJ, Baldwin MA, Cohen FE (2000) A synthetic peptide initiates Gerstmann-Straussler-Scheinker (GSS) disease in transgenic mice. J Mol Biol 295:997–1007
90. Tremblay P, Ball HL, Kaneko K et al (2004) Mutant PrPSc conformers induced by a synthetic peptide and several prion strains. J Virol 78:2088–2099
91. Cohen FE, Pan K-M, Huang Z et al (1994) Structural clues to prion replication. Science 264:530–531
92. Tzaban S, Friedlander G, Schonberger O et al (2002) Protease-sensitive scrapie prion protein in aggregates of heterogeneous sizes. Biochemistry 41:12868–12875
93. Korth C, Stierli B, Streit P et al (1997) Prion (PrP^{Sc})-specific epitope defined by a monoclonal antibody. Nature 389:74–77
94. Nazor KE, Kuhn F, Seward T et al (2005) Immunodetection of disease-associated mutant PrP, which accelerates disease in GSS transgenic mice. EMBO J 24:2472–2480
95. Uptain SM, Lindquist S (2002) Prions as protein-based genetic elements. Annu Rev Microbiol 56:703–741
96. Piccardo P, King D, Telling G et al (2013) Dissociation of prion protein amyloid seeding from transmission of a spongiform encephalopathy. J Virol 87:12349–12356
97. Legname G, Baskakov IV, Nguyen HO et al (2004) Synthetic mammalian prions. Science 305:673–676
98. Asante EA, Grimshaw A, Smidak M et al (2015) Transmission properties of human PrP 102L prions challenge the relevance of mouse models of GSS. PLoS Pathog 11: e1004953
99. Friedman-Levi Y, Meiner Z, Canello T et al (2011) Fatal prion disease in a mouse model of genetic E200K Creutzfeldt-Jakob disease. PLoS Pathog 7:e1002350
100. Chiesa R, Piccardo P, Ghetti B et al (1998) Neurological illness in transgenic mice expressing a prion protein with an insertional mutation. Neuron 21:1339–1351
101. Chiesa R, Piccardo P, Quaglio E et al (2003) Molecular distinction between pathogenic and infectious properties of the prion protein. J Virol 77:7611–7622
102. Dossena S, Imeri L, Mangieri M et al (2008) Mutant prion protein expression causes motor and memory deficits and abnormal sleep patterns in a transgenic mouse model. Neuron 60:598–609
103. Bouybayoune I, Mantovani S, Del Gallo F et al (2015) Transgenic fatal familial insomnia mice indicate prion infectivity-independent mechanisms of pathogenesis and phenotypic expression of disease. PLoS Pathog 11: e1004796
104. Jackson WS, Borkowski AW, Faas H et al (2009) Spontaneous generation of prion infectivity in fatal familial insomnia knockin mice. Neuron 63:438–450
105. Jackson WS, Borkowski AW, Watson NE et al (2013) Profoundly different prion diseases in knock-in mice carrying single PrP codon substitutions associated with human diseases. Proc Natl Acad Sci U S A 110:14759–14764
106. Hsiao KK, Cass C, Schellenberg GD et al (1991) A prion protein variant in a family with the telencephalic form of Gerstmann-Sträussler-Scheinker syndrome. Neurology 41:681–684
107. Hegde RS, Mastrianni JA, Scott MR et al (1998) A transmembrane form of the prion protein in neurodegenerative disease. Science 279:827–834
108. Yang W, Cook J, Rassbach B et al (2009) A new transgenic mouse model of Gerstmann-Straussler-Scheinker syndrome caused by the A117V mutation of PRNP. J Neurosci 29:10072–10080
109. Asante EA, Linehan JM, Smidak M et al (2013) Inherited prion disease A117V is not simply a proteinopathy but produces prions transmissible to transgenic mice expressing homologous prion protein. PLoS Pathog 9: e1003643
110. Buschmann A, Pfaff E, Reifenberg K et al (2000) Detection of cattle-derived BSE prions using transgenic mice overexpressing bovine PrP(C). Arch Virol Suppl 16:75–86
111. Castilla J, Gutierrez Adan A, Brun A et al (2003) Early detection of PrPres in BSE-infected bovine PrP transgenic mice. Arch Virol 148:677–691
112. Jacobs JG, Langeveld JP, Biacabe AG et al (2007) Molecular discrimination of atypical bovine spongiform encephalopathy strains from a geographical region spanning a wide area in Europe. J Clin Microbiol 45:1821–1829

113. Casalone C, Zanusso G, Acutis P et al (2004) Identification of a second bovine amyloidotic spongiform encephalopathy: molecular similarities with sporadic Creutzfeldt-Jakob disease. Proc Natl Acad Sci U S A 101:3065–3070
114. Richt JA, Hall SM (2008) BSE case associated with prion protein gene mutation. PLoS Pathog 4:e1000156
115. Buschmann A, Groschup MH (2005) Highly bovine spongiform encephalopathy-sensitive transgenic mice confirm the essential restriction of infectivity to the nervous system in clinically diseased cattle. J Infect Dis 192:934–942
116. Beringue V, Bencsik A, Le Dur A et al (2006) Isolation from cattle of a prion strain distinct from that causing bovine spongiform encephalopathy. PLoS Pathog 2:e112
117. Beringue V, Andreoletti O, Le Dur A et al (2007) A bovine prion acquires an epidemic bovine spongiform encephalopathy strain-like phenotype on interspecies transmission. J Neurosci 27:6965–6971
118. Buschmann A, Gretzschel A, Biacabe AG et al (2006) Atypical BSE in Germany—proof of transmissibility and biochemical characterization. Vet Microbiol 117:103–116
119. Capobianco R, Casalone C, Suardi S et al (2007) Conversion of the BASE prion strain into the BSE strain: the origin of BSE? PLoS Pathog 3:e31
120. Wadsworth JD, Asante EA, Desbruslais M et al (2004) Human prion protein with valine 129 prevents expression of variant CJD phenotype. Science 306:1793–1796
121. Beringue V, Le Dur A, Tixador P et al (2008) Prominent and persistent extraneural infection in human PrP transgenic mice infected with variant CJD. PLoS One 3:e1419
122. Espinosa JC, Herva ME, Andreoletti O et al (2009) Transgenic mice expressing porcine prion protein resistant to classical scrapie but susceptible to sheep bovine spongiform encephalopathy and atypical scrapie. Emerg Infect Dis 15:1214–1221
123. Westaway D, Dearmond SJ, Cayetano-Canlas J et al (1994) Degeneration of skeletal muscle, peripheral nerves, and the central nervous system in transgenic mice overexpressing wild-type prion proteins. Cell 76:117–129
124. Rubenstein R, Bulgin MS, Chang B et al (2012) PrP(Sc) detection and infectivity in semen from scrapie-infected sheep. J Gen Virol 93:1375–1383
125. Crozet C, Flamant F, Bencsik A et al (2001) Efficient transmission of two different sheep scrapie isolates in transgenic mice expressing the ovine PrP gene. J Virol 75:5328–5334
126. Crozet C, Bencsik A, Flamant F et al (2001) Florid plaques in ovine PrP transgenic mice infected with an experimental ovine BSE. EMBO rep 2:952–956
127. Vilotte JL, Soulier S, Essalmani R et al (2001) Markedly increased susceptibility to natural sheep scrapie of transgenic mice expressing ovine prp. J Virol 75:5977–5984
128. Tamguney G, Miller MW, Giles K et al (2009) Transmission of scrapie and sheep-passaged bovine spongiform encephalopathy prions to transgenic mice expressing elk prion protein. J Gen Virol 90:1035–1047
129. Goldmann W, Hunter N, Smith G et al (1994) PrP genotype and agent effects in scrapie: change in allelic interaction with different isolates of agent in sheep, a natural host of scrapie. J Gen Virol 75:989–995
130. Cordier C, Bencsik A, Philippe S et al (2006) Transmission and characterization of bovine spongiform encephalopathy sources in two ovine transgenic mouse lines (TgOvPrP4 and TgOvPrP59). J Gen Virol 87:3763–3771
131. Saijo E, Kang HE, Bian J et al (2013) Epigenetic dominance of prion conformers. PLoS Pathog 9:e1003692
132. Foster JD, Dickinson AG (1988) The unusual properties of CH 1641, a sheep-passaged isolate of scrapie. Vet Rec 123:5–8
133. Baron T, Biacabe AG (2007) Molecular behaviors of "CH1641-like" sheep scrapie isolates in ovine transgenic mice (TgOvPrP4). J Virol 81:7230–7237
134. Kang HE, Weng CC, Saijo E et al (2012) Characterization of conformation-dependent prion protein epitopes. J Biol Chem 287 (44):37219–37232
135. Bencsik A, Philippe S, Debeer S et al (2007) Scrapie strain transmission studies in ovine PrP transgenic mice reveal dissimilar susceptibility. Histochem Cell Biol 127:531–539
136. Baron T, Crozet C, Biacabe AG et al (2004) Molecular analysis of the protease-resistant prion protein in scrapie and bovine spongiform encephalopathy transmitted to ovine transgenic and wild-type mice. J Virol 78:6243–6251
137. Le Dur A, Beringue V, Andreoletti O et al (2005) A newly identified type of scrapie agent can naturally infect sheep with resistant

PrP genotypes. Proc Natl Acad Sci U S A 102:16031–16036

138. Simmons MM, Konold T, Simmons HA et al (2007) Experimental transmission of atypical scrapie to sheep. BMC Vet Res 3:20
139. Wadsworth JD, Joiner S, Linehan JM et al (2013) Atypical scrapie prions from sheep and lack of disease in transgenic mice over-expressing human prion protein. Emerg Infect Dis 19:1731–1739
140. Cassard H, Torres JM, Lacroux C et al (2014) Evidence for zoonotic potential of ovine scrapie prions. Nat Commun 5:5821
141. Mathiason CK, Powers JG, Dahmes SJ et al (2006) Infectious prions in the saliva and blood of deer with chronic wasting disease. Science 314:133–136
142. Browning SR, Mason GL, Seward T et al (2004) Transmission of prions from mule deer and elk with chronic wasting disease to transgenic mice expressing cervid PrP. J Virol 78:13345–13350
143. Bartz JC, Marsh RF, Mckenzie DI et al (1998) The host range of chronic wasting disease is altered on passage in ferrets. Virology 251:297–301
144. Angers RC, Browning SR, Seward TS et al (2006) Prions in skeletal muscles of deer with chronic wasting disease. Science 311:1117
145. Meade-White K, Race B, Trifilo M et al (2007) Resistance to chronic wasting disease in transgenic mice expressing a naturally occurring allelic variant of deer prion protein. J Virol 81:4533–4539
146. Lafauci G, Carp RI, Meeker HC et al (2006) Passage of chronic wasting disease prion into transgenic mice expressing Rocky Mountain elk (Cervus elaphus nelsoni) PrPC. J Gen Virol 87:3773–3780
147. Tamguney G, Giles K, Bouzamondo-Bernstein E et al (2006) Transmission of elk and deer prions to transgenic mice. J Virol 80:9104–9114
148. Kong Q, Huang S, Zou W et al (2005) Chronic wasting disease of elk: transmissibility to humans examined by transgenic mouse models. J Neurosci 25:7944–7949
149. Sigurdson CJ, Manco G, Schwarz P et al (2006) Strain fidelity of chronic wasting disease upon murine adaptation. J Virol 80:12303–12311
150. Raymond GJ, Raymond LD, Meade-White KD et al (2007) Transmission and adaptation of chronic wasting disease to hamsters and transgenic mice: evidence for strains. J Virol 81:4305–4314
151. Meyerett C, Michel B, Pulford B et al (2008) In vitro strain adaptation of CWD prions by serial protein misfolding cyclic amplification. Virology 382(2):267–276
152. Green KM, Castilla J, Seward TS et al (2008) Accelerated high fidelity prion amplification within and across prion species barriers. PLoS Pathog 4:e1000139
153. Kurt TD, Perrott MR, Wilusz CJ et al (2007) Efficient in vitro amplification of chronic wasting disease PrPRES. J Virol 81:9605–9608
154. Kurt TD, Telling GC, Zabel MD et al (2009) Trans-species amplification of PrP(CWD) and correlation with rigid loop 170N. Virology 387:235–243
155. Haley NJ, Seelig DM, Zabel MD et al (2009) Detection of CWD prions in urine and saliva of deer by transgenic mouse bioassay. PLoS One 4:e4848
156. Tamguney G, Miller MW, Wolfe LL et al (2009) Asymptomatic deer excrete infectious prions in faeces. Nature 461:529–532
157. Haley NJ, Mathiason CK, Zabel MD et al (2009) Detection of sub-clinical CWD infection in conventional test-negative deer long after oral exposure to urine and feces from CWD+ deer. PLoS One 4:e7990
158. Pulford B, Spraker TR, Wyckoff AC et al (2012) Detection of PrPCWD in feces from naturally exposed Rocky Mountain elk (Cervus elaphus nelsoni) using protein misfolding cyclic amplification. J Wildl Dis 48:425–434
159. Henderson DM, Manca M, Haley NJ et al (2013) Rapid antemortem detection of CWD prions in deer saliva. PLoS One 8:e74377
160. Angers RC, Seward TS, Napier D et al (2009) Chronic wasting disease prions in elk antler velvet. Emerg Infect Dis 15:696–703
161. Race B, Meade-White K, Race R et al (2009) Prion infectivity in fat of deer with chronic wasting disease. J Virol 83:9608–9610
162. Johnson C, Johnson J, Clayton M et al (2003) Prion protein gene heterogeneity in free-ranging white-tailed deer within the chronic wasting disease affected region of Wisconsin. J Wildl Dis 39:576–581
163. Raymond GJ, Bossers A, Raymond LD et al (2000) Evidence of a molecular barrier limiting susceptibility of humans, cattle and sheep to chronic wasting disease. EMBO J 19:4425–4430
164. Heaton MP, Leymaster KA, Freking BA et al (2003) Prion gene sequence variation within diverse groups of U.S. sheep, beef cattle, and deer. Mamm Genome 14:765–777

165. O'rourke KI, Spraker TR, Hamburg LK et al (2004) Polymorphisms in the prion precursor functional gene but not the pseudogene are associated with susceptibility to chronic wasting disease in white-tailed deer. J Gen Virol 85:1339–1346
166. Schatzl HM, Wopfner F, Gilch S et al (1997) Is codon 129 of prion protein polymorphic in human beings but not in animals? Lancet 349:1603–1604
167. O'rourke KI, Baszler TV, Miller JM et al (1998) Monoclonal antibody F89/160.1.5 defines a conserved epitope on the ruminant prion protein. J Clin Microbiol 36:1750–1755
168. O'rourke KI, Besser TE, Miller MW et al (1999) PrP genotypes of captive and free-ranging Rocky Mountain elk (Cervus elaphus nelsoni) with chronic wasting disease. J Gen Virol 80(Pt 10):2765–2769
169. Hamir AN, Gidlewski T, Spraker TR et al (2006) Preliminary observations of genetic susceptibility of elk (Cervus elaphus nelsoni) to chronic wasting disease by experimental oral inoculation. J Vet Diagn Investig 18:110–114
170. O'rourke KI, Spraker TR, Zhuang D et al (2007) Elk with a long incubation prion disease phenotype have a unique PrPd profile. Neuroreport 18:1935–1938
171. Green KM, Browning SR, Seward TS et al (2008) The elk PRNP codon 132 polymorphism controls cervid and scrapie prion propagation. J Gen Virol 89:598–608
172. Race B, Meade-White K, Miller MW et al (2011) In vivo comparison of chronic wasting disease infectivity from deer with variation at prion protein residue 96. J Virol 85:9235–9238
173. Keane DP, Barr DJ, Bochsler PN et al (2008) Chronic wasting disease in a Wisconsin white-tailed deer farm. J Vet Diagn Investig 20:698–703
174. Duque Velasquez C, Kim C, Herbst A et al (2015) Deer prion proteins modulate the emergence and adaptation of chronic wasting disease strains. J Virol 89:12362–12373
175. Gossert AD, Bonjour S, Lysek DA et al (2005) Prion protein NMR structures of elk and of mouse/elk hybrids. Proc Natl Acad Sci U S A 102:646–650
176. Sigurdson CJ, Nilsson KP, Hornemann S et al (2009) De novo generation of a transmissible spongiform encephalopathy by mouse transgenesis. Proc Natl Acad Sci U S A 106:304–309
177. Kurt TD, Bett C, Fernandez-Borges N et al (2014) Prion transmission prevented by modifying the beta2-alpha2 loop structure of host PrPC. J Neurosci 34:1022–1027
178. Sigurdson CJ, Nilsson KP, Hornemann S et al (2010) A molecular switch controls interspecies prion disease transmission in mice. J Clin Invest 120:2590–2599
179. Kurt TD, Jiang L, Fernandez-Borges N et al (2015) Human prion protein sequence elements impede cross-species chronic wasting disease transmission. J Clin Invest 125:1485–1496
180. Khan MQ, Sweeting B, Mulligan VK et al (2010) Prion disease susceptibility is affected by beta-structure folding propensity and local side-chain interactions in PrP. Proc Natl Acad Sci U S A 107:19808–19813
181. Jewell JE, Conner MM, Wolfe LL et al (2005) Low frequency of PrP genotype 225SF among free-ranging mule deer (Odocoileus hemionus) with chronic wasting disease. J Gen Virol 86:2127–2134
182. Wolfe LL, Fox KA, Miller MW (2014) "Atypical" chronic wasting disease in PRNP genotype 225FF mule deer. J Wildl Dis 50:660–665
183. Johnson C, Johnson J, Vanderloo JP et al (2006) Prion protein polymorphisms in white-tailed deer influence susceptibility to chronic wasting disease. J Gen Virol 87:2109–2114
184. Bessen RA, Robinson CJ, Seelig DM et al (2011) Transmission of chronic wasting disease identifies a prion strain causing cachexia and heart infection in hamsters. PLoS One 6: e28026
185. Crowell J, Hughson A, Caughey B et al (2015) Host determinants of prion strain diversity independent of prion protein genotype. J Virol 89:10427–10441
186. Bruce ME, Will RG, Ironside JW et al (1997) Transmissions to mice indicate that 'new variant' CJD is caused by the BSE agent. Nature 389:498–501
187. Belay ED, Maddox RA, Williams ES et al (2004) Chronic wasting disease and potential transmission to humans. Emerg Infect Dis 10:977–984
188. Mawhinney S, Pape WJ, Forster JE et al (2006) Human prion disease and relative risk associated with chronic wasting disease. Emerg Infect Dis 12:1527–1535

189. Barria MA, Telling GC, Gambetti P et al (2011) Generation of a new form of human PrP(Sc) in vitro by interspecies transmission from cervid prions. J Biol Chem 286:7490–7495
190. Sandberg M, Al-Doujaily H, Sigurdson C et al (2010) Chronic wasting disease prions are not transmissible to transgenic mice over-expressing human prion protein. J Gen Virol 91(Pt 10):2651–2657
191. Marsh RF, Kincaid AE, Bessen RA et al (2005) Interspecies transmission of chronic wasting disease prions to squirrel monkeys (Saimiri sciureus). J Virol 79:13794–13796
192. Race B, Meade-White KD, Miller MW et al (2009) Susceptibilities of nonhuman primates to chronic wasting disease. Emerg Infect Dis 15:1366–1376
193. Davenport KA, Henderson DM, Bian J et al (2015) Insights into chronic wasting disease and bovine spongiform encephalopathy species barriers by use of real-time conversion. J Virol 89:9524–9531

Chapter 17

In Vivo-Near Infrared Imaging of Neurodegeneration

Victoria A. Lawson, Carolin Tumpach, Cathryn L. Haigh, and Simon C. Drew

Abstract

In vivo near-infrared (NIR) imaging of molecular processes at the preclinical stage promises to provide more valuable mechanistic information about pathological pathways involved in neurodegeneration. NIR imaging has the potential to improve in vivo therapeutic screening protocols by enabling noninvasive monitoring of presymptomatic responses to treatment. We have developed new NIR fluorescent contrast agents conjugated to markers of cell death, and using these agents we have identified molecular pathways associated with prion-induced neurodegeneration and determined the optimal window for meaningful therapeutic intervention in prion disease. This chapter provides a description of the synthesis and purification of our NIR cell Death (NIRD) contrast agent and the application of in vivo NIRD (iNIRD) imaging to a prion model of neurodegeneration.

Key words Prion, In vivo, Near infrared, Fluorescence, Imaging, Caspase, Neurodegeneration

1 Introduction

A key focus of medical imaging for neurodegeneration research and diagnosis is the in vivo measurement of brain amyloid burden [1]; however, other cell signalling pathways, such as activation of caspase enzymes, are found to precede the deposition of protein aggregates [2, 3]. Imaging techniques that can target earlier molecular events are therefore far more attractive since they offer the possibility of detecting physiological changes that precede neuronal dysfunction and death. We have generated a near-infrared contrast agent that can be used to image the molecular pathways that lead to neuronal dysfunction and loss in a live animal model of human prion disease [4]. This contrast agent targets active caspases, the central regulators of apoptosis [5]. Proteolytic caspase activation has been linked with cell death in many neurodegenerative diseases, in particular prion diseases such as Creutzfeldt-Jakob disease (CJD) and Gerstmann-Sträussler-Scheinker syndrome (GSS) [2, 6–12].

Victoria A. Lawson (ed.), *Prions: Methods and Protocols*, Methods in Molecular Biology, vol. 1658, DOI 10.1007/978-1-4939-7244-9_17, © Springer Science+Business Media LLC 2017

We have also demonstrated elevated levels of active caspases in the brain and peripheral organs in a mouse-adapted model of human prion disease, both ex vivo [4, 13] and in vivo [14].

The consensus cleavage site of all known caspases is determined by the sequence of four amino acids N-terminal to the critical aspartic residue at the P1 position [5]. While individual caspases possess differing substrate specificity and tissue localization based upon the identity of the residue at the P4 position (N-terminal to the critical Asp) [15], the Val-Ala-Asp substrate is known to have broad caspase recognition. The Z-VAD-fmk compound (Z = benzyloxycarbonyl; fmk = fluoromethylketone) was developed as an effective general inhibitor of caspase activity and functions by irreversibly binding to the catalytic site of active caspases to prevent further proteolytic activity [16]. Green and red fluorescent analogues of this inhibitor have also been developed. However, their application to in vivo imaging is hampered by the poor tissue penetration depth of visible light [17, 18] and potentially high autofluorescence, especially when imaging diseases of the central nervous system [19, 20]. The use of near-infrared (NIR) wavelengths (650–900 nm) circumvents these problems to a large degree, since NIR light is less attenuated by biological tissue and autofluorescence beyond the visible spectrum is minimal [17, 21]. NIR imaging of amyloid plaques in a mouse model of Alzheimer's disease has been demonstrated [22, 23]; however, it is desirable to be able to monitor other disease pathways associated with neuronal dysfunction and degeneration, such as those mediated by caspases [24].

Below we outline the synthesis of a NIR cell death (NIRD) contrast agent suitable for in vivo imaging by chemically conjugating a NIR dye to the VAD-fmk caspase inhibitor [4]. We then describe its in vivo application (iNIRD) to the noninvasive detection of neuronal apoptosis associated with neurodegeneration in living animals.

2 Materials

2.1 NIRD Probe Synthesis

1. Caspase inhibitor, Z-Val-Ala-Asp(OMe)-fmk (SM Biochemicals LLC, Yorba Linda, USA).
2. NIR dye containing an amino-reactive N-hydroxysuccinimidyl-ester (DY-750-NHS-ester; Dyomics, GmbH, Germany).
3. 33% (v/v) solution of hydrogen bromide (HBr) in acetic acid (AcOH) (Sigma-Aldrich).
4. Anhydrous diethyl ether and petroleum ether (Sigma-Aldrich).

5. Rotary evaporator (BÜCHI, Switzerland) with acetone and dry ice for cold trap.
6. Anhydrous dimethylformamide (DMF) (Sigma-Aldrich).
7. *N,N*-Diisopropylethylamine/triethylamine (Sigma-Aldrich).
8. Electrospray ionization time-of-flight (ESI-TOF) mass spectrometer (Agilent Technologies, USA).
9. Freeze-dryer (CHRIST) with liquid nitrogen as freezing agent.

2.2 NIRD Probe Purification

1. High-performance liquid chromatography (HPLC) instrument, 1200 series (Agilent Technologies,USA).
2. Spherisorb 5 μm cyano 4.6 × 250 mm analytical cartridge (Waters Corporation, USA).
3. HPLC-grade acetonitrile and formic acid (0.1% v/v).
4. Microcentrifuge (Eppendorf, Germany).
5. ESI-TOF mass spectrometer (Agilent Technologies, USA).
6. UV/Vis Spectrometer SPECTROstar Nano (BMG LABTECH, Germany) with 50 μL quartz cuvette (NSG Precision Cells, USA).
7. Methanol (Merck, Germany).
8. Lyophilizer (CHRIST, Germany) with liquid nitrogen as freezing agent.
9. Centrifugal evaporator/speed-vac for microtubes (N-BIOTEK, Korea).

2.3 Intravenous Injection of Imaging Agent

1. Mice (prion-infected and age-matched sham-inoculated control animals).
2. Sterile-filtered dimethylsulfoxide (DMSO) (Sigma-Aldrich).
3. Sterile phosphate-buffered saline (PBS), with no Mg or Ca (Thermo-Fisher Scientific, USA).
4. 26 G sterile needles and 1 mL syringes.
5. Mouse immobilization apparatus.
6. Mouse monitoring checklist.

2.4 Imaging

1. Ketamine (80 mg kg^{-1})/xylazine (10 mg kg^{-1}) anesthetic.
2. Delicate hair removal cream/hair clippers.
3. Petroleum jelly.
4. Cotton buds/swabs.
5. LAS-3000 (Fujifilm, Japan) or LAS-4000 (GE Healthcare Lifesciences, USA) intelligent dark box, equipped with a 710 nm EPI-illuminator and a 785 nm long-pass filter.
6. Heat pad/warm box for animal care.

2.5 Analysis

1. Multi Gauge software, V3.0 (Fujifilm, Japan) or higher.

3 Methods

3.1 NIRD Probe Synthesis

The following protocol is based upon the pan-caspase inhibitor (VAD-fmk); however, the methods are also applicable to the labeling of inhibitors containing a methyl ester (e.g., Z-VAD(OMe)-fmk) and other inhibitors (e.g., caspase-3; Z-DEVD-fmk). Where practicable, protect solutions containing fluorescent dye from ambient light exposure throughout all procedures.

1. It is recommended to perform a quality control of the Z-VAD-fmk by mass spectrometry before carrying subsequent steps (*see* **Note 1**).
2. Add approximately 0.5 mL 33% (v/v) HBr in AcOH to the vial containing the Z-VAD-fmk (typically 1–10 mg) and transfer the dissolved peptide into a glass round-bottom flask. Wash the original vial with another 0.5 mL acid and transfer the wash solution to the flask.
3. Stir the solution for 1 h at room temperature to cleave the Z group and form the HBr salt of NH_2-VAD-fmk.
4. Add ice cold diethylether and petroleum ether to precipitate the NH_2-VAD-fmk (*see* **Note 2**). The de-protected peptide is insoluble in ether and will be visible as brown oily droplets.
5. Carefully remove the excess HBr/AcOH/ether using a Pasteur pipette, and then attach the round-bottom flask to a rotary evaporator. Dry the peptide by lowering the flask into the pre-warmed water bath (40 °C) and rotate gently in low vacuum for approximately 20 min or until all peptide is spread around the flask in fine droplets.
6. Before labeling the peptide with the NIR dye, verify the de-protection reaction by mass spectrometry (*see* **Note 3**).
7. Resuspend the dry peptide in anhydrous DMF at a concentration of ca. 100 mM.
8. Dissolve the desired amount of DY-750-NHS-ester in anhydrous DMF.
9. Combine the DY-750-NHS-ester and NH_2-VAD-fmk at a nominal molar ratio of 1.1:1, and then add three molar equivalents of *N,N*-diisopropylethylamine or triethylamine.
10. Stir the solution in the dark at room temperature for at least 6 h (preferably overnight) to ensure a high degree of labeling.
11. Verify the labeling reaction by mass spectrometry and freeze-dry the crude product (*see* **Note 4**).

3.2 NIRD Probe Purification

After de-protection and NIR labeling the caspase inhibitor, the desired product must be separated from unreacted peptide and DY-750, hydrolyzed DY-750, and other impurities that were present prior to labeling (now both labeled and unlabeled).

1. Dissolve the peptide in approximately 1/3 v/v acetonitrile before adding 2/3 v/v of deionized H_2O. All peptide needs to be fully in solution with no floating particles present to avoid clogging of the HPLC column. Carry out a short spin in a microcentrifuge to force any particulates to the bottom of the tube.
2. Purify the resuspended NIRD probe by HPLC using a cyano column and the following mobile phase:

Buffer A	99.9% deionized H_2O (MQ water), 0.1% (v/v) formic acid
Buffer B	95% (v/v) acetonitrile, 4.9% (v/v) deionized H_2O, 0.1% (v/v) formic acid

The following parameters are recommended to achieve a satisfactory separation of the desired product (*see* **Note 5**)

Stop time	30 min
Flow rate	1.5 mL/min
Column heating	40 °C
Detector wavelength	500 nm
Injection volume	100 μL

Time (min)	% Buffer B
0	10
5	35
25	50
26	100
28	100
30	10

Collect the peptide fractions in 1.7 mL microtubes, either manually or using an automated fraction collector.

Analyze each fraction by mass spectrometry to determine which contains the desired product (*see* **Note 6**).

3. Pool together the correct fractions from multiple HPLC runs, and then determine the concentration of NIRD by measuring the optical density (OD) at 747 nm:

$$\text{Concentration} = \frac{OD_{747nm}}{\varepsilon \times l} \times \text{dilution factor}.$$

where the molar absorptivity (ε) at 747 nm is 270,000 M^{-1} cm^{-1} (*see* **Note 7**), l is the optical path length, and the dilution factor is chosen such that $OD_{747nm} < 1$.

4. Aliquot the NIRD solution into suitably sized portions, and then freeze-dry (*see* **Notes 8** and **9**). Store the dry portions at −80 °C.

3.3 Intravenous Injection of Imaging Agent

All applicable animal ethics approvals and training should be obtained from the relevant authority prior to beginning animal experimentation (*see* **Note 10**).

1. Resuspend a dry NIRD aliquot in sterile PBS with 10% v/v sterile-filtered DMSO (*see* **Notes 11** and **12**). The volume of solution should be such that the desired dose (*see* **Note 9**) can be delivered in a 100 μL injection. Maintain the solution at 37 °C in the dark prior to injection.
2. Immobilize mouse and administer 100 μL of NIRD or control dye to mice via tail vein injection (*see* **Note 13**).
4. Allow 60 min for the probe to circulate before imaging (*see* **Note 14**).

3.4 Imaging

1. Weigh each mouse before administering the anesthetic.
2. Anesthetize mice by intraperitoneal injection of ketamine (80 mg kg^{-1})/xylazine (10 mg kg^{-1}).
3. To prevent the animals' eyes from drying out, carefully cover them with petroleum jelly using a cotton bud or swab.
4. Carefully remove hair from the top of the head of each mouse. This should be done by careful shaving with small clippers, followed by application of sensitive area hair removal cream (*see* **Note 15**).
5. Position the mouse on the upper tray of the LAS-3000 intelligent dark box and adjust focusing.
6. Acquire a bright-field image using white LED epi-illumination (and no filter) with an exposure time of 10–100 ms, an aperture of F2.8, and "standard" resolution (*see* **Note 16**). Save the image in the LAS image format (*see* **Note 17**).
7. Acquire a fluorescence image using 710 nm LED epi-illumination and the 785 nm long-pass filter, an aperture of F0.85, and "standard" resolution, readjusting the focus as necessary. The optimal exposure time will need to be determined based upon the individual application (*see* **Note 18**). Save the image in the LAS image format.

8. Transfer mouse to a warm container and monitor until awake (*see* **Note 19**).

3.5 Analysis

1. Open the Multi Gauge software.
2. Click the "Compose..." button on the "Images" tab and select "Monochrome/32Color" as the overlay type, and then click "Next".
3. For channel 1 and channel 2, click the "Select..." button and locate the bright-field and NIR image files, respectively, to be overlaid. Select the channel 1 and 2 image colors as positive gray and color, respectively.
4. Save the overlay image in the Fuji Multi-scanned image file format (*.set).
5. Click the "Display" tab and threshold the display range of the images (intensity values between 0 and 65,535). The NIR image will exhibit intensity outside the region of interest (shaved area of the body) due to light scatter from surrounding hair. Much of this can be masked by appropriate choice of the lower intensity threshold.
6. Click on the "Measure" tab and trace a region of interest (ROI) around the shaved area of the head, and then click on the "Save" tab. A result table showing the total intensity, ROI area, and the intensity per unit area will be displayed.
7. Repeat **steps 2–6** for each mouse, ensuring that the same image display range is used for each overlaid image set. Analyze the intensity data using appropriate statistical analysis methods and software.

4 Notes

1. In our experience, the quality of the starting material is highly variable between different vendors. In some instances, especially for caspase-specific Z-XXXD-fmk inhibitors, we have been unable to identify any of the desired starting material in the supplied product.
2. Ice cold ether will help to improve the separation process. CAUTION: Ether presents both an inhalation and explosion risk. It should be handled in a fume hood free from sources of open flame and electrostatic discharge. In case the precipitated peptide is difficult to identify, place the flask on ice, increase the amount of ether, and allow longer settlement periods. Remove the supernatant and repeat the ether washing another two times to ensure a clean separation of peptide. It is recommended to retain each wash solution until one is confident the peptide has been correctly isolated.

3. If the starting material contains a methyl ester (OMe) modification to the aspartic acid side chain (to reduce the hydrophilicity of the peptide and thus enhance cell permeability), this can be cleaved to some degree during the de-protection reaction.
4. DMF has a very low vapor pressure and is therefore difficult to remove by centrifugal evaporation. It can be removed more efficiently by addition of acetonitrile and H_2O prior to freeze-drying.
5. While maximum absorbance of the fluorophore occurs near 750 nm, DY-750 has a modest OD at 500 nm that can be monitored. This avoids overloading the detector when concentrated crude solutions are loaded on the column. If the signal is undetectable at 500 nm, increase the detector wavelength to 650 nm or higher.
6. Combine all fractions containing the desired product. The molecular weight of the methylated (non-methylated) form of NIRD is 1028.3 (1014.3). The molecular weight of the hydrolyzed (carboxylate) form of free DY-750 is 713.0. Fractions containing a mix of different peptide species may need to be repurified. Unbound DY-750 will be hydrolyzed and needs to be separated from the desired peptide species as the determination of NIRD concentration relies on the spectral characteristics of the attached DY-750 fluorophore.
7. The extinction coefficient $\varepsilon_{747\ \mathrm{nm}} = 270{,}000\ \mathrm{M^{-1}cm^{-1}}$ corresponds to the value provided by the supplier (http://www.dyomics.com/en/products/nir-excitation/dy-750.html).
8. It is recommended to aliquot slightly more NIRD than required in order to allow for the unusable dead volume in the syringe used for administration of the probe.
9. NIRD dose should be optimized by the user for their specific end application. A reasonable starting value of 10 nmole per mouse may be trialled.
10. All animal procedures described in this protocol conform to the National Health and Medical Research Council of Australia guidelines and were approved by the University of Melbourne Animal Experimentation Ethics Committee.
11. Do not add a 10% v/v DMSO solution directly to the dry portion. Fully solubilize in DMSO first, and then add PBS drop-wise, with thorough mixing.
12. Due to lower solubility of DY750-VAD(OMe)-fmk in aqueous media, it may be preferable to utilize the non-methylated DY750-VAD(OH)-fmk product for in vivo NIR imaging experiments. In our experience, DY750-VAD(OH)-fmk remains sufficiently hydrophobic for cell and blood-brain

barrier permeability. However, the choice will depend upon the desired concentration of NIRD to be administered.

13. It is important that the NIRD is fully solubilized prior to tail vein injection. If not, consider administering a lower dose (*see* also **Notes 11** and **12**).
14. Washout time may need to be optimized depending on the application. This may be empirically estimated by repeated imaging at various times after NIRD administration (but *see* **Note 19**).
15. Care must be taken when removing hair from the top of the head. NIRD may bind to inflammatory caspases activated by injury of the skin by prolonged exposure to the hair removal cream, which will interfere with the signal from deep tissue within the brain. If this is a concern, the hair removal can be performed prior to the day of NIR imaging.
16. Set the image display color to "Positive Gray" and ensure the "Invert pixels" check box is unchecked.
17. The images can be exported into a nonproprietary format (e.g., tiff) using the Multi Gauge software.
18. To ensure maximum image contrast, the exposure time should be chosen such that most of the 16-bit dynamic range of the CCD camera is utilized when imaging mice from which the highest fluorescent signal is expected.
19. Earlier models of the LAS-4000 imaging system were sold with an optional "In vivo Assist" kit, comprising a temperature-controlled anesthesia manifold that enables the simultaneous imaging of three mice. The LAS-3000 and LAS-4000 models not equipped with the In vivo Assist kit are not heated; therefore, imaging should be performed as rapidly as possible and the animal immediately returned to a warm environment to prevent loss of body temperature and failure to recover from anesthetic. If imaging problems are encountered, the animal should be kept warm until rectified.

Acknowledgments

VAL, CLH, and SCD were funded by project grant (APP1044264) administered by the National Health and Medical Research Council (NHMRC) of Australia, and a grant from VCF-George Perry Fund, the Arthur and Mary Osborn Charitable Trust, and the William Buckland Foundation, managed by ANZ Trustees. SCD was supported by a Fellowship (FT110100199) administered by the Australian Research Council.

References

1. Saidlitz P, Voisin T, Vellas B et al (2014) Amyloid imaging in Alzheimer's disease: a literature review. J Nutr Health Aging 18:723–740
2. Jamieson E, Jeffrey M, Ironside JW, Fraser JR (2001) Activation of Fas and caspase 3 precedes PrP accumulation in 87V scrapie. Neuroreport 12:3567–3572
3. de Calignon A, Fox L, Pitstick R et al (2010) Caspase activation precedes and leads to tangles. Nature 464:1201–1204
4. Lawson VA, Haigh CL, Roberts B et al (2010) Near infra-red fluorescence imaging of apoptotic neuronal cell death in a live animal model of prion disease. ACS Chem Neurosci 1:720–727
5. Alnemri ES, Livingston DJ, Nicholson DW et al (1996) Human ICE/CED-3 protease nomenclature. Cell 87:171
6. Liberski PP, Sikorska B, Bratosiewicz-Wasik J et al (2004) Neuronal cell death in transmissible spongiform encephalopathies (prion diseases) revisited: from apoptosis to autophagy. Int J Biochem Cell Biol 36:2473–2490
7. Giese A, Groschup MH, Hess B, Kretzschmar HA (1995) Neuronal cell death in scrapie-infected mice is due to apoptosis. Brain Pathol 5:213–221
8. Jamieson E, Jeffrey M, Ironside JW, Fraser JR (2001) Apoptosis and dendritic dysfunction precede prion protein accumulation in 87V scrapie. Neuroreport 12:2147–2153
9. Lucassen PJ, Williams A, Chung WC, Fraser H (1995) Detection of apoptosis in murine scrapie. Neurosci Lett 198:185–188
10. Jesionek-Kupnicka D, Buczynski J, Kordek R et al (1997) Programmed cell death (apoptosis) in Alzheimer's disease and Creutzfeldt-Jakob disease. Folia Neuropathol 35:233–235
11. Puig B, Ferrer I (2001) Cell death signaling in the cerebellum in Creutzfeldt-Jakob disease. Acta Neuropathol 102:207–215
12. Hetz C, Russelakis-Carneiro M, Maundrell K et al (2003) Caspase-12 and endoplasmic reticulum stress mediate neurotoxicity of pathological prion protein. EMBO J 22:5435–5445
13. Haigh CL, Lawson VA, Drew SC (2014) Blood vessel cell death during prion disease: implications for disease management and infection control. Exp Hematol 42:939–940
14. Drew SC, Haigh CL, Klemm HMJ et al (2011) Optical imaging of apoptosis in the brain and peripheral organs of prion-infected mice. J Neuropathol Exp Neurol 11:143–150
15. Talanian RV, Quinlan C, Trautz S et al (1997) Substrate specificities of caspase family proteases. J Biol Chem 272:9677–9682
16. Van Noorden CJ (2001) The history of Z-VAD-FMK, a tool for understanding the significance of caspase inhibition. Acta Histochem 103:241–251
17. Hawrysz DJ, Sevick-Muraca EM (2000) Developments toward diagnostic breast cancer imaging using near-infrared optical measurements and fluorescent contrast agents. Neoplasia 2:388–417
18. Mahmood U, Weissleder R (2003) Near-infrared optical imaging of proteases in cancer. Mol Cancer Ther 2:489–496
19. Jung T, Bader N, Grune T (2007) Lipofuscin: formation, distribution, and metabolic consequences. Ann N Y Acad Sci 1119:97–111
20. Schnell SA, Staines WA, Wessendorf MW (1999) Reduction of lipofuscin-like autofluorescence in fluorescently labeled tissue. J Histochem Cytochem 47:719–730
21. Weissleder R, Pittet M (2008) Imaging in the era of molecular oncology. Nature 452:580–589
22. Raymond SB, Skoch J, Hills ID et al (2008) Smart optical probes for near-infrared fluorescence imaging of Alzheimer's disease pathology. Eur J Nucl Med Mol Imaging 35(Suppl 1):S93–S98
23. Ran C, Xu X, Raymond S et al (2009) Design, synthesis, and testing of difluoroboron-derivatized curcumins as near-infrared probes for in vivo detection of amyloid-β deposits. J Am Chem Soc 131:15257–15261
24. Mattson MP, Duan W (1999) "Apoptotic" biochemical cascades in synaptic compartments: roles in adaptive plasticity and neurodegenerative disorders. J Neurosci Res 58:152–166

Chapter 18

Strain Typing of Prion Diseases Using In Vivo Mouse Models

Aileen Boyle, Kris Hogan, Jean C. Manson, and Abigail B. Diack

Abstract

Transmissible spongiform encephalopathies (TSE) or prion diseases exhibit strain variation, a phenomenon that has been studied extensively in mouse bioassays. Despite the introduction of many rapid in vitro systems, bioassays remain a key tool in defining prion strains and their ability to transmit disease in vivo. Prion strains can be characterized by a range of phenotypic characteristics such as incubation period, vacuolar pathology, and distribution of the abnormal form of PrP following experimental transmission of the agent into a panel of mice (transgenic or wild type). Interpretation of these characteristics requires considerable experience and an understanding of the procedures used to define them. This chapter reviews the techniques used in strain typing of prion diseases from inoculum preparation and pathological studies to data interpretation alongside an extensive troubleshooting guide.

Key words Transmissible spongiform encephalopathy (TSE), Prion diseases, Strain typing, Bioassay, Histopathology, Vacuolation profile

1 Introduction

Transmissible spongiform encephalopathies (TSE) or prion diseases are a group of fatal neurodegenerative diseases which affect animals and humans. They include scrapie in sheep and goats, bovine spongiform encephalopathy (BSE) in cattle, chronic wasting disease (CWD) in deer, and Creutzfeldt-Jakob disease (CJD) in humans. A key event in all prion diseases is the abnormal misfolding of the endogenous prion protein PrP^{C} to an oligomeric, β-sheet rich form, PrP^{Sc} [1]. Although they share many characteristics such as spongiform vacuolation, neuronal loss, deposition of PrP^{Sc}, and gliosis, the disease phenotype can vary greatly between the different prion diseases due to differences in the prion strain infecting the host [2, 3].

Prion strains are traditionally characterized by their biological properties following their passage to panels of mice. These properties include the incubation time of each mouse line, the order in which each strain of mouse dies of disease (incubation period ranking), and the distribution of vacuolation and PrP^{Sc} in the

Victoria A. Lawson (ed.), *Prions: Methods and Protocols*, Methods in Molecular Biology, vol. 1658,
DOI 10.1007/978-1-4939-7244-9_18, © Springer Science+Business Media LLC 2017

CNS [4, 5]. Together these biological properties provide a "signature" of the prion agent. Although in vitro assays also provide important information regarding TSE strains, bioassay remains the gold standard approach to defining prion strains and determining whether an agent is infectious.

Traditionally strain typing has been carried out in wild-type mouse panels [2, 5]; however some agents do not readily transmit to wild-type mice, i.e., sCJD, GSS, and, in these cases, transgenic mouse panels have proved to be useful. The first transgenic mouse available for TSE research was a transgenic mouse which overexpressed hamster PrP via random insertion of the hamster *Prnp* gene into the murine genome [6]. This approach has been used to produce a multitude of models in which PrP from different species is expressed ectopically at varying levels giving rise to different phenotypes and making direct comparisons between mouse lines difficult. Alternatively, the gene targeting approach in which homologous recombination is used to alter the endogenous gene gives rise to mice which are genetically identical apart from the replaced PrP gene and which have normal expression patterns allowing comparison between them and wild-type control lines [7, 8]. Increasingly, transgenic mouse models expressing human, bovine, ovine, and cervid PrP are being used for strain typing [8–10]. By overcoming the species barrier and allowing not only strain characterization but generation of specific species-adapted strains, they can provide further information on the prion strain. In addition to transgenic models, species such as the bank vole [11, 12] and hamsters [13] may also be used for strain typing by adapting the techniques used in mice.

In traditional wild-type strain typing panels, the inbred lines used are RIII, C57BL, and VM. This combination includes two different *Prnp* genotypes *Prnp*-a (Leu-108, Thr-189) (RIII, C57BL) and *Prnp*-b (Phe 108, Val-189) (VM) [14]. Genetic analysis has revealed that the *Prnp* gene is a major factor in determining disease phenotypes with a single TSE agent resulting in two distinct mouse-passaged strains dependent upon the murine genotype as typified by BSE giving rise to the 301C and 301V strains [3, 15, 16].

Strain typing involves serial passage of an isolate through the RIII or C57BL and VM strains until the strain has stabilized, characterized by shortened and stabilized incubation period (IP) [17]. This can be a lengthy and time-consuming process as primary passage generally involves a long IP and low attack rate due to the species barrier. However a single passage can be adequate to identify major differences between strains if comparative data is available and there are distinctive characteristics. The same approach and techniques are used for transgenic mice; however, we will refer to wild-type mice throughout this chapter unless there are specific differences.

This chapter will review the different techniques used in the strain typing of prion diseases from preparation of the inoculum to the analysis of brain material and how to define a strain.

2 Materials

Prepare all solutions using deionized water unless otherwise stated. All solutions and reagents are prepared and stored at room temperature unless otherwise stated. The appropriate local waste disposal regulations should be followed. The use of personal protective equipment (PPE) is recommended and should follow local regulations.

2.1 Preparation of Inocula

1. 99% ethanol.
2. Sodium chloride for injection 0.9% (Henry Schein Animal Health).
3. Torsion or electronic balance.
4. Glass or Teflon tissue grinder (Griffith's tube tissue grinder).
5. Syringes and needles (23G, 26G).
6. Sterile tubes (screw top).
7. Microbiological Class II hood.
8. Sterile glass bijou.

2.2 Inoculation

1. Anesthesia rig.
2. Isoflurane (inhalant anesthetic).
3. 1 mL syringe and needles (26G 10 mm).

2.3 Collection of Tissues

1. Dissection instruments.
2. Formal saline.
3. Liquid nitrogen.
4. Sample tubes.

2.4 Preparation of Samples

1. Pots of tissues fixed in 10% formal saline.
2. 98% formic acid.
3. 10% formal saline fixative.
4. Waste container to collect formic acid and formal saline.
5. Disposable "trimming boards" prepared using tin foil and a paper towel.
6. Polythene bag for collection of empty polypots.
7. Processing cassettes pre-labeled with coded mouse number.

8. Separate holding containers of formal saline for formic and non-formic-treated tissues.
9. Disposable trimming blades.
10. 70, 90, 95, and 99% IMS (industrial methylated spirits).
11. Xylene.
12. Paraffin wax.
13. Glass slides.

2.5 Hematoxylin and Eosin Staining

1. Racks of slides.
2. Xylene.
3. 70, 95, and 99% IMS.
4. Harris hematoxylin solution.
5. Eosin.
6. Scott's Tap Water Substitute: 8.75 g $NaHCO_3$, 50 g $MgSO_4$ in 2.5 L tap water.
7. Mounting medium.
8. Coverslips.

2.6 Lesion Profiling

1. Light microscope.
2. x4 objective.

2.7 Immunohistochemistry

1. Xylene.
2. 70, 95, and 99% IMS.
3. Autoclave (programmable to 121 °C).
4. 98% formic acid.
5. Hydrogen peroxide.
6. Methanol.
7. TBS/BSA buffer: 0.01 M Tris, 0.15 M NaCl, pH 7.6; add BSA at 1 g/L.
8. Normal blocking serum: normal serum from the species the secondary antibody was raised in, diluted 1 in 20 in TBS/BSA.
9. Secondary biotinylated antibody (anti-mouse or anti-rabbit IgG) dependent upon the species the primary antibody was raised in.
10. Avidin/biotin complex (ABC): Vectastain® Elite® ABC system (*see* **Note 22**).
11. DAB chromogen: 1 g 3,3-diaminobenzidine tetrahydrochloride hydrate in 40 mL water. Filter before storage at −20 °C. Before use, add 2 mL to 200 mL PBS. Filter and add 100 μL hydrogen peroxide.
12. Hematoxylin Z stain solution.

13. Scott's Tap Water Substitute: dissolve 3.5 g $NaHCO_3$ and 20 g $MgSO_4$ substitute in 1 L tap water.
14. Pertex mounting medium.
15. Coverslips.
16. Primary antibody: 6H4 (Prionics).

3 Methods

Carry out all procedures at room temperature unless otherwise stated.

3.1 Safety Measures

Prion-infected material is categorized as either Hazard Group 2 or Hazard Group 3 material and requires appropriate handling. Transmission and laboratory studies may expose workers to hazardous procedures, and strict adherence to safety measures is required. Transmission and laboratory studies should only be undertaken in laboratories and animal facilities with the appropriate level of microbiological containment. In the UK, all safety measures for working with prion-infected material are documented by the Advisory Committee on Dangerous Pathogens (ACDP) (https://www.gov.uk/government/groups/advisory-committee-on-dangerous-pathogens) [18].

3.2 Preparation of Inocula

Prions are resistant to most physical and chemical treatments [19]. Infectivity can persist on surfaces and instruments even after thorough decontamination in some cases. In order to prepare inocula, standard sterile procedures are required, procedures to avoid cross contamination should be employed, and the use of disposable labware is recommended. Due to the nature of samples inoculated and the risks of bacterial contamination, measures should be taken to avoid death of the mice due to contaminated inocula (*see* **Note 1**). All procedures should be carried out within a microbiological Class II hood.

1. Cover tissue in 99% ethanol. Agitate and pour off ethanol. Repeat (*see* **Note 2**).
2. Cover tissue with sterile saline. Agitate and pour off saline. Repeat.
3. Weigh sample using torsion or electronic balance.
4. Using weight of tissue, calculate amount of saline to add to create inocula at required dilution (*see* **Note 3**).
5. Add saline to tissue in Griffith's tube and grind together until smooth homogenate is achieved.
6. Using a syringe fitted with a suitable gauge needle (*see* **Note 4**), transfer inoculum from grinder to sterile glass bijou. Pull

inoculum through a series of needles until it moves freely through the smallest gauge without resistance.

7. Inoculum is now ready for use or storage at −80 °C (*see* **Notes 1** and **5**).

3.3 Inoculation

Each animal in a study should have a unique identification this can be done through ear notches or the use of transponders. Individual animals should have a personal data file maintained throughout the study with any husbandry issues, notes, and clinical scoring recorded.

The route of inoculation is determined by the study protocol and will take into account the pathogenesis of disease according to route and the inoculum used. Prions can replicate both in the CNS and in lymphoid tissues, and the route of inoculation can influence clinical and pathological course of the disease.

Before inoculation, animals must be anesthetized appropriately for the route chosen (*see* **Note 6**). A common route of inoculation in our laboratory for strain typing is intracerebral (i.c.) (*see* **Note** 7). Inoculation can be performed freehand or using a stereotaxic rig if a specific area of the brain is being targeted. In our laboratory, 20 μL of inoculum is used for an i.c. inoculation. Mice of 6–8 weeks are preferred for inoculation due to the relative softness of the skulls; however this can vary with sex and size of the animal. Animals can be i.c. inoculated using a 1 mL syringe and 26G 10 mm needle. Intraperitoneal (i.p.) inoculation is commonly used as a peripheral route either alone or in conjunction with an i.c. route (*see* **Note 6**). 100 μL of inoculum is administered i.p. using a 1 mL syringe and 26G 10 mm needle through the lower left or lower right quadrant of the abdomen avoiding all vital organs.

Following inoculation, animals should be monitored until they are fully recovered from the anesthesia (*see* **Note** 7).

3.4 Clinical Scoring

Prion diseases manifest with behavioral changes and clinical signs of disease. Clinical scoring used in conjunction with pathological examination can help build a complete picture of infection. Clinical scoring should be carried out by an experienced technician from approximately 50 days (d) before the expected clinical onset. If clinical onset is unknown, 50 days post inoculation (dpi) is advised; however in the case of transgenic overexpression models, weekly assessments from inoculation may be advisable. Mice are clinically assessed on a weekly basis without knowledge of previous scores to ensure lack of bias.

Early clinical signs may be nonspecific, dull ruffled coats, and hyperactive behavior. As the disease progresses, clinical signs may include ataxia, kyphosis, rapid weight loss, piloerection, and excessive urination (*see* **Note 8**). Clinical signs vary according to prion strain/mouse line combinations (*see* **Note 8**) and technicians

Table 1
Clinical assessment of prion disease in mice. Technicians should perform scoring on coded mice and with no knowledge of previous scores. These scores may also be accompanied by notes describing novel phenotypes or welfare issues such as scratched eyes, broken skin, etc.

Clinical score	Description
1	Normal mouse – No evidence of disease
2	Early clinical disease symptoms
g	Early symptoms accompanied by alteration of gait
+	Terminal/end point of disease
G	Terminal/end point of disease accompanied by severe deterioration of gait

should also note any novel behavioral phenotypes. In our laboratory, clinical scores are assigned (Table 1) depending on the varying physical and social attributes.

Clinical scoring allows disease progress to be ascertained and establishment of a humane end point and ensures animal welfare standards are maintained in line with the UK Home Office Project Licence severity limits (UK only) or the appropriate legislation for the country of residence.

3.5 Collection of Tissues

Euthanasia of the animals should be carried out using appropriate techniques (*see* **Note 9**). All samples should be labeled clearly and with mouse identification numbers, tissue, and hazard group category. Aseptic technique may be required depending on further use of tissues, e.g., potential for use in transmission studies (*see* **Note 10**). To ensure maximum use of brain material, brains are sagittally cut (*see* **Note 11**). One third should be placed into liquid nitrogen for biochemical analysis or future inoculation. The remaining two thirds should be placed into formal saline for a minimum of 48 h (h). Hazard Group 3 tissues should also undergo treatment with 98% formic acid as per ACDP recommendations (*see* Preparation of Samples and **Note 13**). Other tissues can be collected as required. More detailed morphological examinations of tissue may require perfusion (*see* **Note 12**).

3.6 Preparation of Samples

The aim of tissue processing is to preserve the tissue and its components in a solid medium that is firm enough to support the tissue and allow the preparation of thin, stained sections. The most widely used medium for routine histology is paraffin wax. Tissue processing involves three main steps: dehydration through increasing concentrations of alcohol, clearing in hydrocarbon solvents which are miscible with both the dehydrating agents and the paraffin wax, and embedding in paraffin wax. These steps can be performed by hand or more commonly and efficiently by automated tissue processors.

In order to guarantee that the critical areas of brain tissue are present on the sections, trimming of the brain is required. Rodent brains are trimmed into four specific pieces to allow scoring of vacuolation in very specific areas of the brain from which a lesion profile is generated (Fig. 1). This also allows comparative immunohistochemistry to be carried out.

1. Remove formal saline and replace with approximately 5 mL of 98% formic acid and leave for 1.5 h (*see* **Note 13**).
2. Remove formic acid and replace with approximately 10 mL of 10% formal saline for 1–3 h and then trim tissues as appropriate.
3. Decant fixative into a waste container.
4. Trim the brain into four specific slices as shown in Fig. 1.

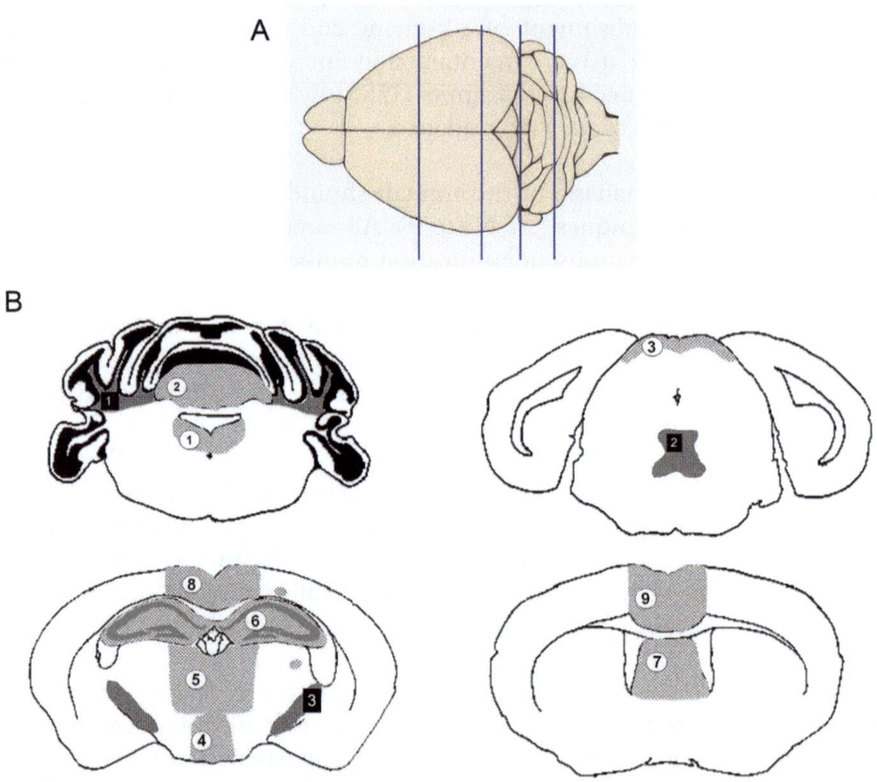

Fig. 1 Schematic of the brain showing standard cutting levels and coronal sections for lesion profiling. (**a**) Standard cutting levels of the brain. (**b**) Coronal brain sections in which the standard scoring regions are shown. *Gray* matter areas (represented in *circle* and *gray*): 1, medulla; 2, cerebellum; 3, superior colliculus; 4, hypothalamus; 5, thalamus; 6, hippocampus; 7, septum; 8, retrospinal cortex; and 9, cingulate and motor cortex. *White* matter areas (represented in *box* and *dark gray*): 1, cerebellar peduncle (CP); 2, decussation of superior CP; 3, basal CP

5. Place in a pre-labeled processing cassette with the printed label. An "extra" piece may be taken from the cervical spinal cord if present.
6. Store the cassettes in 10% formal saline according to hazard group category.
7. Wash all tissues that have been treated with formic acid in running tap water for approximately 1.5–2 h prior to processing.
8. Paraffin embedding schedule, performed by hand or by automatic tissue processor:
   (a) 10% formal saline, one change, 5 min.
   (b) 70% IMS, one change, 40 min.
   (c) 90% IMS, one change, 40 min.
   (d) 95% IMS, one change, 40 min.
   (e) 99% IMS, three changes, 2 × 40 min followed by 1 × 30 min.
   (f) Xylene, three changes, 30 min each.
   (g) Paraffin wax at 61 °C, three changes, 25 min each.
9. Embed tissues in paraffin blocks. Orientate tissues in mold and fill with molten wax.
10. Cut sections at desired thickness (normally 6 μm) at the standard coronal levels and mount onto glass slides.
11. Leave "upright" overnight at room temperature.
12. Incubate at 35–45 °C for a minimum of two nights.

3.7 Hematoxylin and Eosin Staining

The normal staining method used to demonstrate general morphology of tissues is the hematoxylin and eosin (H&E) method. Brain sections stained by this method are required to perform lesion profiling. This procedure can be carried out by hand or using an autostainer.

1. Deparaffinize in three changes of xylene for 5 min each. Hydrate in two changes of 99% IMS for 2 min each, 95% IMS for 1 min. Then rinse in running water for 5 min.
2. Stain in hematoxylin solution for 3 min.
3. Wash in running water for 6 min.
4. Wash in Scott's Tap Water Solution for 2 min.
5. Wash in running water for 5 min.
6. Stain in eosin for 1.5 min.
7. Wash in running water for 30 s.
8. Dehydrate through one change of 70% IMS for 30 s, two changes of 95% IMS for 30 s each, two changes of 99% IMS

for 1 min each, one change of equal proportions xylene/99% IMS for 1 min, and three changes of xylene for 1 min each.

9. Mount and coverslip.

3.8 Lesion Profiling

Lesion profiling is a semiquantitative method of assessing TSE-specific vacuolation in defined regions of the brain [5]. The distribution and severity of vacuolation across the brain is dependent on both prion strain and the host [2]. This allows the construction of a characteristic "lesion profile" for each prion strain/mouse line combination.

Scoring is performed on H&E sections using nine gray matter areas and three white matter areas throughout the brain (Fig. 1) using a x4 objective on a light microscope.

The nine regions of gray matter are:

(1) Medulla, dorsal half of medulla, principally the medial vestibular and repositus nuclei at the level of the cochlear; (2) cerebellum, cerebellar cortex of the folia adjacent to the fourth ventricle; (3) superior colliculus; (4) hypothalamus; (5) medial thalamus; (6) hippocampus; (7) septum, septal nuclei of paraterminal body; (8) cingulate cortex at level of 4, 5, and 6; and (9) motor cortex at level of 7.

White matter regions are:

(1) Cerebellar white matter at the same level as gray matter 1, (2) tegmentum/superior cerebral peduncle at same level as gray matter 3, and (3) pyramidal tract at same level as gray matter 4, 5, 6, and 8.

These are scored as a numerical value (0–5 for gray matter, 0–3 for white (Table 2, Fig. 2; *see* **Notes 14** and **15**)) to represent the degree of vacuolation in that area [5]. Using the mean of these scores, a lesion profile is generated (Fig. 3). The minimum number of mouse brains required to produce a reliable lesion profile is six clinically positive animals. In studies where no clinical signs are observed but vacuolation is present, lesion profiles can still be generated, but it is good practice to state whether this is the case alongside group size.

The presence of vacuolation not related to TSE infection must also be taken into account during examination. Vacuolation outside the scoring regions, aging vacuolation (*see* **Note 15**), and pathological features such as amyloid plaques should also be noted (*see* **Note 16**).

3.9 Incubation Period

The period of time between inoculation and death of an animal can be described in two ways: incubation period or survival period. Incubation period (IP) refers to an animal in which both clinical signs are present and TSE-associated vacuolation is present in the brain. Survival period(s) is used to refer to animals in which no clinical signs and/or vacuolation is evident. When preparing

Table 2
Vacuolation scoring of prion disease in mice. Technicians should perform scoring on coded samples and with no knowledge of TSE agents to allow objective scoring. Pathology outside the scoring region, non-TSE-related vacuolation, and any non-vacuolar pathology should also be noted

Score	Description
Gray matter: 0	No TSE vacuolation or signs of TSE in that area
1	An occasional hole possibly indicative of aging or TSE vacuolation, but diagnosis inconclusive
2	Definite signs of TSE vacuolation, i.e., oval-shaped holes and clumping, *but* no even spread of TSE vacuolation
3	Even spread of TSE vacuolation throughout the area
4	Heavier, even spread of TSE vacuolation throughout the area and the vacuoles are beginning to merge forming large vacuoles
5	Very heavy even spread of TSE vacuolation throughout the area and the vacuoles have frequently merged to make large vacuoles. Lace-like appearance
White matter: 0	No TSE vacuolation or signs of TSE in that area
1	Definite signs of vacuolation in white matter
2	Even spread of vacuolation throughout the white matter
3	Very heavy even spread of vacuolation throughout the white matter with vacuolation joining up to make bigger vacuoles

incubation graphs, the mean value (in dpi) $\pm$ SEM are plotted for each mouse line against inocula. This allows both length of IP and the order in which each the mouse line succumb (ranking of incubation time) to be compared as shown in Fig. 4. It is good practice to define the categories of animals that are included in incubation or survival graphs. Some researchers may wish to use the presence of PrP deposition as an alternative (or additional) to vacuolation in incubation graphs; however this should be noted in the figure legend.

3.10 Immunohistochemistry

Immunohistochemistry (IHC) utilizes the specific binding between the antigen of interest in the tissue sections and an antibody. Binding is identified and enhanced by attaching an avidin/biotin complex in conjunction with a chromogen which allows visualization. Immunohistochemistry is routinely used as a diagnostic technique in humans and in large animals for the detection of PrP^{Sc} in the brain and other organs. Proteinase K treatment is not routinely used in mouse studies (dependent upon exact protocol; *see* **Note 21**), and although pretreatments such as autoclaving and formic acid are presumed to denature PrP^{C}, appropriate controls

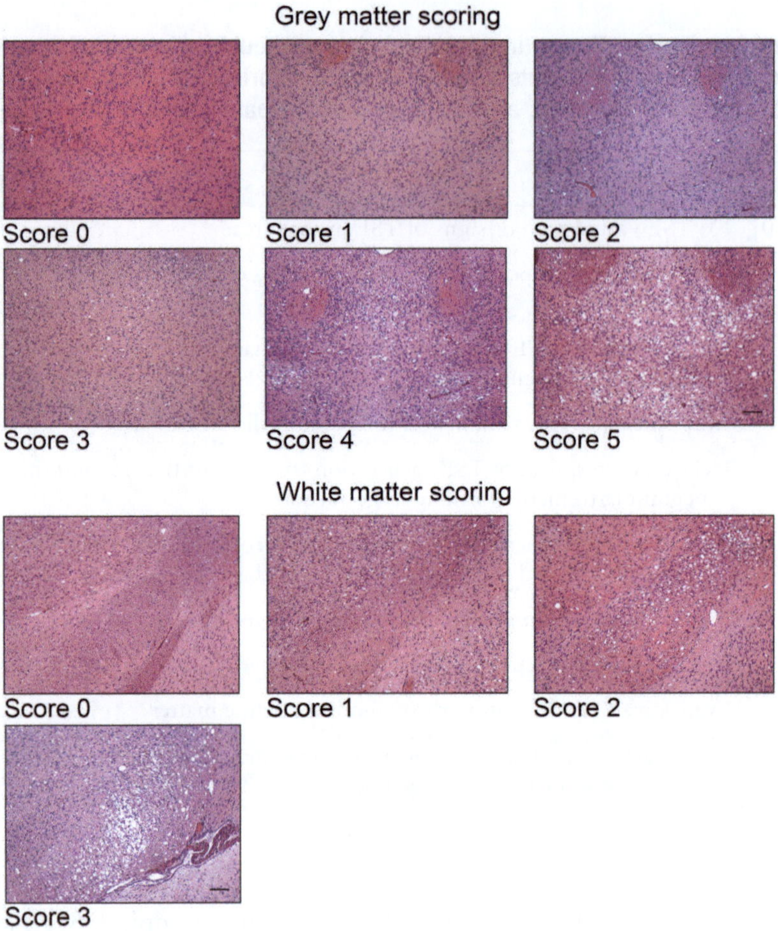

Fig. 2 Representative images of vacuolation scores in *gray* and *white* matter brain regions

should be included (*see* **Note 17**). The choice of antibody should be made dependent upon the PrP epitope of interest and species from which the inoculums is derived (*see* **Note 18**).

In mice, IHC reveals distinct distribution patterns of PrP^{Sc} staining dependent upon the mouse line/prion agent combination. PrP^{Sc} distribution can be described in a combination of ways: location within the brain and type of deposition, and qualitative or quantitative scoring using programs such as ImageJ. Deposition types include but are not limited to granular, punctate, aggregates, plaques, and stellate [20]. A comprehensive description of the PrP^{Sc} staining can allow strain characterization both between inocula, where vacuolation is not present or where numbers are too low for lesion profiling to be used. Additionally, although less common, IHC can be carried out on sagittal sections if required.

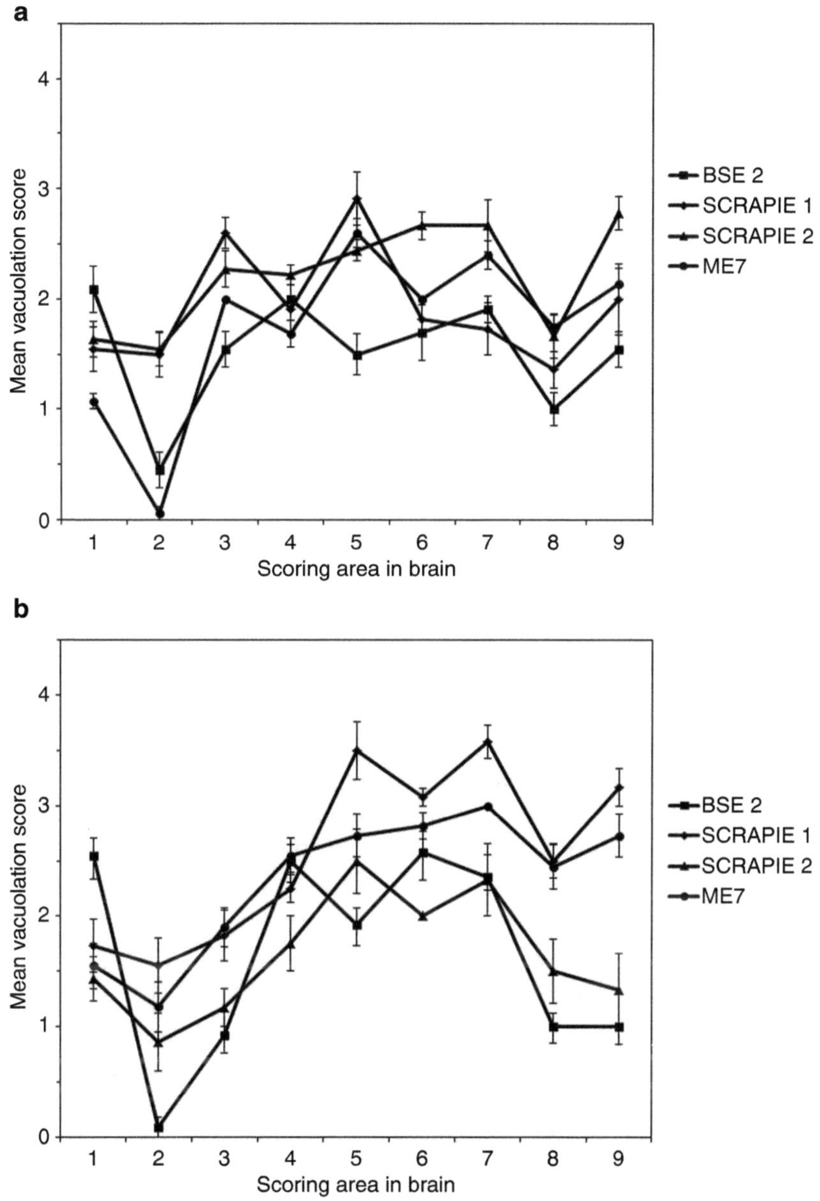

Fig. 3 Lesion profile of BSE and scrapie isolated in C57BL in (**a**) C57BL and (**b**) RIII mice (±SEM). ME7 is included as a reference strain. Note that in this example, lesion profile alone could not be used to distinguish the strain of disease

1. Dewax the sections in four changes of xylene, 2 min each. Hydrate in one change of equal proportions of xylene/99% ethanol for 2 min, followed by three changes of 99% ethanol, two changes of 95% ethanol, and one change of 70% alcohol for 2 min each. Rinse in running water for 5 min. Agitate in each reagent and allow to drain between reagents.

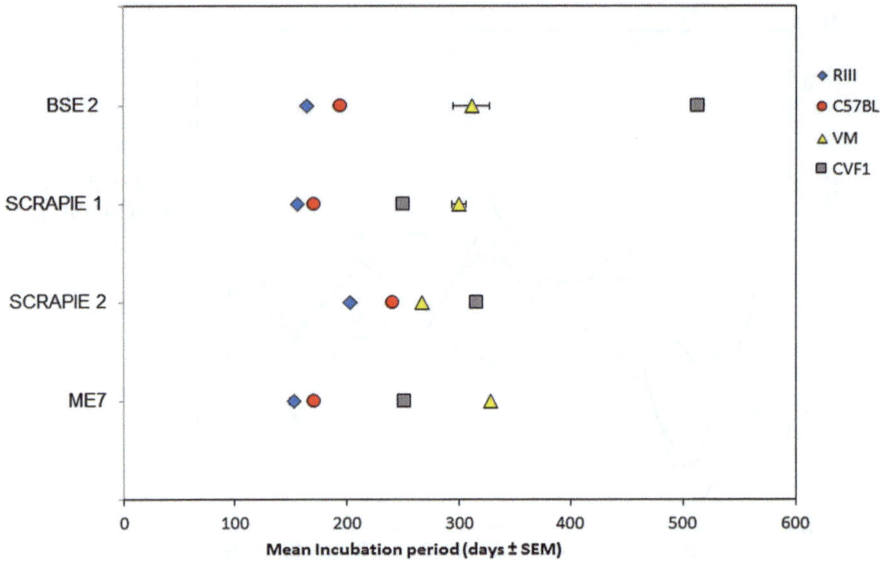

Fig. 4 Incubation period of BSE and scrapie isolated in C57BL mice in C57BL, RIII, VM, and CVF1 mice (±SEM). In this example, when combined with the lesion profile data in Fig. 3, incubation period ranking would be used to distinguish the strain of disease

2. Antigen retrieval (**step 1**): Immerse sections in distilled water in an uncovered container (metal or glass) (*see* **Note 19**). Autoclave at a holding temperature of 121 °C for 15 min. Once the autoclave cycle has completed and is safe to open, remove container and place in slowly running water to cool to room temperature. Once cool, replace distilled water with fresh distilled or tap water.
3. Antigen retrieval (**step 2**): Immerse the sections in 98% formic acid for 5 min.
4. Immerse sections in tap water. Tip out tap water and place sections in slowly running tap water for 10 min.
5. Peroxidase blocking: To reduce nonspecific background labeling, endogenous peroxidase activity is prevented by immersing the sections in fresh 1% hydrogen peroxidase in methanol for 10 min (*see* **Note 20**).
6. Immerse sections in slowly running tap water for 5 min (*see* **Note 21**). Follow by three washes in TBS/BSA for 5 min each.
7. Normal serum block: To block nonspecific binding, sections can be incubated with appropriate serum for 20 min.
8. Remove normal serum.
9. Primary antibody: Incubate sections with primary antibody diluted in TBS/BSA overnight (*see* **Note 18**).

10. Wash in TBS/BSA for 3 × 5 min.
11. Secondary antibody: Incubate sections with the appropriate diluted biotinylated secondary antibody for 1 h.
12. Wash in TBS/BSA for 3 × 5 min.
13. Detection: Incubate sections in avidin/biotin complex (ABC) for 30 min (*see* **Note 22**).
14. Wash in TBS/BSA for 3 × 5 min.
15. Chromogen: Add DAB substrate and allow to develop for approximately 5 min. This time is variable depending upon the substrate, and it is recommended that the optimal time is determined for each protocol. Check that staining is sufficient before continuing to next step.
16. Rinse sections in slowly running water for 1 min.
17. Counterstain: Stain the nuclei by immersing sections in hematoxylin for 30 s, and rinse in running tap water for 1 min. Immerse in Scott's Tap Water for 1 min, and rinse in running tap water for 1 min.
18. Dehydrate: Immerse sections in two changes of 70% ethanol, two changes of 95% ethanol, three changes of 99% ethanol, one change of equal proportions 99% ethanol/xylene, and four changes of xylene for 30 s each. Agitate in each reagent and allow to drain between reagents.
19. Mount the slides with mounting medium.

3.11 Defining a Strain

Strain typing of prion diseases relies on a combination of the above techniques, incubation period and ranking (order in which the mouse lines succumb to disease), lesion profile, and PrP^{Sc} deposition. The use of one technique is not sufficient to identify a prion strain or isolate as some strains may show similar lesion profiles or incubation period rankings to others as demonstrated in Fig. 3 and Fig. 4. The number of animals used must be large enough to gain a true representation, i.e., a minimum of six per group. Once the data has been compiled, it is necessary to compare it with previous transmissions and/or published data, taking into account differences in inoculation route, titer, and procedures. In particular titer of the inoculum can affect IP where a higher titer can give a shorter IP but no change in ranking.

In few instances a single primary passage will be sufficient to identify a prion strain if there are distinctive characteristics such as a particular lesion profile pattern or IP ranking, i.e., BSE-related strains show a specific lesion profile in RIII mice; however if any differences are noted in incubation ranking, vacuolation distribution, or PrP^{Sc} distribution patterns, it is advisable to carry out a second passage into panels of mice following the same techniques. The majority of strains and isolates will require two or more passages until the strain becomes stabilized, i.e., IP and lesion profiles remain constant. Upon subsequent passages, transmission tends to be more efficient with shorter incubation periods. Once lesion profiles and IP have stabilized, the prion strain can be fully defined.

4 Notes

1. To prevent illness and death of the mice due to bacterial contamination of inoculum, primary tissues, i.e., human, bovine, and ovine, should be checked for bacterial contamination after homogenization, and prior to inoculation. This is particularly important for field samples. Drop 0.05 mL of inoculum into the center of two blood agar plates, and spread inocula to cover the surface area of each plate using a sterilized spreader. Cover plates and place one plate into an anaerobic jar or bag and then place both anaerobic and aerobic plates into a 37 °C incubator for 48 h. Monitor plates for growth and record results. If visible growth is apparent, discussion with an animal unit manager and/or vet is required. In cases of low growth (<10 small diffuse colonies across the plate), inoculation can take place following administration of antibiotic (20 μL subcutaneous injection of Synulox (Pfizer)) to the animals; however this should be subject to discussion with your animal unit manager and/or vet. It is also advisable to inoculate a small number of mice (4–6) and monitor for 1 week for adverse reactions before inoculating an entire cohort.
2. Washing in 99% ethanol will remove surface contamination and is recommended for primary tissues or non-sterile material.
3. To calculate the volume of saline (mL) required for 10% w/v dilution: (weight of tissue (mg)*0.9)/100.
4. Suitable gauge needles: 26G for mouse brains or 23G for other tissues.
5. Dilution of inoculum used can vary according to the study. In our laboratory, 10% w/v is generally used for primary transmissions; however it has been noted that some transgenic mouse strains show signs of toxicity and require a 1% w/v dilution. A 1% w/v dilution is used for serial passages. If further dilutions are required, dilution of the inoculum in sterile saline should occur prior to use to avoid degradation of the inocula.
6. In our laboratory we use general anesthesia for i.c. and i.c./i.p. combination route of inoculation. No anesthetic is used for i.p. inoculation alone. Advice from your animal unit manager and/or vet should be sought for alternative inoculation routes and should comply with United Kingdom Home Office Project Licence requirements' or the appropriate country's legislation.

7. Anesthesia is induced using a vaporized isoflurane and oxygen mix. Mice are placed in a chamber supplied with an O_2 and isoflurane mix at 3 L/min; once the mice are unconscious, this can be reduced to 1.5 L/min – unconsciousness should be confirmed by lack of pedal response to toe pinch. Recovery is usually achieved within 2–3 min; if recovery is taking longer, animals should be placed on heat mats and monitored until fully conscious.
8. Clinical symptoms can vary according to prion strain/mouse line combination. Table 3 gives representative clinical symptoms from a range of transmission studies from our laboratory.
9. Mice are routinely euthanized by cervical dislocation; death is confirmed prior to sample collection by disrupting a main blood vessel. Should a cardiac blood sample be required, mice are exposed to a rising concentration of carbon dioxide and a 21G needle attached to a 1 mL syringe is inserted below the sternum to obtain a sample directly from the heart.
10. The skull, after removing the fur, should be held by the orbital sockets using curved forceps and passed through a Bunsen flame to remove any surface contamination. Using sterile surgical scissors, cut around the skull to expose the brain. The brain is dropped into a sterile Petri dish after severing any remaining nerve and membrane tissue.
11. Place the brain on a Petri dish or board spinal cord end closest. Cut down the full length of the brain parallel to the midline 1 mm from the left side. The right side should be placed in formal saline and the left (non-inoculated side to prevent false-positives in biochemical analysis) into liquid nitrogen.
12. Animal(s) should be terminally anesthetized prior to removing the rib cage to expose the heart. A butterfly needle (with attached capillary tubing) should be inserted in the left ventricle; an incision should be made in the vena cava to allow flow through. Using a 20 mL syringe attached to the tubing, slowly infuse PBS until organs become pale and the runoff is clear. Fixative can be introduced by attaching another 20 mL syringe to the capillary tubing while keeping the needle in place.
13. 98% formic acid has been shown to significantly reduce the levels of TSE infectivity in formal saline-fixed tissues [21]. In our laboratory this is required for all Hazard Group 3 material, i.e., material originating from human or BSE agents or any TSE agent passaged through a transgenic mouse expressing human PrP.
14. If the scorer finds that the vacuolation falls between two scores, it is good practice to record the lower score. It is important to remember that the average score of the group represents the true vacuolation trend – not the individual mouse brain score.

Table 3
Clinical features of common prion strain/mouse line combinations. Description of representative clinical features of common prion strain transmissions in a range of mouse lines. Average incubation periods are also given, and clinical scoring should begin at least 50 d prior to estimated clinical onset

TSE strain	Mouse strain	Mean incubation (d)	Clinical symptoms
ME7	C57Bl6/J	160	Clinical symptoms appearing lasting 8–10 weeks before humane end point (HEP): Accumulation of white discharge around the eyes, lethargy/reluctance to move freely around cage, piloerection of coat, increasing socio-isolation, steady continuous weight loss, muscle ataxia leading to some deterioration of gait, wet anal/genital area, kyphosis
87V	VM	320	Clinical symptoms up to 14+ weeks before HEP: Lack of nest-building instinct (previously very strong), steady weight loss, accompanied deterioration of gait, lethargy, scratching/broken skin on neck/shoulder area, piloerection and loss of coat condition, excessive urination
ME7	RIII	360	Lethargy/unresponsiveness to stimuli preceded by a period of hyperactivity, loss of coat condition, wet genital/anal region, absence of nest building, excessive urination, muscle ataxia
GSS	101LL	110	Very short clinical phase: Rapid weight loss, deterioration of gait which can lead to complete paralysis of rear limbs, head tilt, occasionally wet genital/anal region, kyphosis
ME7	Tg20-multi PrP copy transgenic	75	Extremely short clinical phase requiring multiple welfare checks per day: Circling behavior tracking edge of cage, severe deterioration of gait leading to paralysis within short time span, wet genital/anal area observed in most, no loss of coat condition, kyphosis. Sudden death possible
vCJD	HuVV-human transgenic	400–650 days	Loss of gait, hyperactivity in early stage, cage circling, lack/slowness in motion, wet anal/genital area, kyphosis, "absent" look
BSE	B6-bovine transgenic	450	Lack/slowness in motion, wet anal/genital area, "clasping" of rear limbs when lifted by the tail, loss of gait – Severe in some cases

Also it should never be relied upon as a sole method for strain identification.

15. Some mouse strains exhibit severe white matter vacuolation which may or may not be due to a TSE, e.g., due to aging. For this reason white matter vacuolation should not be relied upon for a positive diagnosis in the absence of positive gray matter vacuolation.
16. In addition to the lesion profile, other associated neuropathology should be recorded. For example, the ME7 strain of scrapie usually demonstrates neuronal loss from pyramidal cell layer and gliosis in the hippocampus. Early vacuolation in the dorsal lateral geniculate is observed in some strains – an area not included in a lesion profile. Both primary- and mouse-passaged BSE strains exhibit vacuolation in the cochlear nucleus, and in VM mice, where incubation period of BSE is greatly reduced on subsequent passage, vacuolation is frequently observed in the CA2 region of the hippocampus. Amyloid plaques may also be present and are a characteristic of some prion strains, i.e., the 87 V strain of scrapie in VM mice. These pathologies help to build a complete picture of the strain and should not be ignored.
17. In IHC, positive controls (known infected mice) are required to test the reactivity of the antigen and antibody and to confirm the procedure has been performed correctly. Negative controls (noninfected mouse) allow assessment of nonspecific immunolabeling and effectiveness of pretreatments to remove PrP^C. Control sections in which the primary antibody was omitted and replaced with normal rabbit or mouse serum or buffer allow assessment of nonspecific binding of the secondary antibody. In addition, age-matched controls may also be necessary if transgenic mice are being utilized in order to determine if age has an effect on PrP expression.
18. There are numerous monoclonal and polyclonal anti-PrP antibodies available for use in IHC. These should be chosen based on their specificity to the target PrP sequence taking into account the mouse model used, i.e., wild type or transgenic, the species from which the inoculum is derived, and the epitope of interest, i.e., within the central or C-terminal domain. Some antibodies may also recognize conformational epitopes. The antibodies most commonly used in our laboratories are 6H4, 3F4, and BH1 [22].
19. Antigen retrieval can sometimes be improved by immersing the sections in 0.1 M citric acid buffer (pH 6.4) during the autoclave process.

20. The peroxidase blocking step can be increased up to 30 min. This is often dependent on the exact protocol followed and optimization of the technique.
21. This step may be followed by proteinase K (PK) treatment at 10 μg/mL for 5 min. The PK treatment further increases PrP^{Sc} antigenicity and can be assumed to denature PrP^{C}. However, PK treatment is very harsh and can destroy tissues requiring gentle handling of the sections during the procedure.
22. Our laboratory uses the Vectastain® Elite® ABC system. Follow the manufacturer's instructions, preparing 30 min before use.

Acknowledgments

The authors wish to thank our colleagues in the Neurobiology Division, Biomedical Research Facility, and Pathology Services of the Roslin Institute and R(D)SVS for their technical expertise and assistance.

References

1. Prusiner SB (1982) Novel proteinaceous infectious particles cause scrapie. Science 216 (4542):136–144
2. Fraser H, Dickinson AG (1973) Scrapie in mice: agent-strain differences in the distribution and intensity of grey matter vacuolation. J Comp Pathol 83(1):29–40. doi:10.1016/0021-9975(73)90024-8
3. Bruce ME, McConnell I, Fraser H et al (1991) The disease characteristics of different strains of scrapie in Sinc congenic mouse lines: implications for the nature of the agent and host control of pathogenesis. J gen Virol 72(Pt 3):595–603
4. Bruce ME (2003) TSE strain variation. Br med Bull 66(1):99–108. doi:10.1093/bmb/66.1.99
5. Fraser H, Dickinson AG (1968) The sequential development of the brain lesion of scrapie in three strains of mice. J Comp Pathol 78 (3):301–311
6. Scott M, Foster D, Mirenda C et al (1989) Transgenic mice expressing hamster prion protein produce species-specific scrapie infectivity and amyloid plaques. Cell 59(5):847–857. doi:0092-8674(89)90608-9 [pii]
7. Moore RC, Hope J, McBride PA et al (1998) Mice with gene targetted prion protein alterations show that Prnp, Sinc and Prni are congruent. Nat Genet 18(2):118–125
8. Bishop MT, Hart P, Aitchison L et al (2006) Predicting susceptibility and incubation time of human-to-human transmission of vCJD. Lancet Neurol 5(5):393–398
9. Browning SR, Mason GL, Seward T et al (2004) Transmission of prions from mule deer and elk with chronic wasting disease to transgenic mice expressing cervid PrP. J Virol 78(23):13345–13350. doi:10.1128/JVI.78.23.13345-13350.2004
10. Vilotte JL, Soulier S, Essalmani R et al (2001) Markedly increased susceptibility to natural sheep scrapie of transgenic mice expressing ovine prp. J Virol 75(13):5977–5984. doi:10.1128/JVI.75.13.5977-5984.2001
11. Di Bari MA, Chianini F, Vaccari G et al (2008) The bank vole (Myodes glareolus) as a sensitive bioassay for sheep scrapie. J gen Virol 89(Pt 12):2975–2985. doi:10.1099/vir.0.2008/005520-0
12. Di Bari MA, Nonno R, Castilla J et al (2013) Chronic wasting disease in bank voles: characterisation of the shortest incubation time model for prion diseases. PLoS Pathog 9(3): e1003219. doi:10.1371/journal.ppat.1003219

13. Kimberlin RH, Cole S, Walker CA (1986) Transmissible mink encephalopathy (TME) in Chinese hamsters: identification of two strains of TME and comparisons with scrapie. Neuropathol Appl Neurobiol 12(2):197–206
14. Westaway D, Goodman PA, Mirenda CA et al (1987) Distinct prion proteins in short and long scrapie incubation period mice. Cell 51 (4):651–662
15. Dickinson AG, Meikle VMH, Fraser H (1968) Identification of a gene which controls the incubation period of some strains of scrapie agent in mice. J Comp Pathol 78(3):293–299
16. Bruce ME, Boyle A, Cousens S et al (2002) Strain characterization of natural sheep scrapie and comparison with BSE. J gen Virol 83 (3):695–704
17. Bruce ME (1993) Scrapie strain variation and mutation. Br Med Bull 49(4):822–838
18. ACDP (2013) Minimise transmission risk of CJD and vCJD in healthcare settings. Prevention of CJD and vCJD by Advisory Committee on Dangerous Pathogens' Transmissible Spongiform Encephalopathy (ACDP TSE) Risk Management Subgroup. https://www.gov.uk/government/publications/guidance-from-the-acdp-tse-risk-management-subgroup-formerly-tse-working-group. Accessed 24 June 2014
19. Taylor DM (1999) Inactivation of prions by physical and chemical means. J Hosp Infect 43(Suppl):S69–S76
20. Corda E, Beck KE, Sallis RE et al (2012) The interpretation of disease phenotypes to identify TSE strains in mice: characterisation of BSE using PrPSc distribution patterns in the brain. Vet Res 43:86. doi:10.1186/1297-9716-43-86
21. Taylor DM (2003) Preventing accidental transmission of human transmissible spongifom encephalopathies. Br Med Bull 66:293–303
22. McCutcheon S, Langeveld JP, Tan BC et al (2014) Prion protein-specific antibodies that detect multiple TSE agents with high sensitivity. PLoS One 9(3):e91143. doi:10.1371/journal.pone.0091143

Chapter 19

Preparation and Immunostaining of the Myenteric Plexus of Prion-Infected Mice

Laura J. Ellett and Victoria A. Lawson

Abstract

Intracerebral inoculation of mice with the M1000 strain of mouse-adapted human prions results in the consistent accumulation of PrP^{Sc} in the ileum of the gastrointestinal tract (GIT) of mice with clinical signs of prion disease. The accumulation of PrP^{Sc} in the ileum is accompanied by caspase activation and loss of immunoreactivity in subpopulations of neurons in the enteric nervous system. This suggests that like neurons in the central nervous system, cells in the enteric nervous system are also susceptible to prion-induced toxicity. In this chapter we describe the immunostaining of cells in myenteric plexus preparations of whole mounts prepared from the gastrointestinal tract of prion-infected mice.

Key words Fluorescence microscopy, Gastrointestinal tract, Ileum, Microdissection, Myenteric plexus, Neuron, Prion infection, Wholemount

1 Introduction

Prion disease is predominantly associated with degeneration of neurons in the central nervous system. However, in many cases, prions can also be found in tissue external to the central nervous system where they may contribute to clinical signs of disease and present as a potential source of disease transmission [1–6]. Following ingestion, prions accumulate in the gut-associated lymphoid tissues and enteric nervous system [7–10] from where they may spread to the central nervous system via the autonomic nervous system [11–13] or blood [14]. In sporadic forms of disease, where prions are believed to arise in the central nervous system, PrP^{Sc} has been detected in peripheral tissues including the spleen, muscle, and peripheral nerves [1, 3, 4, 15], and there have now been a number of reports of PrP^{Sc} accumulation in the gastrointestinal tract of patients with familial forms of prion disease [5, 16].

In animals the presence of prion infectivity in the feces of natural models of prion disease and in experimental models is a potential source of environmental contamination and spread of

Victoria A. Lawson (ed.), *Prions: Methods and Protocols*, Methods in Molecular Biology, vol. 1658,
DOI 10.1007/978-1-4939-7244-9_19,

prions between animals [17–19]. The source of this infectivity is unknown, but it may arise from the shedding of infected cells from the gastrointestinal tract.

The enteric nervous system (ENS) has been described as the second brain to the central nervous system. The ENS is the division of the autonomic nervous system that provides intrinsic control of the gastrointestinal system [20]. The neurons of the ENS are organized into two major sets of ganglia: the myenteric plexus (MP) located between the longitudinal and smooth muscle layers and the submucosal plexus (SMP) found in the submucosa. The function of the gastrointestinal tract is also influenced by extrinsic innervations that arise from the dorsal motor nucleus of the vagus to promote increased gut motility and sympathetic innervations from the spinal ganglia to inhibit gastric motility. The neuronal types in the ENS include primary afferent neurons, interneurons, and motor neurons (inhibitory or excitatory) that can be identified by their chemical signatures and structure [20, 21]. Most neurons involved with gastrointestinal motility are found in the myenteric plexus.

Prion accumulation in the enteric nervous has been observed to be associated with the activation of caspases [22] and to cause a specific decrease in neuronal populations within the myenteric plexus [23], which may be the cause of gastrointestinal dysfunction reported in some patients with prion disease [5, 16] and may, as has been observed in patients with Parkinson's disease, be an early marker of neurodegeneration [24]. This chapter describes the method for preparing the myenteric plexus of prion-infected mice for immunofluorescence detection of neuronal populations and enteric glial cells.

2 Materials

1. PBS: 0.15 M NaCl in 0.01 M sodium phosphate buffer, pH 7.2.
2. Nicardipine: 0.01% v/v in PBS.
3. Instruments for gross dissection (1× coarse forceps, scissors).
4. Instruments for fine tissue preparation (1× coarse forceps, 1× angled iris sharp scissors).
5. Blunt-ended needle/cannula and syringe.
6. Zamboni's fixative: 2% formaldehyde (or paraformaldehyde from formaldehyde), 0.2% picric acid, pH 7.0 (*see* **Note 1**).
7. Glass dish containing Zamboni's fixative within a secondary container.
8. Transfer pipettes.
9. 1.5 mm balsa wood boards cut to ~6 × 3 cm.

10. Entomological pins.
11. Containers for washing prepared tissue and storage, e.g., small beakers.
12. 0.1% sodium azide prepared in PBS.
13. Dimethyl sulfoxide (DMSO).
14. Waste container for Zamboni's fixative and DMSO (hazardous chemical disposal).
15. Small vials for storing wholemounts.
16. Instruments for fine tissue preparation (1× coarse dissection forceps for handling pins, 2× fine forceps).
17. Glass slide.
18. Black board.
19. Dissecting microscope.
20. Light source.
21. Scalpel or razor blade.
22. Silicone-covered Petri dish (we use Sylgard, Dow Corning).
23. Staining trays within a humid, lidded container.
24. 1.7 mL snap cap tubes.
25. Glass slides and coverslips.
26. Non-fluorescent mounting medium (we use Dako S3023).
27. Fine paintbrush.
28. Lint-free tissue.
29. BSA/TX-100: 1% bovine serum albumin (BSA), 0.1% v/v Triton-X-100 (TX-100) in PBS.
30. Primary antibodies directed to markers of various classes of enteric neurons (Table 1).
31. Fluorescently labeled secondary antibodies directed to the relevant primary antibody species.
32. Fluorescence microscope with appropriate filters.

3 Methods

3.1 Wholemount Preparation

1. Remove a 4–5 cm length of ileum terminal to the caecum from mouse/rat and place in container of PBS/nicardipine (*see* **Note 2**).
2. Using coarse forceps, pick up section of intestine holding one side of lumen so that the blunt-ended needle can be inserted into the lumen to flush out remaining contents using PBS/nicardipine.
3. Lay the tissue on balsa board and wet with PBS/nicardipine.

Table 1
Primary anti-sera used for immunohistochemical detection of neurons/cells in murine myenteric plexus

Tissue antigen	Class of neuron/cell detected in the murine myenteric plexus	Source[a]/reference for anti-sera
Hu (RNA-regulating protein)	All neurons	[25]
GFAP (glial fibrillary acidic protein)	Enteric glial cell	Chemicon [26]
Calretinin	Interneurons and excitatory muscle motor neuron	Swant
NFM (neurofilament medium) 145 kDa	Intrinsic primary afferent neurons	Chemicon—AB1897
nNOS (neuronal nitric oxide synthase)	Inhibitory muscle motor neuron	[27]

[a]Supply companies: Chemicon International, Dako, Carpinteria, CA, USA; Swant, Bellinzona, Switzerland

4. Locate mesenteric boarder (where mesentery attaches and blood vessels meet) and place facing up.
5. Pin section to board using an entomological pin through the back edge of the lumen (*see* **Note 3**).
6. Insert sharp, angled iris scissor tip into the lumen and cut up the mesenteric boarder using long straight cuts.
7. Gently open section and turn it over so that the mucosa is surface down on the balsa wood.
8. Gently pull edge and pin out. Work down the length of the tissue from side to side, pulling and pinning out as you go. The tissue should be taunt and square when finished.
9. Once the section has been stretched out, push the pins into the balsa so that the section can float in fixative section side down.
10. Float section tissue side down in lidded glass dish of Zamboni's fixative.
11. Place dish containing tissue in a securely lidded secondary container and place at 4 °C overnight.
12. Remove sections from Zamboni's fixative and unpin from balsa wood.
13. Place in container of DMSO and wash for exactly 10 min.
14. Repeat 10 min DMSO washes twice more.
15. Collect waste in appropriately labeled container for disposal (*see* **Note 4**).
16. Wash three times for 10 min in PBS.

17. Transfer section to fresh container of PBS/ 0.1% sodium azide and store at 4 °C.

3.2 Dissection of Wholemount Preparations

1. Place tissue mucosa up on a glass slide on a black board and cover with PBS and then cut off a small square section of about 0.5–1 cm^2 with razor blade (*see* **Note 5**). Orient the tissue to help you remember the course/grain of the circular muscle.
2. Place tissue section mucosa up into silicone-covered Petri dish filled with PBS and pin down slightly stretched with a pin at each corner and with the circular muscle running from left to right.
3. Place the light flat on the black board so that light shines horizontally into dish and tissue (try to find best angle at which different layers of tissue are visible under the dissecting microscope).
4. Place fine forceps beneath one side of tissue and gently lift so that the tissue is stretched on one side (mucosa and muscular layers beneath should now be distinguishable), and then separate mucosa, connective tissue, and submucosa from layers below using the second pair of forceps to tear at the mucosa (*see* **Note 6**).
5. Strip mucosa, connective tissue, and submucosa completely off by carefully pushing tissue down while pulling up the mucosa with the second pair of forceps. Strip mucosa off piece by piece if necessary (*see* **Note** 7).
6. Next, peel off strips of circular muscle beginning on left or right side by first loosening it and then tearing it with the fine forceps while pushing tissue down a little further up with the second pair of forceps. Continue until all circular muscle is stripped off. If circular muscle remains, staining will be impaired in that area.
7. Completed myenteric plexus preparations tissues can be stored at 4 °C in PBS-azide in snap cap 1.7 mL tubes or further processed (*see* **Note 8**).

3.3 Immunostaining of Dissected Sections

1. Place a 20 μL droplet of 1% BSA with 0.1% v/v Triton-X-100 into a well of a staining tray (*see* **Note 9**). Incubate for 30mins at room temperature.
2. Place tissue in 20 μL droplet of desired primary antibody diluted in PBS (*see* **Note 10**). Ensure that tissue section is covered by the antibody solution and doesn't float on top of it. Incubate at 4 °C, overnight.
3. Wash tissue section three times 10 min with PBS (~ 500 μL).
4. Incubate in 20 μL droplet secondary antibody diluted in PBS for 1 h at room temperature.
5. Wash tissue section three times 10 min in PBS.

6. Place a droplet of non-fluorescent mounting medium on glass slide and remove bubbles with a lint-free tissue. Place tissue in this droplet and smooth out wrinkles with fine paintbrush, then carefully lower coverslip on top and let it dry in a darkened place for at least 1 h at 4 °C to set.
7. Analyze via fluorescence microscopy.
8. Store analyzed tissue sections at 4 °C away from light.

4 Notes

1. Combine 50 mL formaldehyde and 150 mL filtered, saturated, aqueous, picric acid solution. Make up to 1 L with 300 mL water and 500 mL 0.2 M phosphate buffer, pH 7, with 10 M NaOH. Store in glass bottle at 4 °C.
2. It is important that the tissue remains moist during the procedure but also that it is not immersed in PBS/nicardipine for too long (no more than 30 min).
3. Keep track of the fine entomological pins. It is a good idea to pin out a regular amount of pins into the balsa wood before you start dissecting. Sixteen pins are normally enough for each section of ileum. They are very easy to lose and easy to sustain a needle stick injury from. Only handle pins with forceps not hands.
4. Zamboni's fixative and DMSO are hazardous chemicals and should be disposed of through appropriate hazardous waste routes. Zambonis is also explosive when dry so either rinse any item that has come into contact with the fixative or keep it wet. For example, tissues which have absorbed Zamboni's can be placed in a waste bottle containing water.
5. Handle gut tissue at the edge and not in the middle to avoid damaging the tissue.
6. Never use extra fine forceps for picking up pins as this will damage the delicate tips. Only use them for dissecting.
7. Often all these three layers will come off together. Avoid ripping the tissue and be careful at corners and the areas where pins are holding the tissue in place. Make sure to strip off all connective tissue (which looks like cobwebs). Any areas which still have connective tissue remaining will be near impossible to remove the circular muscle from.
8. If you have the light hitting the tissue at the right angle, you should be able to see the ganglia of the myenteric plexus. Always handle dissected tissues carefully and only lift tissue sections by letting it attach to forceps; don't grab tissue with forceps to avoid bruising and/or ripping of tissue.

9. Put staining tray into a lidded container with some water at the bottom—this humid environment prevents the small droplets of block/antibody solution from evaporating. Keep track of which wells have been used and what they have been used for.
10. You can double or triple stain. Just ensure you have the required number of different species of secondary antibodies all conjugated to different fluorophores.

References

1. Glatzel M, Abela E, Maissen M et al (2003) Extraneural pathologic prion protein in sporadic Creutzfeldt-Jakob disease. N Engl J Med 349:1812–1820
2. Wadsworth JD, Joiner S, Hill AF et al (2001) Tissue distribution of protease resistant prion protein in variant Creutzfeldt-Jakob disease using a highly sensitive immunoblotting assay. Lancet 358:171–180
3. Peden AH, Ritchie DL, Head MW et al (2006) Detection and localization of PrPSc in the skeletal muscle of patients with variant, iatrogenic, and sporadic forms of Creutzfeldt-Jakob disease. Am J Pathol 168:927–935
4. Lee CC, Kuo LT, Wang CH et al (2005) Accumulation of prion protein in the peripheral nervous system in human prion diseases. J Neuropathol Exp Neurol 64:716–721
5. Mead S, Gandhi S, Beck J et al (2013) A novel prion disease associated with diarrhea and autonomic neuropathy. N Engl J Med 369:1904–1914
6. Wadsworth JD, Joiner S, Fox K et al (2007) Prion infectivity in variant Creutzfeldt-Jakob disease rectum. Gut 56:90–94
7. Herzog C, Riviere J, Lescoutra-Etchegaray N et al (2005) PrPTSE distribution in a primate model of variant, sporadic, and iatrogenic Creutzfeldt-Jakob disease. J Virol 79:14339–14345
8. Kruger D, Thomzig A, Lenz G et al (2009) Faecal shedding, alimentary clearance and intestinal spread of prions in hamsters fed with scrapie. Vet Res 40:4
9. Beekes M, Mcbride PA (2000) Early accumulation of pathological PrP in the enteric nervous system and gut-associated lymphoid tissue of hamsters orally infected with scrapie. Neurosci Lett 278:181–184
10. Seelig DM, Mason GL, Telling GC et al (2011) Chronic wasting disease prion trafficking via the autonomic nervous system. Am J Pathol 179:1319–1328
11. Mcbride PA, Beekes M (1999) Pathological PrP is abundant in sympathetic and sensory ganglia of hamsters fed with scrapie. Neurosci Lett 265:135–138
12. Sigurdson CJ, Spraker TR, Miller MW et al (2001) PrP(CWD) in the myenteric plexus, vagosympathetic trunk and endocrine glands of deer with chronic wasting disease. J Gen Virol 82:2327–2334
13. Kaatz M, Fast C, Ziegler U et al (2012) Spread of classic BSE prions from the gut via the peripheral nervous system to the brain. Am J Pathol 181:515–524
14. Siso S, Jeffrey M, Gonzalez L (2009) Neuroinvasion in sheep transmissible spongiform encephalopathies: the role of the haematogenous route. Neuropathol Appl Neurobiol 35:232–246
15. Guiroy DC, Shankar SK, Gibbs CJ Jr et al (1989) Neuronal degeneration and neurofilament accumulation in the trigeminal ganglia in Creutzfeldt-Jakob disease. Ann Neurol 25:102–106
16. Fong JC, Rojas JC, Bang J et al (2017) Genetic prion disease caused by PRNP Q160X mutation presenting with an orbitofrontal syndrome, cyclic diarrhea, and peripheral neuropathy. J Alzheimers Dis 55:249–258
17. Tamguney G, Miller MW, Wolfe LL et al (2009) Asymptomatic deer excrete infectious prions in faeces. Nature 461:529–532
18. Safar JG, Lessard P, Tamguney G et al (2008) Transmission and detection of prions in feces. J Infect Dis 198:81–89
19. Pritzkow S, Morales R, Moda F et al (2015) Grass plants bind, retain, uptake, and transport infectious prions. Cell Rep 11:1168–1175
20. Furness JB (2012) The enteric nervous system and neurogastroenterology. Nat Rev Gastroenterol Hepatol 9:286–294
21. ZD Q, Thacker M, Castelucci P et al (2008) Immunohistochemical analysis of neuron types in the mouse small intestine. Cell Tissue Res 334:147–161

22. Drew SC, Haigh CL, Klemm HM et al (2011) Optical imaging detects apoptosis in the brain and peripheral organs of prion-infected mice. J Neuropathol Exp Neurol 70:143–150
23. Lawson VA, Furness JB, Klemm HM et al (2010) The brain to gut pathway: a possible route of prion transmission. Gut 59:1643–1651
24. Braak H, De Vos RA, Bohl J et al (2006) Gastric alpha-synuclein immunoreactive inclusions in Meissner's and Auerbach's plexuses in cases staged for Parkinson's disease-related brain pathology. Neurosci Lett 396:67–72
25. Fairman CL, Clagett-Dame M, Lennon VA et al (1995) Appearance of neurons in the developing chick gut. Dev Dyn 204:192–201
26. Albanese V, Lawson VA, Hill AF et al (2008) Evidence for prion protein expression in enteroglial cells of the myenteric plexus of mouse intestine. Auton Neurosci 140:17–23
27. Williamson S, Pompolo S, Furness JB (1996) GABA and nitric oxide synthase immunoreactivities are colocalized in a subset of inhibitory motor neurons of the guinea-pig small intestine. Cell Tissue Res 284:29–37

Part V

Methods for the Detection and Treatment of Prion Diseases

Chapter 20

Cell Culture Methods for Screening of Prion Therapeutics

Hilary E. McMahon

Abstract

It is currently difficult to predict the number of asymptomatic prion carriers who will potentially go on to develop a prion disease or who will unknowingly transmit the prion agent to another individual. As prion therapeutic clinical trials have lacked success, there is a continuous need for novel therapeutics that have the potential to prevent, as for inherited prion disorders; slow, as for all prion disorders; and ultimately stop disease progression. Prion-infected cell models provide an ideal tool to search for new treatment avenues. This chapter describes the use of prion cell culture systems in the identification of prion therapeutics. It also deals with the methods required to validate the potential of an antiprion agent through cell viability and impact on cell growth rate.

Key words Therapeutics, Prion, Scrapie, Viability, Screening, Growth rate

1 Introduction

A recent study of appendix samples has indicated that potentially 1 in 2000 people in the United Kingdom may be asymptomatic carriers of the prion infectious agent that causes variant Creutzfeldt-Jakob disease (vCJD) [1]. This finding adds to concerns for the transmission of prion diseases through silent prion carriers [2, 3]. It also strengthens the need for the development of successful prion therapeutics, of which there is currently none. A number of compounds have been evaluated for their antiprion abilities. They have been directed toward interfering with the prion conversion process, altering prion trafficking, and enhancing prion degradation [4–7]. Compounds that have been tested clinically have failed to show activity against the clinical signs of prion disease [8, 9]; this stresses the requirement for studies that source novel therapeutic approaches to treat these disorders.

The advent and development of prion cell culture systems have been fundamental to the understanding of prion biology and to providing us with an essential screening tool to identify new prion therapeutics. Their development has required persistence, not all

Victoria A. Lawson (ed.), *Prions: Methods and Protocols*, Methods in Molecular Biology, vol. 1658,
DOI 10.1007/978-1-4939-7244-9_20,

cells are susceptible to infection [10], and prion strains have been proposed to have potential cell-type specificity and infectability [11]. Vorberg [11] noted that although initial prion infection can occur in many cells, acute infection does not guarantee persistence of infection. The key to prion cell culture as a therapeutic tool is the establishment of cell models that can maintain a persistent infection. In spite of the hurdles experienced in the development of persistently infected cell models, a number of prion cell models have been established [12]. For our work, we employ the neuroblastoma N2a cell line overexpressing mouse PrP^C, known as N2a58, which is highly infectible [5, 13] and is persistently infected with scrapie strain 22L. These cells were originally described by Nishida et al., [13] and we have subcloned these to obtain the line N2a22L20 [4], which expresses a high level of PrP^{Sc}. In our work, we also use the hypothalamic cell line GT1–7 infected with 22L and Chandler [13]. This work presents the methodologies for screening for prion therapeutics in the prion-infected N2a cell culture models. It outlines the significance of toxicity testing through the MTT (3,[4,5 dimethylthiazol-2-yl]-2,5 diphenyltetrazolium bromide) cell viability assay. It is important that a potential therapeutic does not impact significantly on the growth of the infected cell lines. As such this work also presents protocols that enable the monitoring of the effect of potential therapeutics on cell growth rates. The methods outlined here use the N2a22L20 cell model, but they can be adapted to other prion-infected cell models.

2 Materials

2.1 Screening for Antiprion Agents in Prion-Infected Cell Culture

1. "Category 2" biological safety cabinet.
2. Water bath.
3. CO_2 cell culture incubator.
4. Disposable cell culture dishes, pipettes, and tubes.
5. Dulbecco's Modified Eagle's Medium (DMEM).
6. Fetal bovine serum (FBS).
7. Penicillin/streptomycin.
8. Geneticin (G418 sulfate) (SIGMA).
9. 0.2/0.1 μM filter device.
10. Phosphate-buffered saline (PBS).
11. 10× trypsin (25 g/l porcine trypsin in 0.9% (w/v) NaCl). Prepare 1× stock in PBS and store in the freezer.
12. The N2a22L20 cell line in Subheading 3.1 is maintained in DMEM supplemented with 10% (v/v) FBS, penicillin/streptomycin solution, and 300 μg/ml geneticin (G418).

2.2 Short But Precise Protocol for the Detection of PrP^{Sc} in the Cell Lines

1. PBS.
2. Refrigerated centrifuge for Eppendorfs.
3. Lysis buffer: Triton/DOC lysis buffer (0.5% (w/v) sodium deoxycholate, 0.5% (v/v) Triton X-100, 150 mM NaCl, and 50 mM Tris–HCl pH 7.5. Store at 4 °C. When you are ready to use the lysis buffer, add proteinase inhibitors pepstatin A (1 μg/ml), leupeptin (1 μg/ml), and EDTA (2 mM) to your aliquot (*see* **Note 1**).
4. Eppendorfs.
5. Bicinchoninic acid assay (BCA) solutions (SIGMA).
6. Proteinase K. Stock solution is prepared at 1 mg/ml in sterile ddH_20. Aliquot as 20 μl and store in the freezer (*see* **Note 2**).
7. Pefabloc (SIGMA). Stock solution is prepared by dissolving 24 mg in 2 ml sterile ddH_2O. Aliquot as 20 μl and store in the freezer.
8. Water bath at 37 °C.

2.3 Long Protocol for the Detection of PrP^{Sc} in the Cell Lines

1. Requirements are as per Subheading 2.2.

2.4 Determination of the Effect of Your Treatment on Cell Viability

1. 96-well tissue culture plate.
2. CO_2 cell culture incubator.
3. PBS.
4. 10× trypsin (25 g/l porcine trypsin in 0.9% (w/v) NaCl). Prepare 1× stock in PBS and store in the freezer.
5. Trypan Blue solution (0.4%).
6. Neubauer hemocytometer cell counting slide and coverslip.
7. Aspirator.
8. Petri dish.
9. MTT (3,[4,5 dimethylthiazol-2-yl]-2,5 diphenyltetrazolium bromide). Stock solution is prepared by dissolving 50 mg MTT in 10 ml sterile PBS, aliquot into 2 ml lots and freeze at −20 °C.
10. DMSO (dimethyl sulfoxide).
11. Spectrophotometer.

2.5 Determination of the Effect of the Antiprion Compound on Cell Growth Rate

1. 24-well cell culture plate.
2. CO_2 cell culture incubator.
3. PBS.
4. 10× trypsin (25 g/l porcine trypsin in 0.9% (w/v) NaCl). Prepare 1× stock in PBS and store in the freezer.

5. Trypan Blue solution (0.4%) (SIGMA).
6. Neubauer hemocytometer cell counting slide and coverslip.

3 Methods

3.1 Screening for Antiprion Agents in Prion-Infected Cell Culture

All media preparation and passaging of cells are performed in a "Category 2" biological safety cabinet. Sterilize all media components by filtering through a 0.2 μM filter unit (see **Note 3**). Media are stored at 4 °C and warmed to 37 °C in a water bath prior to use. Cells employed must be free of mycoplasma contamination (*see* **Note 4**). The protocols described can be adapted as appropriate to various cell lines. The following protocol outlines the procedure for screening for potential antiprion agents:

1. Passage cells just before they reach confluence (*see* **Note 5**).
2. Remove media from cells, and rinse with warm PBS with a volume equal to that of the media removed.
3. Remove PBS and add 1 ml 1× trypsin in PBS (volume is for a 60 mm dish). Leave the trypsin on for 20 s and remove. In the case of N2a22L20 cells, leave the cells at room temperature for 5 min (*see* **Note 6**), and tap dish to dislodged cells.
4. Place a volume of media onto the cells equal to that used during growth, and slowly pipette the cells up and down to generate a cell suspension (*see* **Note 7**).
5. Seed the cells at 2×10^5 cells for a 6-welled plate. At this time of passage, add your sterile potential therapeutic compound under test to the passaged cells in their media; set this up in triplicate. Also set up a triplicate test for the solvent/carrier used to dissolve your test compound; this is added at the same concentration as used when adding the test compound (*see* **Note 8**). Incubate the cells.
6. Continue to passage your cells for the required period of time, and then assay for the presence of the prion infectious agent.

3.2 Short But Precise Protocol for the Detection of PrP^{Sc} in the Cell Lines

At the appropriate time, cells are tested for their PrP^{Sc} content. We employ two methods for the detection of PrP^{Sc}, one requiring a 45 min centrifugation and one without. This procedure outlines the quicker protocol for detection of PrP^{Sc}; it is a convenient protocol when screening for therapeutic agents.

1. Remove media from cells, and rinse with cold PBS with a volume equal to that of the growth media (*see* **Note 9**). Aspirate the PBS and then add 1 ml cold PBS to the cells. Scrape the cells into the PBS (*see* **Note 10**) and transfer to an Eppendorf.
2. Centrifuge the cells down at 10,621 × *g* for 5 min at 4 °C. Remove the supernatant, and add 60 μl (vary this volume

depending on pellet size) lysis buffer and resuspend the pellet. Leave tubes on ice for 15 min.

3. Centrifuge the tubes for 2 min at 10,621 × *g* at 4 °C; remove the supernatant to a fresh Eppendorf tube.
4. Determine the protein concentration in each sample using the BCA assay. For the N2a22L20 cells, only 50 μg protein is needed to detect PrP^{Sc}. Remove the required volume of lysate to give your required protein level; bring samples to same volume with lysis buffer. The volumes used in this protocol remain low.
5. Place all samples on a rack in ice for at least 5 min.
6. Dilute proteinase K stock at a 1/2.5 dilution in ddH_2O, and then add 2 μl ice-cold proteinase K (16 μg/mg protein) to each tube on ice (*see* **Note 11**).
7. Remove the rack of tubes from the ice once proteinase K has been added to every tube, and place in water bath at 37 °C for 15 min.
8. Remove the rack of tubes from the water bath and place on ice. Add 1 mM Pefabloc to stop the reaction, and leave on ice for 5 min.
9. As the volumes are kept low by virtue of the lysing protocol, 2× loading buffer is added directly to the samples, and samples are heated for 5 min at 90 °C. Run your samples on SDS-PAGE and analyze by Western blotting. If your sample volume increases, use 3× or 5× loading buffer. Your sample size, once loading buffer is added, should be under the size of your wells on the SDS-PAGE, as the complete sample must be loaded into the well. This protocol reproducibly allows for very good detection of PrP^{Sc} (Fig. 1).

3.3 Long Protocol for the Detection of PrP^{Sc} in the Cell Lines

If your protein levels were low after lysis, that may preclude the potential usage of the method laid out in Subheading 3.2. If you think that the final sample volume will exceed the volume of your loading wells, then you can use this protocol.

1. **Steps 1–3** are as per Subheading 3.2.
2. Determine the protein concentration in each sample using the BCA assay. For the N2a22L20 cells, only 50 μg protein is needed to detect PrP^{Sc}. Remove the required volume of lysate to give your required protein level; bring samples to 50 μl (or your desired volume level) with lysis buffer.
3. Place all samples on a rack in ice for at least 5 min.
4. Dilute proteinase K stock at a 1/2.5 dilution in ddH_2O, and then add 2 μl ice-cold proteinase K (16 μg/mg protein) to each tube on ice (*see* **Note 11**).

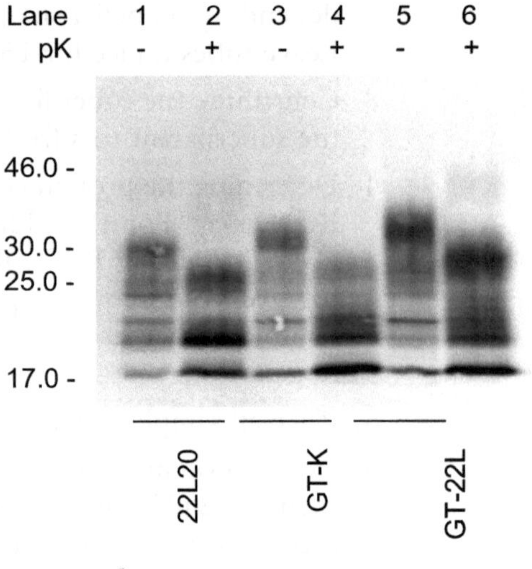

Fig. 1 Detection of PrP^{Sc} in prion-infected cell lines used for therapeutic screening. This figure shows total PrP (*lanes 1, 3,* and *5*) and PrP^{Sc} (*lanes 2, 4,* and *6*) in three prion-infected cell lines: N2a22L20 (*lanes 1* and *2*), GT-K (*lanes 3* and *4*), and GT-22L (*lanes 5* and *6*). These cells were lysed into lysis buffer; the lysate (20 μg) was either loaded directly onto the 12% SDS-PAGE gel without proteinase K digestion (pK–) for total PrP, or 50 μg of respective lysate was treated with pK (pK+) for the detection of PrP^{Sc}. Western blots were then analyzed with the antibody 7A12, which recognizes the epitope 175–185 of mouse PrP [16]

5. Remove the rack of tubes from the ice once proteinase K has been added to every tube, and place in water bath at 37 °C for 15 min.
6. Remove the rack of tubes from the water bath and place on ice. Add 1 mM Pefabloc to stop the reaction, and leave on ice for 5 min.
7. Centrifuge your samples at 20,817.16 × *g* for 45 min at 4 °C.
8. Remove the supernatant and carefully resuspend the pellet in 1× loading buffer, and heat the samples for 5 min at 90 °C. Run your samples on SDS-PAGE and analyze by Western blotting.

3.4 Determination of the Effect of Your Treatment on Cell Viability

The ideal antiprion agent will have little or no effect on cell viability or growth rate. It is important to validate the effect of your selected compound on cell viability, and this can be done by the MTT assay which is used for quantifying metabolically active cells and analyzing cell viability [14].

1. The first part of this protocol requires the cells to be counted to seed your cells at 8000 cells per well in a 96-well plate. For this rinse growing cells with PBS and treat with 1× trypsin (20 s and incubate at room temp for 5 min for N2a22L20 cells).

2. Resuspend cells in an appropriate volume of growth media, and ensure cells are well suspended. Take 50 μl of the cell suspension and add to 50 μl of Trypan Blue (0.4%) mix.
3. Count the number of clear cells in four squares of the Neubauer hemocytometer chamber:
   (a) Set up Neubauer hemocytometer on an inverted microscope; place its coverslip over the counting squares.
   (b) Follow your hemocytometer's instructions to load your sample into the chambers of the hemocytometer.
   (c) Clear cells are counted in a large square under the microscope.
   (d) Ignore cells lying on the top and right margin of the squares.
   (e) Count cells over the large square and repeat your count over four large squares.
4. Once the number of cells per ml has been calculated, determine the volume to be added to each well in order to transfer 8000 cells to each well. Dilute your cells in media so that you have 8000 cells per 100 μl.
5. 100 μl of cell suspension containing 8000 cells must be added to each well in triplicate. Use a multipipette to dispense 100 μl to each well. Incubate for 24 h at 37C in a CO_2 cell culture incubator.
6. Prepare drug dilutions in media; always set up control no drug samples.
7. Remove media from the plate containing the cells using an aspirator. Then transfer 100 μl drug solutions prepared in media to the 96-well plate containing the cells using a multipipette. Incubate for 24 h at 37C in a CO_2 cell culture incubator.
8. Remove the media and drugs using an aspirator. Add 10mls of media to 1 ml of stock MTT in a 50 ml tube and invert to mix, transfer the solution to a petri dish, and then transfer 110 μl from there to each well of the 96-well plate using a multipipette. Incubate at 37 °C for 2 h in a CO_2 cell culture incubator. Remove the medium.
9. Add 100 μl DMSO to each well. Read the results at 570 nm and determine the percentage cell viability.

3.5 Determination of the Effect of the Antiprion Compound on Cell Growth Rate

Work has indicated that growth rate of cells can affect PrP^{Sc} levels [15]. In testing a potential antiprion agent, you must examine its effect on growth rate relative to untreated cells. The procedure below demonstrates how this can be done using the Trypan Blue exclusion assay.

1. Passage the cells according to Subheading 3.1; except use a 24-well plate when passaging the cells.
2. At various time points post passage, monitor the cell number; the best time to start this is once the cells have adhered to the cell plate. Each time point should be analyzed in triplicate.
3. At your dedicated time point, remove the media from your triplicate wells; rinse the cells with warm PBS. Aspirate the PBS and treat with 1× trypsin for 10 min; do not remove the trypsin (*see* **Note 12**). Add media to the wells; the serum in the media will inactivate the trypsin.
4. Ensure cells are well suspended; if they aren't, the assay will be distorted. Take 50 μl of cells and add to 50 μl of Trypan Blue (0.4%) mix. Count the number of cells in four squares of the Neubauer hemocytometer chamber (as per step 3 in Subheading 3.4).
5. Calculate your cells per ml of sample.
6. Determine the cell count at appropriate time points until the cells are in the stationary phase of growth. Compare the growth curves between treatments and controls.

4 Notes

1. As this protocol is for the detection of PrP^{Sc} which requires proteinase K, and as proteinase K is a serine proteinase, PMSF is not added as a proteinase inhibitor.
2. Proteinase K does not like to be frozen and thawed multiple times; therefore, store stock solution in small aliquots.
3. If you are adding powder components to your media, sterilize by filtering through a 0.1 μM filter system. This is to eliminate any mycoplasma entrance into the medium.
4. A number of procedures are available for testing cells for mycoplasma contamination, including molecular, PCR-based, and colorimetric techniques (PlasmoTest from InvivoGen).
5. Cells should be passaged while they are in the log phase of growth to maintain the health of the cells.
6. Treatment to detach cells from the cell culture dish varies with cell type. CHO cells require incubation for 5 min at 37 °C. GT1–7 cells benefit from using trypsin EDTA.
7. Avoid generating bubbles during the resuspension as this can result in damage to the cells.
8. The effect of the solvent used to dissolve your test compound must be itself tested in case the solvent induces prion clearance.

9. Rinsing of the cells is essential to prevent interference in Western blots.
10. To scrape the cells into PBS from cell dishes, use the top of a 1 ml pipette tip, the widest part, and in a circular motion, scrape the cells from the dish into the PBS. If you are carrying out a large test, this is a very convenient method of detaching the cells.
11. All procedures here are performed on ice to prevent the proteinase K from starting digestion in the tubes. This is especially important when you are dealing with large numbers of samples.
12. It is important not to remove the trypsin at this stage. If the trypsin is removed, you risk lifting some of the cells in the well, which will negatively give impact on your analysis. This is very significant if you are dealing with many samples at the same time.

Acknowledgments

We are grateful to Professor Man-Sun Sy for the use of antibody 7A12.

References

1. Gill ON, Spencer Y, Richard-Loendt A et al (2013) Prevalent abnormal prion protein in human appendixes after bovine spongiform encephalopathy epizootic: large scale survey. BMJ Clin Res Ed 347:f5675
2. Thomas JG, Chenoweth CE, Sullivan SE (2013) Iatrogenic Creutzfeldt-Jakob disease via surgical instruments. J Clin Neurosci 20:1207–1212
3. O'dowd A (2013) Variant CJD still poses threat to public health, warn experts. BMJ Clin Res Ed 347:f7142
4. Prior M, Lehmann S, Sy MS et al (2007) Cyclodextrins inhibit replication of scrapie prion protein in cell culture. J Virol 81:11195–11207
5. Molloy B, Mcmahon HE (2014) A cell-biased effect of estrogen in prion infection. J Virol 88:1342–1353
6. Caughey B, Ernst D, Race RE (1993) Congo red inhibition of scrapie agent replication. J Virol 67:6270–6272
7. Mange A, Nishida N, Milhavet O et al (2000) Amphotericin B inhibits the generation of the scrapie isoform of the prion protein in infected cultures. J Virol 74:3135–3140
8. Tsuboi Y, Doh-Ura K, Yamada T (2009) Continuous intraventricular infusion of pentosan polysulfate: clinical trial against prion diseases. Neuropathology 29:632–636
9. Geschwind MD, Kuo AL, Wong KS et al (2013) Quinacrine treatment trial for sporadic Creutzfeldt-Jakob disease. Neurology 81:2015–2023
10. Krauss S, Vorberg I (2013) Prions ex vivo: what cell culture models tell us about infectious proteins. Int J Cell Biol 2013:704546
11. Vorberg I, Raines A, Priola SA (2004) Acute formation of protease-resistant prion protein does not always lead to persistent scrapie infection in vitro. J Biol Chem 279:29218–29225
12. Solassol J, Crozet C, Lehmann S (2003) Prion propagation in cultured cells. Br Med Bull 66:87–97
13. Nishida N, Harris D, Vilette D et al (2000) Successful transmission of three mouse-adapted scrapie strains to murine neuroblastoma cell lines overexpressing wild-type mouse prion protein. J Virol 74:320–325
14. Mosmann T (1983) Rapid colorimetric assay for cellular growth and survival: application to

proliferation and cytotoxicity assays. J Immunol Methods 65:55–63

15. Ghaemmaghami S, Phuan PW, Perkins B et al (2007) Cell division modulates prion accumulation in cultured cells. Proc Natl Acad Sci U S A 104:17971–17976

16. Pham N, Yin S, Yu S et al (2008) Normal cellular prion protein with a methionine at position 129 has a more exposed helix 1 and is more prone to aggregate. Biochem Biophys Res Commun 368:875–881

Chapter 21

Real-Time Quaking-Induced Conversion for Diagnosis of Prion Disease

Katsuya Satoh, Ryuichiro Atarashi, and Noriyuki Nishida

Abstract

Sporadic human prion diseases are defined on the basis of clinical features, with periodic sharp discharge (PSD) on electroencephalograms (EEG), a positive 14-3-3 protein assay of CSF samples, and abnormal signals on cerebral cortex on diffusion-weighted (DWI) MR images. It is essential to detect the abnormal prion protein in neuropathological or immunochemical detection of brain tissues when we diagnose definite cases for human prion disease. We performed definite diagnosis of sporadic human prion disease in alive patients. Recently, testing of CSF with a new in vitro abnormal prion protein amplification technology, designated real-time quaking-induced conversion (RT-QUIC), has shown considerable promise as a highly specific diagnostic test for human prion disease.

Key words 14-3-3 protein, CSF, RT-QUIC method

1 Introduction

The posttranscriptional conversion of normal prion protein (PrP^c) into abnormal form (PrP^{Sc}) is thought to be a pathogenic mechanism of human prion diseases. Caughey et al. [1] applied the seeding theory for amyloid-fibril formation to demonstrate the cell-free conversion of PrP in vitro. The original cell-free system has produced in vitro PrP^{Sc} amyloid with low efficiency; recently, Soto et al. [2] described a method for the amplification of PrP^{Sc} called "protein misfolding cyclic amplification" (PMCA) and suggested that PMCA could have tremendous implications for research and diagnosis. Unfortunately, PrP^{Sc} in human samples is, however, not detected by the PMCA method, except for the variant in Creutzfeldt-Jakob disease (CJD). Atarashi et al. developed a new method called the "quaking-induced conversion" (QuIC), which improves the speed and practicality by using bacterially expressed recombinant PrP^C ($rPrP^c$) and by substituting the shaking for sonication [3]. This seeding assay of PrP-amyloid formation in vitro could be monitored using thioflavin T (ThT) in real time,

Victoria A. Lawson (ed.), *Prions: Methods and Protocols*, Methods in Molecular Biology, vol. 1658,
DOI 10.1007/978-1-4939-7244-9_21, © Springer Science+Business Media LLC 2017

so we call this new method as a real-time QuIC (RT-QuIC) [4]. With recent advancements, we could detect one femtogram of human PrP^{Sc} in affected brain using the RT-QuIC [5], indicating that the method is the most sensitive and highly specific test for prion and useful for diagnosis of human prion diseases at this time.

2 Material

2.1 Recombinant Human PrP Expression and Purification

1. Human PrP residues 23–231 (129 M) ligated into an appropriate recombinant protein expression vector. We use the pET41 vector (EMD Biosciences, San Diego, CA, USA) using NdeI-HindIII PCR inserts. Verify the accuracy of the sequences.
2. *E. coli* BL21 (DE3).
3. Overnight Express™ Autoinduction System (EMD Biosciences).
4. BugBuster® and Lysonase™ kit (EMD Biosciences).
5. EDTA-free protease inhibitors (Roche Diagnostics Japan, Tokyo, Japan).
6. Ni-NTA SuperFlow™ resin (Qiagen, Hamburg, Germany).
7. Denaturation buffer: 100 mM sodium phosphate, pH 8.0, 10 mM Tris, 6 M guanidine-HCl.
8. Refolding buffer: 100 mM sodium phosphate, pH 8.0, 10 mM Tris.
9. Elution buffer: 100 mM sodium phosphate, pH 5.8, 500 mM imidazole, 10 mM Tris.
10. AKTA Explorer System (GE Healthcare, New York, NY, USA).
11. 10 mM phosphate buffer (pH 6.8).

2.2 RT-QuIC Using CSF or Brain Homogenates

1. CSF samples stored at −80 °C prior to analysis. Divide samples into 5 μL aliquots for use with the RT-QuIC method. Do not use samples that have been subjected to repeated freezing and thawing, as they show poor results using the QuIC method.
2. 96-well optical-bottom black microplate (Thermo Fisher Scientific, Waltham, MA, USA).
3. Reaction buffer: 500 mM NaCl, 50 mM PIPES, pH 7.0, 1 mM EDTA, and 10 μM ThT.
4. Microplate reader (Infinite M200 or F200 fluorescence plate reader; TECAN, Männedorf, Switzerland).

3 Method

When attempting to detect PrP^{Sc} from human tissue samples with the RT-QuIC method, we are using recombinant human PrP, although recombinant animal PrP [6, 7] can also be used as substrates instead of human PrP (*see* **Notes 1** and **2**).

3.1 Recombinant Human PrP Expression and Purification

1. Transform plasmids into *E. coli* BL21 (DE3). Express soluble recombinant human PrP (rHuPrP-sen) using the Overnight Express™ Autoinduction System.
2. Lyse cell pellets using a BugBuster® and Lysonase™ kit in the presence of EDTA-free protease inhibitors.
3. Wash rHuPrP-sen inclusion bodies twice with 0.1× BugBuster® and pellet by centrifugation.
4. Load Ni-NTA SuperFlow™ resin with denatured protein from the inclusion bodies, and refold the protein using a linear gradient (100 mM sodium phosphate (pH 8.0), 10 mM Tris, 6 M guanidine-HCl) over 6 h at a flow rate of 1 mL/min using an AKTA Explorer System (GE Healthcare, New York, NY, USA).
5. Elute protein with 100 mM sodium phosphate (pH 5.8), 500 mM imidazole, and 10 mM Tris. Dilute pooled fractions to 0.2 mg/mL with water.
6. To determine the concentration of rHuPrP-sen, measure the absorbance at 280 nm. The purity of the final protein preparation should be ≥99%, as estimated by SDS-PAGE immunoblotting.
7. After purification, store protein aliquots at −80 °C in 10 mM phosphate buffer (pH 6.8).

3.2 RT-QuIC Using CSF or Brain Homogenates

1. RT-QuIC reactions are prepared in a 96-well optical-bottom black microplate (Thermo Fisher Scientific, Waltham, MA, USA) to a final volume of 100 μL.
2. Set up reactions in quadruplicate or sextuplicate in reaction buffer in microplate wells.
3. Add freshly thawed rHuPrP-sen to reaction buffer. Mild variations in rHuPrP-sen concentrations (0.06–0.15 mg/mL) may be observed between rHuPrP-sen samples.
4. Add CSF to 5 μL/well for use as seeds for RT-QuIC reactions. CSF samples from healthy subjects should be added as negative controls. This type of control is referred to as a no-seed control.
5. Cover the 96-well microplate with sealing tape, and incubate at 37 °C using a microplate reader (Infinite M200 or F200 fluorescence plate reader; TECAN, Männedorf, Switzerland).
6. Shake the microplate intermittently by circular shaking for 30 s at the highest speed, followed by no shaking for 30 s, with a 2 min pause to measure fluorescence. Perform 300 cycles per microplate.
7. The microplate reader measures ThT fluorescence in relative fluorescence units (arbitrary units), which reflect the kinetics of fibril formation.

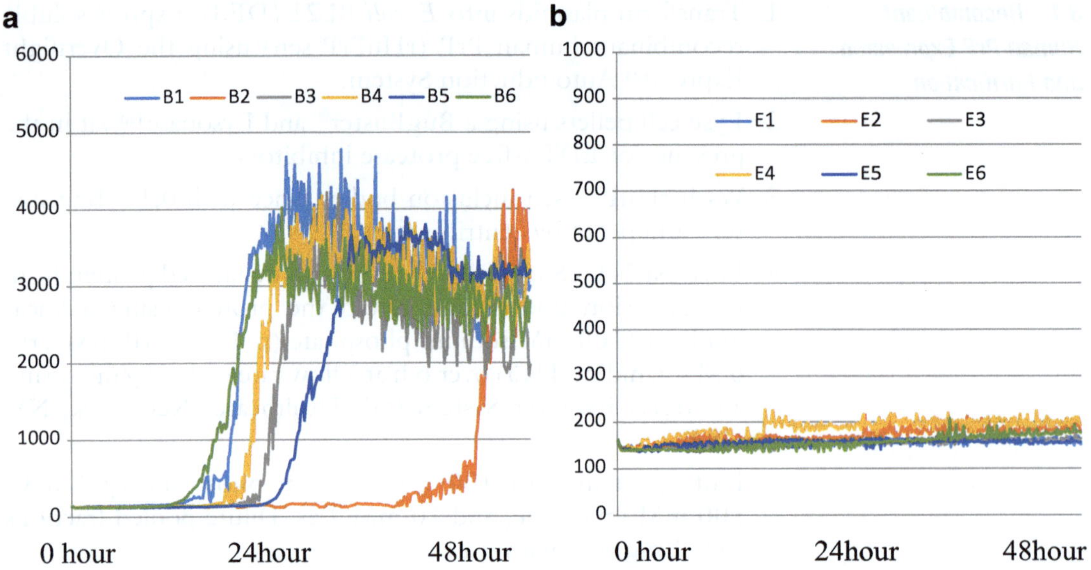

Fig. 1 Analysis of RT-QuIC reaction data of CSF samples from sporadic CJD patients using the RT-QuIC method. An example of a kinetic graph from an RT-QuIC reaction using CSF samples from sporadic CJD patients is shown in (**a**). The baseline kinetic graph from an RT-QuIC reaction using CSF samples from non-prion disease patients shows no change (**b**). When analyzing RT-QuIC reactions using CSF samples from sporadic CJD patients, the baseline kinetics of ThT in the RT-QuIC reactions are not changed if they fail to exceed 5% of the baseline

8. Monitor the fluorescence using the mild fluorescence intensity level (100 ~ 1000 arbitrary unites) reading every 10 min at 440 nm excitation and 485 nm emission wavelengths using a monochromator (Infinite M200) or filter (Infinite F200).
9. It is important that the samples are not contaminated. First prepare non-prion material inside a biological safety cabinet in a prion-free laboratory, and use aerosol-resistant tips. After preparing non-prion material in a 96-well optical-bottom black microplate, prepare prion-infected material inside the biological safety cabinet.

3.3 RT-QuIC Analysis

A positive reaction is regarded as an increase in the level of fluorescence that exceeds 200% of the baseline fluorescence level of the buffer and healthy no-seed controls over six consecutive readings (Fig. 1).

4 Notes

1. The detection of PrP^{Sc} in human materials using recombinant hamster PrP protein

   Recombinant hamster PrP expression and purification

(a) Amplify DNA sequences encoding hamster PrP residues 23–231 and 90–231 (K02234) using PCR, and ligate the PCR products into the pET41 vector (EMD Biosciences) as NdeI-HindIII inserts. Verify the accuracy of the sequences.

(b) Transform plasmids into *E. coli* BL21 (DE3). Express the rPrP-c using the Overnight Express™ Autoinduction System (EMD Biosciences).

(c) Treat the cell pellets from 0.25 L of cultures with two liquid nitrogen freeze-thaw cycles. Lyse cells with BugBuster® master mix (EMD Biosciences) to isolate the inclusion bodies.

(d) Wash inclusion bodies twice with 0.1× BugBuster®, pellet by centrifugation, and store at −20 °C for later use.

(e) Denature pellets using 8 M guanidine-HCl at pH 8.0 on a rotator for 30 min at room temperature and then centrifuge at 16,000 × *g* for 5 min. Add the denatured proteins to Ni-NTA SuperFlow resin (Qiagen) that has been equilibrated in denaturing buffer (100 mM sodium phosphate (pH 8.0), 10 mM Tris, 6 M guanidine-HCl).

(f) Wash the resin in the column using a linear gradient of refolding buffer (100 mM sodium phosphate (pH 8.0) and 10 mM Tris) and the Bio-Rad system (Bio-Rad, Hercules, CA, USA). Refold the denatured protein on the column using a linear gradient of refolding buffer.

(g) Elute bound protein using a linear gradient of elution buffer (100 mM sodium phosphate (pH 5.8), 10 mM Tris, 500 mM imidazole) at 1 mL/min for 60 min.

2. RT-QuIC method using CSF or brain homogenate

(a) Prepare reactions in a 96-well optical-bottom black microplate (Thermo Fisher Scientific) to a final volume of 100 μL.

(b) Make up reaction buffer components to final concentrations of 170 mM NaCl, 10 mM PBS (pH 7.4), 10 μM EDTA, 10 μM ThT, and 0.1 mg/mL recombinant hamster PrP.

(c) Add 15 μL of CSF or 2 μL of diluted brain homogenate (10% brain homogenate) to the RT-QuIC reaction buffer in a final volume of 100 μL.

(d) Cover the 96-well microplate with sealing tape and incubate at 42 °C using a microplate reader (FLUOstar OPTIMA plate reader; BMG LABTECH, Offenburg, Germany) for 90–120 h.

(e) Shake the microplate intermittently by circular shaking for 1 min at the highest speed (600 rpm), followed by no shaking for 1 min, with a 2 min pause to measure fluorescence.

(f) Incubate and shake the microplate for ~10 h for sporadic CJD brain homogenate and >30 h for CSF samples from CJD patients.

(g) After 90 h, the majority of CSF samples in the RT-QuIC reaction should show an increase in fluorescence.

(h) The microplate reader measures ThT fluorescence in relative fluorescence units (arbitrary units), which reflect the kinetics of fibril formation.

(i) Monitor the fluorescence using the lowest fluorescence intensity level every 10 min at 450 nm excitation and 485 nm emission wavelengths using a biochromator (FLUOstar OPTIMA plate reader).

References

1. Kocisko DA, Come JH, Priola SA et al (1994) Cell-free formation of protease-resistant prion protein. Nature 370:471–474
2. Saborio GP, Permanne B, Soto C (2001) Sensitive detection of pathological prion protein by cyclic amplification of protein misfolding. Nature 411:810–813
3. Atarashi R (2009) Recent advances in cell-free PrPSc amplification technique. Protein Pept Lett 16:256–259
4. Wilham JM, Orru CD, Bessen RA et al (2010) Rapid end-point quantitation of prion seeding activity with sensitivity comparable to bioassays. PLoS Pathog 6:e1001217
5. Atarashi R, Satoh K, Sano K et al (2011) Ultrasensitive human prion detection in cerebrospinal fluid by real-time quaking-induced conversion. Nat Med 17:175–178
6. McGuire LI, Peden AH, Orru CD et al (2012) Real time quaking-induced conversion analysis of cerebrospinal fluid in sporadic Creutzfeldt-Jakob disease. Ann Neurol 72:278–285
7. Cramm M, Schmitz M, Karch A et al (2015) Characteristic CSF prion seeding efficiency in humans with prion diseases. Mol Neurobiol 51:396–405

Chapter 22

Methods for Molecular Diagnosis of Human Prion Disease

Jonathan D.F. Wadsworth, Gary Adamson, Susan Joiner, Lara Brock, Caroline Powell, Jacqueline M. Linehan, Jonathan A. Beck, Sebastian Brandner, Simon Mead, and John Collinge

Abstract

Human prion diseases are associated with a range of clinical presentations, and they are classified by both clinicopathological syndrome and etiology, with subclassification according to molecular criteria. Here, we describe updated procedures that are currently used within the MRC Prion Unit at UCL to determine a molecular diagnosis of human prion disease. Sequencing of the *PRNP* open reading frame to establish the presence of pathogenic mutations is described, together with detailed methods for immunoblot or immunohistochemical determination of the presence of abnormal prion protein in the brain or peripheral tissues.

Key words Bovine spongiform encephalopathy, Creutzfeldt-Jakob disease, Fatal familial insomnia, Gerstmann-Sträussler-Scheinker disease, Kuru, Prion, Prion disease, Prion protein, Transmissible spongiform encephalopathy, Variant Creutzfeldt-Jakob disease

1 Introduction

Prion diseases are fatal neurodegenerative disorders that include Creutzfeldt-Jakob disease (CJD), Gerstmann-Sträussler-Scheinker disease (GSS), fatal familial insomnia (FFI), kuru, and variant CJD (vCJD) in humans [1–4]. Their central feature is the posttranslational conversion of host-encoded, cellular prion protein (PrP^C), to an abnormal isoform, designated PrP^{Sc} [1, 5]. Human prion diseases are biologically unique in that the disease process can be triggered through inherited germline mutations in the human prion protein gene (*PRNP*) and infection (by inoculation or in some cases by dietary exposure) with prion-infected tissue or by rare sporadic events that generate PrP^{Sc} [1–3]. Substantial evidence indicates that abnormal PrP isoforms are the principal, if not the sole, components of the transmissible infectious agent or prion [1, 2, 5]. The existence of multiple strains or isolates of prions has been difficult to accommodate within a protein-only model of prion propagation; however, considerable experimental evidence

Victoria A. Lawson (ed.), *Prions: Methods and Protocols*, Methods in Molecular Biology, vol. 1658,
DOI 10.1007/978-1-4939-7244-9_22, © Springer Science+Business Media LLC 2017

suggests that prion strain diversity is encoded within PrP itself and that phenotypic diversity in human prion diseases relates to differing physicochemical properties of abnormal PrP isoforms [5–12]. Furthermore, the propagation of distinct abnormal PrP isoforms may be determined by the host genome [13–15].

Human prion diseases are associated with a range of clinical presentations, and they are classified by both clinicopathological syndrome and etiology, with subclassification according to molecular criteria [3, 4, 16–18]. Approximately 85% of cases occur sporadically as Creutzfeldt-Jakob disease (sporadic CJD) at a rate of one to two cases per million population per year across the world, with an equal incidence in men and women [2, 19, 20]. Approximately 15% of human prion disease is associated with autosomal dominant pathogenic mutations in *PRNP*, and to date, more than 40 mutations have been described [2, 18, 21, 22]. These include insertions of between 4 and 12 extra repeats within the octapeptide repeat region between codons 51 and 91, a two-octapeptide repeat deletion (OPRD), and various point mutations causing missense or stop substitutions (Fig. 1).

Although human prion diseases are transmissible diseases, acquired forms have, until recently, been confined to rare and unusual situations. The most frequent causes of iatrogenic CJD occurring through medical procedures have arisen as a result of treatment with growth hormone derived from human cadavers or implantation of dura mater grafts [23–25]. Less frequent incidences of iatrogenic human prion disease have resulted from transmission of CJD prions during corneal transplantation, contaminated electroencephalographic electrode implantation, and surgical operations using contaminated instruments or apparatus [25]. The most well-known example of acquired prion disease in humans is kuru, which was transmitted by cannibalism among the Fore and neighboring linguistic groups of the Eastern Highlands in Papua New Guinea [17, 26–28]. Remarkably, kuru demonstrates that incubation periods of infection with human prions can exceed 50 years [28]. The appearance of vCJD in the United Kingdom from 1995 onward [29], and the experimental confirmation that this is caused by the same prion strain as that causing bovine spongiform encephalopathy (BSE) in cattle [7, 14, 30, 31], has led to a widespread concern that exposure to the epidemic of BSE poses a distinct threat to public health in the United Kingdom and other countries [3, 32]. The extremely prolonged and variable incubation periods of these diseases, particularly when crossing a transmission barrier [3, 33], mean that it will be many years before the full consequences of human exposure to bovine prions can be predicted with confidence. In the meantime, significant numbers in the population may be incubating this disease [34] and pose a risk for transmitting infection to others via blood transfusion, blood products, tissue and organ transplantation, and other iatrogenic routes [32, 35–39].

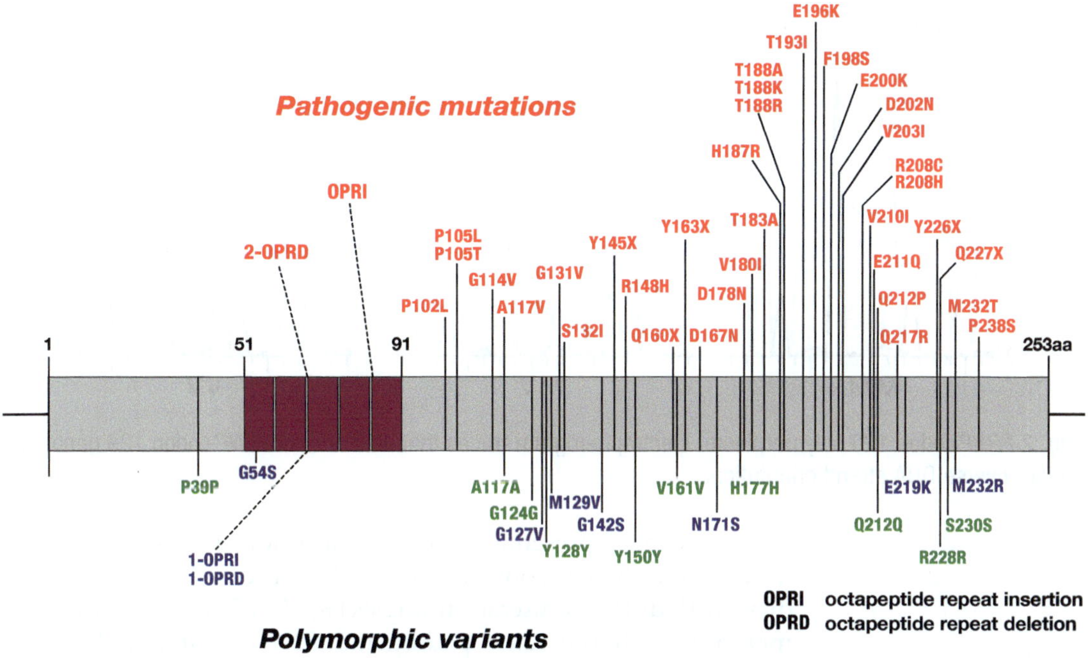

Fig. 1 Pathogenic mutations and polymorphisms in human prion protein. The pathogenic mutations associated with human prion disease are shown above the human PrP coding sequence. These consist of 4–12 OPRI within the octapeptide repeat region between codons 51 and 91, a two OPRD, and various point mutations causing missense or stop codon substitutions. Some of these changes have been observed in individual patients only and should be considered as possible pathogenic mutations that require confirmation. Point mutations are designated by the wild-type amino acid preceding the codon number, followed by the mutant residue, using single-letter amino acid nomenclature (X denotes stop). Polymorphic variants are shown below the PrP coding sequence (synonymous changes, *green*; nonsynonymous changes, *blue*). Codon 129 and 219 polymorphisms have profound susceptibility, disease-modifying effects, or both. Deletion of one octapeptide repeat is not associated with prion disease in humans

Polymorphism at codon 129 of *PRNP* (Fig. 2) (encoding either methionine [M] or valine [V]) powerfully affects susceptibility to human prion diseases [20, 27, 40–44]. About 38% of Northern Europeans are homozygous for the more frequent methionine allele, 51% are heterozygous, and 11% homozygous for valine. Homozygosity at *PRNP* codon 129 predisposes to the development of sporadic and acquired CJD [20, 27, 40–43] and is most strikingly observed in vCJD. At the time of writing all but one pathologically proven cases of vCJD have been homozygous for codon 129 methionine of *PRNP* [3, 38, 44–46] with one heterozygous patient confirmed in 2016 [47]. Heterozygosity at codon 129 is thought to confer resistance to prion disease by inhibiting homologous protein-protein interactions essential for efficient prion replication, while the presence of methionine or valine at residue 129 controls the propagation of distinct human prion strains via conformational selection [3, 5, 32, 45, 46].

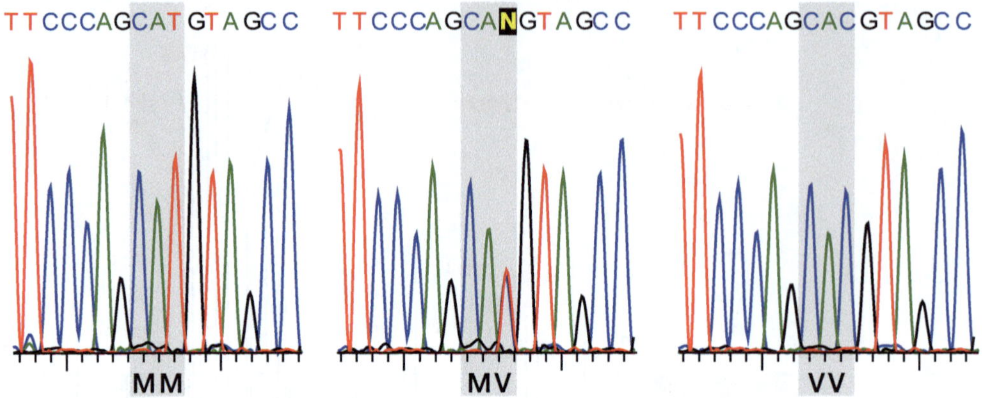

Fig. 2 *PRNP* codon 129 polymorphism. Electropherogram traces that illustrate all *PRNP* codon 129 genotypes in the reverse DNA strand orientation

The clinical presentation of human prion disease varies enormously, and there is considerable overlap observed between individuals with different disease etiologies [3, 19, 22] and even in family members with the same pathogenic *PRNP* mutation [21, 22, 48–52]. Progressive dementia, cerebellar ataxia, pyramidal signs, chorea, myoclonus, extrapyramidal features, pseudobulbar signs, seizures, and amyotrophic features can be seen in variable combinations. Criteria used for diagnosis of human prion disease are available [3, 53], and definite diagnosis of sporadic and acquired prion disease relies upon neuropathological examination and the demonstration of abnormal PrP deposition in the central nervous system by either immunoblotting or immunohistochemistry. In the appropriate clinical setting, identification of a pathogenic *PRNP* mutation provides clear diagnosis of inherited prion disease and subclassification according to mutation; *PRNP* analysis is also used for presymptomatic genetic testing in affected families [3, 17, 18, 22, 48, 54, 55]. Because of the extensive phenotypic variability seen in inherited prion disease and its ability to mimic other neurodegenerative conditions, notably Alzheimer's disease, *PRNP* analysis should be considered in all patients with undiagnosed dementing and ataxic disorders [3, 22, 49]. In addition, we have recently described a novel clinical and pathological phenotype associated with a Y163X mutation in *PRNP*, a prion protein systemic amyloidosis, characterized by slow disease progression, diarrhea, autonomic failure, and neuropathy [56]. This newly recognized novel disease phenotype indicates that *PRNP* analysis should also be considered in the investigation of unexplained chronic diarrhea associated with a neuropathy or an unexplained syndrome similar to familial amyloid polyneuropathy.

Clinicians are fortunate that a range of sensitive and specific investigations are available to secure a diagnosis in life. Principal among these is magnetic resonance imaging (MRI) of the brain which shows increased signal on T2- or diffusion-weighted images in the cerebral cortex, basal ganglia, and thalamus in >80% of patients with sporadic CJD [57, 58], although these are sometimes overlooked [59]. Cerebrospinal fluid (CSF) analysis is also helpful for differential diagnosis as the absence of a raised cell count or oligoclonal bands is reassuring from the point of view of inflammatory etiologies. Raised CSF proteins 14-3-3, S100b, and NSE are also found in ~90% sporadic CJD patients although their presence is not specific for the disease. CSF PrP amplification technologies show promise as specific and sensitive confirmatory tests [60, 61].

The brains of patients with prion disease frequently show no recognizable abnormalities on gross examination at necropsy; however, microscopic examination of the brain at either necropsy or in antemortem biopsy specimens typically reveals characteristic histopathologic changes, consisting of neuronal vacuolation and degeneration, which give the cerebral gray matter a microvacuolated or "spongiform" appearance accompanied by a reactive proliferation of astroglial cells [62–64]. Although spongiform degeneration is frequently detected, it is not an obligatory neuropathologic feature of prion disease; astrocytic gliosis, although not specific to the prion diseases, is more constantly seen. The lack of a lymphocytic inflammatory response is also an important characteristic. Demonstration of abnormal PrP immunoreactivity or more specifically biochemical detection of PrP^{Sc} in brain material by immunoblotting techniques is diagnostic of prion disease, and some forms of prion disease are characterized by deposition of amyloid plaques composed of insoluble aggregates of PrP [62–65]. Amyloid plaques are a notable feature of kuru and GSS [62, 65, 66], but they are less frequently found in the brains of patients with sporadic CJD, which typically show a diffuse pattern of abnormal PrP deposition [9, 62, 65]. The histopathologic features of vCJD are remarkably consistent and distinguish it from other human prion diseases, with large numbers of PrP-positive amyloid plaques that differ in morphology from the plaques seen in kuru and GSS in that the surrounding tissue takes on a microvacuolated appearance, giving the plaques a florid appearance [29, 67]. The tissue distribution of PrP^{Sc} in vCJD differs strikingly from that in classical CJD with uniform involvement of lymphoreticular tissues [35, 68–72]. Depending upon the density of lymphoid follicles, PrP^{Sc} concentrations in vCJD peripheral tissues can vary enormously, with levels relative to the brain as high as 10% in tonsil [35, 68] or as low as 0.002% in rectum [35, 39]. Tonsil biopsy is used for diagnosis of vCJD, and to date, it has shown 100% sensitivity and specificity for diagnosis of vCJD at an early clinical stage [3, 35, 38, 68], although some patients show scanty deposition of abnormal PrP and a large number of follicles may have to be examined by immunohistochemistry [73].

In this chapter, we update our previous contribution to *Prion Protein Protocols* [74] and describe the procedures that are currently used within the MRC Prion Unit at UCL to provide a molecular diagnosis of human prion disease. Methods for sequencing the *PRNP* open reading frame to establish the presence of pathogenic mutations and to determine *PRNP* polymorphic codon 129 genotype are described together with procedures used for immunoblot or immunohistochemical determination of the presence of abnormal PrP in the brain or peripheral tissues.

2 Materials

2.1 Molecular Genetics

1. BACC3 DNA extraction kit from GE Healthcare Life Sciences.
2. TE buffer: 10 mM Tris and 1 mM EDTA, pH 8.0.
3. MegaMix-Royal or MegaMix-Blue (Microzone, Haywards Heath, UK).
4. HyperLadder 100 bp (Bioline, London, UK).
5. Micro-Clean (Microzone).
6. BigDye version 1.1 Cycle Sequencing Kit (Applied Biosystems, Foster City, CA).
7. BetterBuffer (Microzone).
8. 0.5 M EDTA, pH 8.0, diluted fourfold in water.
9. Hi-Di formamide (Applied Biosystems).
10. Performance-optimized polymer 7 (Applied Biosystems).
11. DdeI restriction endonuclease, including NEBuffer 3 (New England Biolabs, Ipswich, MA).
12. MetaSieve agarose (Flowgen, Ashby, Leicestershire, UK).
13. PflFI restriction endonuclease, including NEBuffer 4 (New England Biolabs).
14. BsaI restriction endonuclease, including NEBuffer 3 (New England Biolabs).
15. TOPO TA Cloning Kit for Sequencing (Invitrogen, Paisley, UK).
16. Luria-Bertani (LB) broth.
17. TaqMan GTXpress Master Mix (Life Technologies, Paisley, UK).
18. TaqMan MGB probes (Life Technologies, Paisley, UK).

2.2 Immunoblotting

1. Dulbecco's sterile phosphate-buffered saline (PBS) lacking Ca^{2+} and Mg^{2+} ions.
2. Duall tissue grinders (Anachem Ltd., Luton, Bedfordshire, UK).

3. Proteinase K (specific enzymatic activity ~30 Anson units/g) prepared as a stock solution of 1 mg/ml in water.
4. Sodium dodecyl sulfate (SDS) sample buffer.
   (a) A stock concentrate of 2× SDS sample buffer [142 mM Tris, 22.72% (v/v) glycerol, 4.54% (w/v) SDS, and 0.022% (w/v) bromophenol blue] is prepared in water and titrated to pH 6.8 with HCl.
   (b) This solution requires adjustment with reducing agent and proteinase K inhibitor immediately before use to produce 2× working SDS sample buffer.
   (c) For preparation of 0.5 ml of 2× working SDS sample buffer, mix the following: 440 μl of stock concentrate of 2× SDS sample buffer plus 20 μl of 2-mercaptoethanol plus 40 μl of 100 mM 4-(2-aminoethyl)-benzenesulfonyl fluoride prepared in water.
   (d) This produces 2× working SDS sample buffer of the following final composition: 125 mM Tris–HCl, 20% (v/v) glycerol, pH 6.8, containing 4% (w/v) SDS, 4% (v/v) 2-mercaptoethanol, 8 mM 4-(2-aminoethyl)-benzenesulfonyl fluoride, and 0.02% (w/v) bromophenol blue.
5. Novex® 16% Tris-glycine SDS polyacrylamide gel electrophoresis (PAGE) gels (Life Technologies Ltd., Paisley, UK).
6. SeeBlue prestained molecular mass markers (Life Technologies Ltd).
7. SDS-PAGE electrophoresis buffer: 100 ml of 10× Tris-glycine, SDS concentrate [0.25 M Tris, 1.92 M glycine, 1% (w/v) SDS (National Diagnostics, USA)] plus 900 ml water.
8. Immobilon-P transfer membrane.
9. Electroblotting buffer: 100 ml of 10× Tris-glycine concentrate [0.25 M Tris and 1.92 M glycine (National Diagnostics, USA)], 700 ml of water, and 200 ml of methanol.
10. PBST: 100 ml of 10× PBS concentrate (low in phosphate) (VWR, Lutterworth, UK), 900 ml of water, and 0.5 ml of Tween 20.
11. Anti-PrP monoclonal antibody 3F4 (Covance, Princeton, New Jersey).
12. Goat anti-mouse IgG (fab-specific) alkaline phosphatase conjugate (absorbed with human serum proteins) (Sigma-Aldrich Poole, Dorset, UK).
13. CDP-Star chemiluminescent substrate (Life Technologies Ltd).
14. Carestream BioMax MR film (Anachem Ltd).
15. AttoPhos chemifluorescent substrate (Promega, Madison, WI).

   (a) Mix 36 mg of AttoPhos substrate in 60 ml of AttoPhos buffer.
   (b) Store as 3 ml aliquots at −20 °C.
16. Sodium lauroylsarcosine (Merck Chemicals Ltd., Nottingham, UK).
17. Benzonase (Benzon nuclease purity 1 [25 U/μl], Merck Chemicals Ltd).
18. Sodium phosphotungstic acid stock solution.
   (a) Stock solution is 4% (w/v) sodium phosphotungstic acid containing 170 mM $MgCl_2$ prepared in water, pH 7.4.
   (b) For preparation of 10 ml of stock solution, add 0.4 g of sodium phosphotungstic acid and 0.35 g $MgCl_2{\cdot}6H_2O$ to a 50 ml polypropylene tube, and make to ~9 ml with water.
   (c) The pH of this solution is acidic and needs to be titrated with 5 M NaOH to pH 7.4 before adjusting to a final volume of 10 ml with water.
   (d) On addition of NaOH, immediate formation of insoluble $MgOH_2$ occurs that will redissolve on vortexing.
   (e) Addition of 5 M NaOH followed by vortexing and measurement of pH needs to be done repetitively.
   (f) For 10 ml of stock solution, addition of 360 μl of 5 M NaOH will generate pH 7.4.

2.3 Immunohistochemistry

2.3.1 Procurement

1. 10% buffered formal-saline.
2. Biopsy cassettes (R. A. Lamb, Eastbourne, UK).

2.3.2 Prion Deactivation with Formic Acid

1. Biopsy cassettes (R. A. Lamb).
2. 98% formic acid.
3. 2 M sodium hydroxide: 80 g of sodium hydroxide pellets in water to 1 liter.
4. 10% buffered formal-saline.

2.3.3 Tissue Processing

1. 10% buffered formal-saline.
2. Industrial methylated spirits (J.M. Loveridge Ltd., Southampton, UK), diluted in water to desired concentration.
3. Xylene.
4. Pure paraffin wax (R. A. Lamb).

2.3.4 Tissue Sectioning

1. Microtome (Leica, Wetzlar, Germany).
2. SuperFrost microscope slides (VWR, West Chester, PA).

2.3.5 Tissue Staining

1. Xylene.
2. Absolute ethanol diluted in water to desired concentration.

3. Harris hematoxylin (BDH).
4. Acid alcohol: 1% HCl in absolute ethanol.
5. Eosin Y solution 0.5%, aqueous (VWR).
6. Pertex mounting medium (Cox Scientific Ltd., Kettering, UK).
7. Benchmark staining machine (Ventana Medical Systems, Illkirch Cedex, France).
8. Protease 1 (Ventana Medical Systems).
9. Rabbit anti-glial fibrillary protein (Dako UK Ltd., Ely, Cambridgeshire, UK); antibody diluent (Ventana Medical Systems).
10. iView DAB Detection Kit (Ventana Medical Systems), containing an inhibitor solution (3% hydrogen peroxide) (4 min), universal biotinylated secondary antibody (10 min), streptavidin-horseradish peroxidase solution (10 min), 3,3-diaminobenzidine and hydrogen peroxide (20 min), and copper solution (4 min).
11. Hematoxylin (Ventana Medical Systems).
12. Bluing reagent (Ventana Medical Systems).
13. Tris-EDTA-citrate buffer, pH 7.8: 2.1 mM Tris, 1.3 mM EDTA, and 1.1 mM sodium citrate.
14. 98% formic acid.
15. 10 mM sodium citrate buffer, pH 6.0: solution A, 10.5 g of citric acid in 500 ml of deionized water and solution B, 29.41 g of sodium citrate in 1000 ml of water.
    (a) Add 18 ml of solution A to 82 ml of solution B, and then adjust to 1 liter final volume with water.
16. Anti-PrP monoclonal antibody ICSM35 (D-Gen Ltd., London, UK).
17. Protease 3 (Ventana Medical Systems).
18. SuperBlock (Pierce Chemical, Rockford, IL).
19. Prepared with methanol GPR and 30% hydrogen peroxide.
20. Tris-buffered saline (TBS): 50 mM Tris, 145 mM NaCl, pH 7.6.
    (a) For 1 liter of 10× stock, add 60.5 g of Trizma base and 84.7 g of sodium chloride to 800 ml of water.
    (b) Adjust to pH 7.6 with 32.00 ml of concentrated HCl.
    (c) Make to 1 liter final volume with water.
    (d) Dilute 10× TBS to 1× concentration with water.
21. 4 M guanidine thiocyanate.
    (a) Add 472.64 g guanidine thiocyanate in water to a final volume of 1 liter.
22. Normal rabbit serum (Dako UK Ltd.).

23. Biotinylated rabbit anti-mouse immunoglobulins (Dako UK Ltd.).
24. Strept AB complex/HRP Duet Kit (Dako UK Ltd.).
25. 3,3-diaminobenzidine tetrachloride.
    (a) Use at a final concentration of 25 mg in 100 ml of 1× TBS.

3 Methods

3.1 Molecular Genetics

3.1.1 Isolation of Genomic DNA from Blood

1. All procedures are performed within a class 1 microbiological safety cabinet situated within an ACDP level II containment laboratory with strict adherence to local rules of safe working practice. Informed consent for the analysis of a sample must be established before investigation. This may be obtained from the patient (or next of kin or other advocate in accordance with the Mental Capacity Act 2005 in the UK). For predictive testing, we expect evidence of appropriate genetic counseling before analysis.
2. Genomic DNA is extracted from whole anticoagulated blood (typically from a 5 ml EDTA tube) by using the Nucleon BACC3 DNA extraction kit (*see* Subheading 2.1, **item 1**) following the manufacturer's instructions. DNA concentrations are determined using a NanoDrop ND-1000 spectrophotometer and adjusted to 200–250 ng/μl in TE buffer (*see* Subheading 2.1, **item 2**). Concentrations are remeasured before dilution of DNA in TE buffer to a final concentration of 20 ng/μl and storage at 4 °C.

3.1.2 Sequencing of PRNP Open Reading Frame

PCR of *PRNP* Open Reading Frame

1. Prepare a premix of MegaMix-Royal (a 2× concentrate, *see* Subheading 2.1, **item 3**) containing primers at 0.5 μM sufficient for 25 μl reactions on a 96-well plate. PCR primers used to amplify the open reading frame are 5′-CTA TGC ACT CAT TCA TTA TGC-3′ (forward) and 5′-GTT TTC CAG TGC CCA TCA GTG-3′ (reverse). Use 12.5 μl MegaMix-Royal, and make each reaction up to 25 μl using water and primers.
2. Add 1 μl of 20 ng/μl genomic DNA.
3. Thermal cycling is performed on an MJ Research (Watertown, MA) Tetrad 1 PCR machine or similar using the following cycling parameters:
    (a) 95 °C for 5 min.
    (b) 95 °C for 30 s.
    (c) 58 °C for 40 s.
    (d) 72 °C for 1 min.
    (e) Repeat **steps b–d** an additional 34 times.

4. Assess polymerase chain reaction (PCR) by electrophoresis of 5 μl of product on a 2% ethidium bromide-stained agarose gel with 5 μl of HyperLadder IV (*see* Subheading 2.1, **item 4**) size standard. The gel is viewed using a Gel Doc 1000 transilluminator (Bio-Rad, Hemel Hempstead, UK) and Quantity One 4.5.1 software or similar.

PCR Product Cleanup

1. An equal volume of Micro-Clean (*see* Subheading 2.1, **item 5**) is added to the PCR product and mixed well by pipetting or vortexing.
2. The mixture is left at room temperature for 15 min.
3. The plate is centrifuged at 2000–4000 × *g* for 40 min at room temperature.
4. The supernatant is removed by centrifuging the plate at 40 × *g* for 30 s in an inverted position on tissue paper by using centrifuge plate holders.
5. Resuspend the cleaned PCR product in 150 μl of water.

Sequencing Reactions

1. For each sequencing reaction, prepare a premix of 1 μl of BigDye (*see* Subheading 2.1, **item 6**), 5 μl of BetterBuffer (*see* Subheading 2.1, **item 7**), 0.75 μl of sequencing primer at 5 pmol/μl, approximately 2.5 ng of cleaned PCR product, and water to a final volume of 15 μl. The amount of PCR product is estimated using visual comparison with known amounts of HyperLadder IV (*see* Subheading 2.1, **item 4**) size standard.
2. Sequencing primers are 5′-GAC GTT CTC CTC TTC ATT TT-3′ (forward 1), 5′-CCG AGT AAG CCA AAA ACC AAC-3′ (forward 2), 5′-CAC CAC CAC TAA AAG GGC TGC-3′ (reverse 1), and 5′-TTC ACG ATA GTA ACG GTC C-3′ (reverse 2).
3. Sequencing reactions are thermally cycled on an MJ Research Tetrad 1 PCR machine or similar using the following cycling parameters:
   (a) 96 °C for 1 min.
   (b) 96 °C for 10 s.
   (c) 50 °C for 5 s.
   (d) 60 °C for 4 min.
   (e) Repeat **steps b–d** 24 times.

Sequencing Product Cleanup

1. To each sequencing reaction, add 3.75 μl of 0.125 M EDTA, pH 8.0 (*see* Subheading 2.1, **item 8**).
2. Add 45 μl of 100% ethanol to each reaction and mix by pipetting.

3. Leave reactions at room temperature for 15 min.
4. Centrifuge the plate at 3000 × *g* for 30 min at 4 °C.
5. The supernatant is removed by centrifuging the plate at 185 × *g* for 1 min in an inverted position on tissue paper.
6. Add 50 μl of 70% ethanol in water.
7. Centrifuge the plate at 1650 × *g* for 15 min at 4 °C.
8. The supernatant is removed by centrifuging the plate at 185 × *g* for 1 min in an inverted position on tissue paper.
9. Place the plate on the PCR block held at 37 °C for 5 min to remove final traces of ethanol.

Electrophoresis

1. Add 10 μl of Hi-Di formamide-loading solution (*see* Subheading 2.1, **item 9**), and vortex the plate for 30 s.
2. Denature the samples by placing on the PCR block held at 95 °C for 2 min, and then immediately transfer to ice.
3. Standard run conditions are applied to electrophoresis of sequencing products on an Applied Biosystems 3730 × l, using polymer POP7 (*see* Subheading 2.1, **item 10**), 50 cm arrays, and a standard run module with a sample injection time of 15 s.

Data Analysis

1. Data analysis is performed using Applied Biosystems SeqScape software version 2.5.
2. Analysis filter settings are adjusted to allow assembly of poor data due to insertions or deletions (maximum mixed bases 95%, maximum Ns 95%, minimum clear length bp of 1, and minimum sample score of 1).
3. Poor data or failed reactions are removed from projects by visual inspection of data.

3.1.3 PCR Size Fractionation to Investigate Insertion or Deletion Variants

1. Prepare a premix of MegaMix Blue (*see* Subheading 2.1, **item 3**) containing primers at 0.5 μM sufficient for 25 μl reactions on a 96-well plate. PCR primers are 5′-GAC CTG GGC CTC TGC AAG AAG CGC-3′ (forward) and 5′-GGC ACT TCC CAG CAT GTA GCC G-3′ (reverse).
2. Add 1 μl of 20 ng/μl genomic DNA.
3. Thermal cycling is performed on an MJ Research Tetrad 1 PCR machine or similar using the following cycling parameters:
   (a) 94 °C for 5 min.
   (b) 94 °C for 30 s.
   (c) 65 °C for 30 s.
   (d) 72 °C for 1 min.
   (e) Repeat **steps b–d** 34 times.
   (f) 72 °C for 5 min.

1 2 3 4 5 6 7 8 9

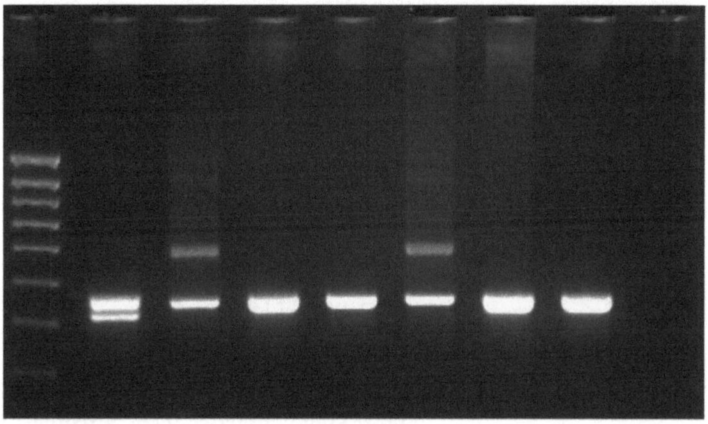

Fig. 3 Analysis of *PRNP* OPRI mutations. Image from agarose gel electrophoresis of *PRNP* amplicons illustrating the presence of an insertional mutation. Lane 1, HyperLadder IV; lane 2, 1-OPRD control; lane 3, 6-OPRI control; lane 4–8, patient samples; lane 6 demonstrates amplification of a heterozygous insertional mutation of 144 base pairs (6-OPRI mutation positive); lanes 4, 5, 7, and 8 are wild-type alleles only; lane 9, no-template control

4. Assess PCR by electrophoresis of 5 μl of product on a 2% ethidium bromide-stained agarose gel with 5 μl of HyperLadder IV (*see* Subheading 2.1, **item 4**) size standard. The gel is viewed using a Bio-Rad Gel Doc 1000 transilluminator and Quantity One 4.5.1 software or similar.
5. 1-OPRD and 6-OPRI controls are run on each gel (Fig. 3).

3.1.4 Mutation Confirmation

A second assay is performed to confirm the presence or absence of missense or stop mutations when a predictive genetic test is being carried out. PCR size fractionation, as described above, is sufficient in addition to sequencing when testing for insertion mutants in a predictive setting; however, unexpected or unknown insertion mutations may require cloning to confirm exact base pair composition. Examples of confirmatory assays used to detect the more common *PRNP* missense mutations and cloning methodology are described in Subheading "Confirmation of P102L."

Confirmation of P102L

1. Prepare a premix of MegaMix Blue (*see* Subheading 2.1, **item 3**) containing primers at 0.5 μM sufficient for 25 μl reactions on a 96-well plate. PCR primers are 5′-GAC CTG GGC CTC TGC AAG AAG CGC-3′ (forward) and 5′-GGC ACT TCC CAG CAT GTA GCC G-3′ (reverse).
2. Add 1 μl of genomic DNA at 20 ng/μl.
3. Thermal cycling is performed on an MJ Research Tetrad 1 PCR machine or similar using the following cycling parameters:
   (a) 94 °C for 5 min.

(b) 94 °C for 30 s.
(c) 65 °C for 30 s.
(d) 72 °C for 1 min.
(e) Repeat **steps b–d** 34 times.
(f) 72 °C for 5 min.

4. Prepare restriction endonuclease reaction by adding 10 μl of PCR product, 1 μl of DdeI (*see* Subheading 2.1, **item 11**), 2.5 μl of 10× NEBuffer 3 (*see* Subheading 2.1, **item 11**), and 11.5 μl of H_2O.
5. Incubate reaction at 37 °C for 3 h.
6. Electrophorese 10 μl of digested PCR product on a 3% 2:1 MetaSieve agarose (*see* Subheading 2.1, **item 12**) ethidium bromide-stained gel using 5 μl of HyperLadder IV (*see* Subheading 2.1, **item 4**) size standard. The gel is viewed using a Bio-Rad Gel Doc 1000 transilluminator and Quantity One 4.5.1 software.
7. Digested positive and negative controls are run on each gel to visualize mutant DNA (95, 101, and 152 bp) and wild-type DNA (101 and 247 bp) fragment patterns.

D178N

1. Prepare a premix of MegaMix Blue (*see* Subheading 2.1, **item 3**) containing primers at 0.5 μM sufficient for 25 μl reactions on a 96-well plate. PCR primers are 5′-CTA TGC ACT CAT TCA TTA TGC-3′ (forward) and 5′-GTT TTC CAG TGC CCA TCA GTG-3′ (reverse).
2. Add 1 μl of genomic DNA at 20 ng/μl.
3. Thermal cycling is performed on an MJ Research Tetrad 1 PCR machine or similar using the following cycling parameters:
   (a) 94 °C for 5 min.
   (b) 94 °C for 30 s.
   (c) 55 °C for 40 s.
   (d) 72 °C for 45 s.
   (e) Repeat **steps b–d** 34 times.
   (f) 72 °C for 5 min.
4. Prepare restriction endonuclease reaction by adding 10 μl of PCR product, 1 μl of PflFI (*see* Subheading 2.1, **item 13**), 2.5 μl of 10× NEBuffer 4 (*see* Subheading 2.1, **item 13**), and 11.5 μl of H_2O.
5. Incubate reaction at 37 °C for 3 h.
6. Electrophorese 10 μl of digested PCR product on a 2% ethidium bromide-stained agarose gel by using 5 μl of HyperLadder IV (*see* Subheading 2.1, **item 4**) size standard. The gel is

viewed using a Bio-Rad Gel Doc 1000 transilluminator and Quantity One 4.5.1 software.

7. Digested positive and negative controls are run on each gel to visualize mutant DNA (1015 bp) and wild-type DNA (386 and 629 bp) fragment patterns.

E200K

1. Prepare a premix of MegaMix Blue (*see* Subheading 2.1, **item 3**) containing primers at 0.5 μM sufficient for 25 μl reactions on a 96-well plate. PCR primers are 5′-CTA TGC ACT CAT TCA TTA TGC-3′ (forward) and 5′-GTT TTC CAG TGC CCA TCA GTG-3′ (reverse).
2. Add 1 μl of genomic DNA at 20 ng/μl.
3. Thermal cycling is performed on an MJ Research Tetrad 1 PCR machine or similar using the following cycling parameters:
   (a) 94 °C for 5 min.
   (b) 94 °C for 30 s.
   (c) 55 °C for 40 s.
   (d) 72 °C for 45 s.
   (e) Repeat **steps b–d** 34 times.
   (f) 72 °C for 5 min.
4. Prepare restriction endonuclease reaction by adding 10 μl of PCR product, 1 μl of BsaI (*see* Subheading 2.1, **item 14**), 2.5 μl of 10× NEBuffer 3 (*see* Subheading 2.1, **item 14**), and 11.5 μl of H_2O.
5. Incubate reaction at 50 °C for 3 h.
6. Electrophorese 10 μl of digested PCR product on a 2% ethidium bromide-stained agarose gel by using 5 μl of HyperLadder IV (*see* Subheading 2.1, **item 4**) size standard, and view using a Bio-Rad Gel Doc 1000 transilluminator and Quantity One 4.5.1 software.
7. Digested positive and negative controls are run on each gel to visualize mutant DNA (1015 bp) and wild-type DNA (318 and 697 bp) fragment patterns.

3.1.5 Characterization of Insertion Mutations

Generation of Amplicon to Be Cloned

1. Prepare a premix of MegaMix Blue (*see* Subheading 2.1, **item 3**) containing primers at 0.5 μM sufficient for 25 μl reactions on a 96-well plate. PCR primers are 5′-GAC CTG GGC CTC TGC AAG AAG CGC-3′ (forward) and 5′-GGC ACT TCC CAG CAT GTA GCC G-3′ (reverse). (Note that MegaMix Blue contains an enzyme that has 3-prime terminal adenosine triphosphate transferase activity that ensures that the amplicon has "A" overhangs to anneal to the "T" overhangs of the cloning vector.)
2. Add 1 μl of genomic DNA at 20 ng/μl.

3. Thermal cycling is performed on an MJ Research Tetrad 1 PCR machine or similar using the following cycling parameters:
   (a) 94 °C for 5 min.
   (b) 94 °C for 30 s.
   (c) 65 °C for 30 s.
   (d) 72 °C for 1 min.
   (e) Repeat **steps b–d** 34 times.
   (f) 72 °C for 10 min.
4. Electrophorese 5 μl of PCR product on a 2% ethidium bromide-stained agarose gel by using 5 μl of HyperLadder IV (*see* Subheading 2.1, **item 15**) size standard, and view using a Bio-Rad Gel Doc 1000 transilluminator and Quantity One 4.5.1 software.
5. To preserve the "A" overhangs, before cloning, carry out as little manipulation as possible, and use fresh product.

Ligation and Cloning

1. Perform TOPO TA cloning as described in the online user manual version O, April 10, 2006 (25-0276). Updates of this protocol are available at www.invitrogen.com. Use Invitrogen cat. no. K4575-01 (*see* Subheading 2.1, **item 15**) (TOP10, Chemically Competent *E. coli*, 20 reactions).

Analysis and Sequencing of Recombinant Clones

1. There is no need to miniprep possible positive clones. Pick white and light blue clones (color enhancement can be obtained by leaving the plate at 4 °C overnight if this is preferred), and inoculate wells of a prewarmed 96-well tissue culture or storage plate containing 150–200 μl of LB broth (*see* Subheading 2.1, **item 16**) containing the appropriate antibiotic.
2. Incubate at 37 °C for about 5–6 h or until the wells are opaque.
3. Transfer 50 μl aliquots to a 96-well PCR plate.
4. Seal the plate and place on a PCR machine for 10 min at 99 °C to lyse the bacteria. Aliquots (1 μl) of these crude DNA preparations can then be used to produce amplicons by using the same methods used to produce the original amplicon by simple transfer to a fresh 96-well PCR plate containing the appropriate PCR mix.
5. Electrophorese 5 μl of PCR product on a 2% ethidium bromide-stained agarose gel by using 5 μl of HyperLadder IV (*see* Subheading 2.1, **item 4**) size standard. The gel is viewed using a Bio-Rad Gel Doc 1000 transilluminator and Quantity One 4.5.1 software.
6. Amplicons from positive wells can be purified and sequenced according to the automated sequencing protocol for PCR

products (*see* Subheadings "PCR Product Cleanup," "Sequencing Reactions," "Sequencing Product Cleanup," "Electrophoresis," and "Data Analysis").

7. Amplicons will contain Taq polymerase artifacts preserved in the cloning process. Therefore, at least three clones should be sequenced to obtain a consensus sequence, preferably on both strands.

3.1.6 Alternative Confirmation of Missense or Stop Mutations Using Real-Time PCR

P102L

1. For each reaction, prepare a premix of 10 μl of GTXpress Master Mix (*see* Subheading 2.1, **item 17**); 2 μl each of forward primer, reverse primer, wild-type probe, and mutant probe; and 1 μl of water (Primer stocks at 9 μM, probe stocks are at 2 μM). Primer and probes are 5′-GGA GGT GGC ACC CAC AGT C-3′ (forward primer), 5′-GCC ATG TGC TTC ATG TTG GTT-3′ (reverse primer), 5′-FAM-CTT ACT CGG CTT GTT C-3′ (wild-type probe), and 5′-VIC-CTT ACT CAG CTT GTT CC-3′ (mutant probe) (*see* Subheading 2.1, **item 18**).
2. Add 1 μl of genomic DNA at 20 ng/μl. Positive, negative, and no-template controls are also run on each plate.
3. Thermal cycling and data capture are performed on a QuantStudio 12 K Flex machine (Life Technologies) using a fast 96-well block, choosing genotyping mode and a fast instrument run.
4. On completion of the run, inspecting the allele discrimination plot will show the unknown sample clustering with either the positive controls or the negative controls.

D178N

1. For each reaction, prepare a premix of 10 μl of GTXpress Master Mix (*see* Subheading 2.1, **item 17**), and 2 μl each of forward primer, reverse primer, wild-type probe, and mutant probe and 1 μl of water. Primer and probes are 5′-CAG GCC CAT GGA TGA GTA CA-3′ (forward primer), 5′-CGT GTG CTG CTT GAT TGT GA-3′ (reverse primer), 5′-FAM-TTG ACG CAG TCG TGC A-3′ (wild-type probe), and 5′-VIC-TTG ACG CAG TTG TGC A-3′ (mutant probe) (*see* Subheading 2.1, **item 18**).
2. Add 1 μl of genomic DNA at 20 ng/μl. Positive, negative, and no-template controls are also run on each plate.
3. Thermal cycling and data capture are performed on a QuantStudio 12 K Flex machine (Life Technologies) using a fast 96-well block, choosing genotyping mode and a fast instrument run.
4. On completion of the run, inspecting the allele discrimination plot will show the unknown sample clustering with either the positive controls or the negative controls.

E200K

1. For each reaction, prepare a premix of 10 μl of GTXpress Master Mix (*see* Subheading 2.1, **item 17**); 2 μl each of forward primer, reverse primer, wild-type probe, and mutant probe; and 1 μl of water. Primer and probes are 5′-CGG TCA CCA CAA CCA CCA A-3′ (forward primer), 5′-CAC GCG CTC CAT CAT CTT AA-3′ (reverse primer), 5′-FAM-AAC TTC ACC GAG ACC GA-3′ (wild-type probe), and 5′-VIC-AGA ACT TCA CCA AGA CC-3′ (mutant probe) (*see* Subheading 2.1, **item 18**).
2. Add 1 μl of genomic DNA at 20 ng/μl. Positive, negative, and no-template controls are also run on each plate.
3. Thermal cycling and data capture are performed on a Quant-Studio 12 K Flex machine (Life Technologies) using a fast 96-well block, choosing genotyping mode and a fast instrument run.
4. On completion of the run, inspecting the allele discrimination plot will show the unknown sample clustering with either the positive controls or the negative controls.

3.2 Immunoblotting

3.2.1 Biosafety

1. All procedures are performed within a class 1 microbiological safety cabinet situated within an ACDP level III containment laboratory with strict adherence to local rules of safe working practice. Informed consent for the analysis of samples must be in place before investigation.
2. No unsealed biological material (tissue or derivative sample thereof) is manipulated outside of the class 1 microbiological safety cabinet. Disposable gloves, safety gown, and safety glasses are worn at all times.
3. 1.5 ml screw-top microfuge tubes containing a rubber O-ring are used.
4. Guidelines for decontamination of human prions are available [75]. All disposable plasticware (e.g., tubes, tips, and so on) and solutions containing biological material are decontaminated in 50% (v/v) sodium hypochlorite solution (containing >20,000 ppm available chlorine prepared in water) for at least 1 h before disposal of the liquid phase down the designated laboratory sinks within the containment laboratory. Sharps (needles and scalpels) are disposed of immediately after use into a sharps bin and autoclaved at 136 °C for 20 min before incineration.
5. Decontaminated plasticware is transferred to a sharps bin and autoclaved at 136 °C for 20 min before incineration.

3.2.2 Preparation of Tissue Homogenate

1. Tissue specimens, stored frozen in sealed pots within the ACDP level III containment laboratory, are transferred into a class 1 microbiological safety cabinet and partially thawed and placed on a petri dish.

2. A suitable quantity of tissue is excised using a scalpel and sealed in a disposable plastic pot and weighed. The tissue is then prepared as a 10% (w/v) homogenate in Dulbecco's sterile PBS lacking Ca^{2+} and Mg^{2+} ions (*see* Subheading 2.2, **item 1**). The amount of PBS to add in microliters is equal to nine times wet weight of tissue in milligrams. This calculation will produce a homogenate very close to a true 10% (w/v) ratio without the necessity of having to accurately measure the total volume of tissue in PBS before the homogenization process.
3. Homogenization of brain tissue is achieved by serial passage of the tissue through syringe needles of decreasing diameter (needle gauges 19, 21, 23, and 25).
4. Homogenization of peripheral tissue is achieved through the use of glass Duall tissue grinders (*see* Subheading 2.2, **item 2**).
5. The homogenate is stored as aliquots in 1.5 ml screw-top microfuge tubes at −80 °C.

3.2.3 Proteinase K Digestion and Electrophoresis

1. 10% brain homogenate is thawed, thoroughly vortexed, and then centrifuged at 100 × *g* (800 rpm) for 1 min in a microfuge (*see* **Note 1**).
2. 20 μl aliquots of the resultant supernatant are adjusted to a final concentration of 50 μg/ml proteinase (*see* Subheading 2.2, **item 3**) by addition of 1.05 μl of a 1 mg/ml proteinase K stock solution (*see* **Note 2**).
3. Samples are incubated at 37 °C for 1 h, followed by centrifugation at 16,100 × *g* (13,200 rpm) for 1 min in a microfuge.
4. The digestion is terminated by resuspension of the sample with an equal volume (21 μl) of 2× working SDS sample buffer (*see* Subheading 2.2, **item 4**) and *immediate* transfer to a 100 °C heating block for 10 min.
5. Samples for analysis in the absence of proteinase K treatment are treated directly with an equal volume of 2× working SDS sample buffer (*see* Subheading 2.2, **item 4**) and heated similarly.
6. All samples are centrifuged at 16,100 × *g* (13,200 rpm) for 1 min in a microfuge, thoroughly vortexed, and then recentrifuged 16,100 × *g* for 1 min before electrophoresis of the supernatant.
7. 10 μl of the supernatant is loaded on Novex® 16% Tris-glycine polyacrylamide mini gel (*see* Subheading 2.2, **item 5**) (*see* **Note 3**). The remainder of the sample can be stored at −80 °C. Then, 10 μl of 1× working SDS sample buffer (prepared by mixing 2× working SDS sample buffer with an equal volume of water) should be added in any blank lane. Ten microliters of SeeBlue prestained molecular mass markers (*see* Subheading 2.2, **item 6**) is used to calibrate the gel.

8. Gels are run at a constant voltage of 200 V for 80 min in SDS-PAGE running buffer (*see* Subheading 2.2, **item** 7) (*see* **Note 3**).
9. Gels are electroblotted (one gel per XCell II™ Blot Module (Novex®)) on to polyvinylidene difluoride membrane (*see* Subheading 2.2, **item 8**) in electroblotting buffer (*see* Subheading 2.2, **item 9**) at a constant voltage of 35 V for 2 h or 15 V overnight. Immobilon-P membrane is soaked for 2 min in 100% methanol and then rinsed in electroblotting buffer immediately before use.

3.2.4 High-Sensitivity Chemiluminescence (ECL)

1. Blots are blocked with 5% (w/v) nonfat milk powder in PBST (*see* Subheading 2.2, **item 10**) for 1 h followed by brief rinsing with PBST.
2. Blots are incubated with anti-PrP monoclonal antibody 3F4 (*see* Subheading 2.2, **item 11**) at a final concentration of 0.2 μg/ml in PBST containing 0.1% (w/v) sodium azide for either 90 min or overnight.
3. Blots are washed for a minimum of 30 min and up to 60 min with at least six changes of PBST.
4. Blots are incubated for 1 h with a 1:10,000 dilution of goat anti-mouse IgG-phosphatase conjugate (*see* Subheading 2.2, **item 12**) in PBST.
5. Blots are washed for a minimum of 30 min and up to 60 min with at least six changes of PBST.
6. Blots are washed 2 × 5 min with 20 mM Tris, pH 9.8, containing 1 mM $MgCl_2$.
7. Blots are developed with chemiluminescent substrate CDP-Star (*see* Subheading 2.2, **item 13**) and visualized on BioMax MR film (*see* Subheading 2.2, **item 14**) (*see* **Note 4**).

3.2.5 Standard Enhanced Chemifluorescence (ECF)

1. Blots are blocked with 5% (w/v) nonfat milk powder in PBST (*see* Subheading 2.2, **item 10**) for 1 h followed by brief rinsing with PBST.
2. Blots are incubated with anti-PrP monoclonal antibody 3F4 (*see* Subheading 2.2, **item 11**) at a final concentration of 0.2 μg/ml in PBST containing 0.1% (w/v) sodium azide for either 90 min or overnight.
3. Blots are washed for a minimum of 30 min and up to 60 min with at least six changes of PBST.
4. Blots are incubated for 1 h with a 1:5000 dilution of goat anti-mouse IgG-phosphatase conjugate (*see* Subheading 2.2, **item 12**) in PBST.
5. Blots are washed for a minimum of 30 min and up to 60 min with at least six changes of PBST.

6. Blots are washed 2 × 5 min with 20 mM Tris, pH 9.8, containing 1 mM $MgCl_2$.
7. Blots are developed with chemifluorescent substrate AttoPhos (*see* Subheading 2.2, **item 15**) and visualized on a Storm 840 PhosphorImager (GE Healthcare, Little Chalfont, Buckinghamshire, UK). PrP glycoforms are quantified with ImageQuaNT software (GE Healthcare) (*see* **Note 4**).

3.2.6 Sodium Phosphotungstic Acid Precipitation

Methods are adapted from the original procedure of Safar et al. [76] as described by Wadsworth et al. [35].

1. 10% (w/v) homogenates from human brain or peripheral tissues prepared in Dulbecco's PBS lacking Ca^{2+} and Mg^{2+} ions (*see* Subheading 2.2, **item 1**) are centrifuged at 100 × *g* (800 rpm) for 1 min in a microfuge.
2. 500 μl of the resultant supernatant is mixed with an equal volume of 4% (w/v) sodium lauroylsarcosine (*see* Subheading 2.2, **item 16**) prepared in Dulbecco's PBS lacking Ca^{2+} and Mg^{2+} ions (*see* Subheading 2.2, **item 1**) and incubated for 10 min at 37 °C with constant agitation.
3. Samples are adjusted to final concentrations of 50 U/ml Benzonase (*see* Subheading 2.2, **item 17**) (add 2 μl of 25 U/μl Benzon nuclease, purity 1) and 1 mM $MgCl_2$ (add 0.5 μl of 2 M $MgCl_2$ prepared in water) and incubated for 30 min at 37 °C with constant agitation.
4. Samples are adjusted with 81.3 μl of a sodium phosphotungstic acid stock solution (*see* Subheading 2.2, **item 18**) to give a final concentration in the sample of 0.3% (w/v) sodium phosphotungstic acid. This stock solution is prewarmed to 37 °C before use, and both the sample and the stock solution should be at 37 °C upon mixing to avoid formation of insoluble magnesium salts.
5. Samples are incubated at 37 °C for 30 min with constant agitation before centrifugation at 16,100 × *g* (13,200 rpm) for 30 min in a microfuge. The microfuge rotor can be prewarmed to 37 °C before use, because this helps to avoid salt precipitation during centrifugation.
6. After careful isolation of the supernatant, the sample is recentrifuged at 16,100 × *g* (13,200 rpm) for 2 min, and the residual supernatant is discarded. New tops are placed on the microfuge tubes.
7. Pellets are resuspended to a 20 μl final volume with Dulbecco's PBS lacking Ca^{2+} and Mg^{2+} ions containing 0.1% (w/v) sodium lauroylsarcosine and proteinase K digested and processed for immunoblotting as described in Subheading 3.2.3, **steps 2–9** (*see* **Note 5**).

3.3 Immunohistochemistry

3.3.1 Procurement

Biosafety

For samples suspected to contain infectious prions, all procedures are performed within a class 1 microbiological safety cabinet situated within an ACDP level III containment laboratory with strict adherence to local rules of safe working practice. Informed consent for the analysis of samples must be in place before investigation. Samples are kept in a category III laboratory before decontamination with formic acid. Guidelines for decontamination of human prions are available [75]. Liquids that have been in contact with infected samples are decontaminated by mixing with an equal volume of 2 M sodium hydroxide for at least 1 h. For certain reagents, specialist disposal is preferred due to chemical incompatibilities [75].

Whole Brain, Brain Hemispheres, or Whole Internal Organs

1. Large specimens of tissue (whole brain, brain hemispheres, whole internal organs) are suspended in 10% buffered formal-saline (*see* Subheading 2.3.1, **item 1**). The volume added should be approximately five times the volume of the tissue.
2. If there is excess blood within the sample, the 10% buffered formal-saline should be exchanged until it remains clear.
3. Tissue is left for up to 3 weeks to ensure adequate fixation and hardening.
4. After fixation, samples of tissue are excised with dimensions suitable for histology cassettes (*see* Subheading 2.3.1, **item 2**).

Small Specimens of Brain or Peripheral Tissues

1. Smaller pieces of the brain (frontal cortex, temporal cortex, parietal cortex, occipital cortex, cerebellum) or samples of other peripheral tissues (tonsil, spleen, lymph nodes, appendix), with dimensions no larger than approximately 3 cm × 3 cm × 1 cm, are commonly provided for investigation.
2. Tissue samples are immersed in approximately 5 volumes of 10% buffered formal-saline (*see* Subheading 2.3.1, **item 1**).
3. If there is excess blood within the sample, the 10% buffered formal-saline should be exchanged until it remains clear.
4. Fixation of the samples is achieved after 2 days.
5. After fixation, samples of tissue are excised with dimensions suitable for histology cassettes (*see* Subheading 2.3.1, **item 2**).

3.3.2 Prion Deactivation with Formic Acid

1. All brain tissue must be of a size suitable for processing. Generally, this is considered as the size and thickness of the histology cassettes (*see* Subheading 2.3.2, **item 1**). Care must be taken not to overfill the cassettes, because this will result in poor processing and distortion.
2. After being encased in labeled cassettes, the samples are immersed in 98% formic acid (*see* Subheading 2.3.2, **item 2**) for 1 h.

3. Formic acid is decanted into a waste pot half filled with 2 M sodium hydroxide (*see* Subheading 2.3.2, **item 3**).
4. Specimens are treated with approximately 5 volumes of 10% buffered formal-saline (*see* Subheading 2.3.2, **item 4**) for 1 h.
5. The 10% buffered formal-saline (*see* Subheading 2.3.2, **item 4**) is exchanged at least once to ensure any excess of formic acid has been removed before tissue processing.
6. Samples are removed from the ACDP level III containment laboratory.
7. Samples are placed on a tissue processor in an ACDP level II containment laboratory.

3.3.3 Tissue Processing

Ten percent buffered formal-saline (*see* Subheading 2.3.3, **item 1**) is an aqueous fixative; therefore, the samples are treated through a series of processing stages before wax embedding. Each stage needs to be of sufficient length to ensure impregnation. The stages are as follows:

1. Dehydration: The samples are taken through a series of industrial methylated spirits (IMS) (*see* Subheading 2.3.3, **item 2**) (70, 90, 100%) to remove water (Table 1).

Table 1
Protocol for overnight processing of tissue samples for immunohistochemistry

Solution	Time (min)	Temperature
10% buffered formal-saline	30	Ambient
IMS 70%	75	Ambient
IMS 70%	75	Ambient
IMS 70%	75	Ambient
IMS 90%	60	Ambient
IMS 90%	60	Ambient
IMS 100%	75	Ambient
IMS 100%	75	Ambient
Xylene	75	Ambient
Xylene	75	Ambient
Molten paraffin wax	50	60 °C
Molten paraffin wax	50	60 °C
Molten paraffin wax	50	60 °C
Molten paraffin wax	50	60 °C

2. Clearing: The alcohol is replaced by xylene (*see* Subheading 2.3.3, **item 3**), a fluid miscible with IMS and paraffin wax (*see* Subheading 2.3.3, **item 4**) (Table 1).
3. Impregnation: The xylene is replaced with molten paraffin wax (*see* Subheading 2.3.3, **item 4**) (Table 1).
4. Embedding: The samples are embedded in the desired orientation in molten paraffin wax (*see* Subheading 2.3.3, **item 4**). Once the wax has hardened, the samples are ready for sectioning.

3.3.4 Tissue Sectioning

1. The microtome (*see* Subheading 2.3.4, **item 1**) is set at 8 μm for tissue sectioning, although this measure can be varied.
2. The sample block, now in wax, is mounted on to the microtome chuck, and serial sections of the sample are taken.
3. Sections are floated out on a water bath set at 40 °C.
4. The sections are mounted on SuperFrost microscope slides (*see* Subheading 2.3.4, **item 2**) and left to air-dry at 37 °C for approximately 2 h.
5. Slides are dried at 60 °C for a minimum of 2 h, after which they are ready to be stained.
6. Tonsil sections require cutting just before staining. Immunoreactivity is markedly reduced if sections are exposed to air for long periods of time.

3.3.5 Tissue Staining

Staining with Hematoxylin and Eosin (H&E)

1. Rehydrate the sections by removing paraffin in three changes of xylene (*see* Subheading 2.3.5, **item 1**), followed by sequential washing for 1–2 min with graded alcohol (*see* Subheading 2.3.5, **item 2**) (100% × 2, 90%, and 70%) and final washing in running tap water.
2. Place the slides in filtered Harris hematoxylin solution (*see* Subheading 2.3.5, **item 3**) for 5 min.
3. Wash briefly in running tap water, and differentiate in 1% acid alcohol (*see* Subheading 2.3.5, **item 4**) for 30 s.
4. Wash well in running tap water and allow the color to develop. Check microscopically. Nuclei look dark blue, whereas background shows a weak residual hematoxylin coloration.
5. Wash briefly in running water, and stain with Eosin Y solution (*see* Subheading 2.3.5, **item 5**) for 2–3 min.
6. Wash sections sequentially for ~1–2 min with water, 70% ethanol, 90% ethanol, 100% ethanol, and xylene.
7. Mount sections in a xylene-based mounting medium, Pertex (*see* Subheading 2.3.5, **item 6**) (*see* **Note 6**).

Staining for Glial Fibrillary Acidic Protein (GFAP)

1. The sections to be stained are placed in plastic racks, and paraffin was removed, as described in Subheading "Staining with Hematoxylin and Eosin (H&E)."
2. The slides are placed on the Benchmark XT staining machine (*see* Subheading 2.3.5, **item** 7) (Ventana Medical Systems) with a 4 min pretreatment with Protease 1 (*see* Subheading 2.3.5, **item 8**).
3. The slides are incubated with a GFAP antibody (*see* Subheading 2.3.5, **item 9**) diluted 1:1000 in antibody diluent (*see* Subheading 2.3.5, **item 9**).
4. The slides are stained using the staining kit iView DAB (*see* Subheading 2.3.5, **item 10**) and counterstained using hematoxylin (*see* Subheading 2.3.5, **item 11**) and a bluing reagent (*see* Subheading 2.3.5, **item 12**).
5. Once the run is finished, the slides are washed in hot soapy water (diluted washing up liquid) and dehydrated through alcohol and xylene.
6. Mount sections in a xylene-based mounting medium, Pertex (*see* Subheading 2.3.5, **item 6**) (*see* **Note** 7).

Staining for PrP

1. The sections to be stained are placed in plastic racks, and paraffin was removed as described in Subheading "Staining with Hematoxylin and Eosin (H&E)."
2. The pretreatment for detection of abnormal PrP deposition is dependent upon the tissue and the length of fixation. For human brain samples that have been fixed for up to ~2 weeks, the microwave heat retrieval method is preferred. If the brain samples are fixed for longer periods, the pressure cooker method is used. If tonsil or other secondary lymphoid tissue is being examined, the autoclaving heat retrieval method is used.

Microwave Method

3. After removal of paraffin (*see* Subheading "Staining with Hematoxylin and Eosin (H&E)") the slides are placed in 1 liter of Tris-EDTA-citrate buffer (*see* Subheading 2.3.5, **item 13**), and then they were placed in a microwave for 25 min at 800-W power.
4. The slides are washed in running cold tap water for 3 min.
5. The samples are covered with 98% formic acid (*see* Subheading 2.3.5, **item 14**), incubated for 5 min, and then washed in running cold tap water for 5 min to remove excess formic acid.

Pressure Cooker Method

6. After removal of paraffin (*see* Subheading "Staining with Hematoxylin and Eosin (H&E)"), the slides are placed in a pressure cooker containing 1.5 liters of boiling Tris-EDTA-citrate buffer (*see* Subheading 2.3.5, **item 13**) for 5 min at high pressure and 5 min at low pressure.

7. Place the slides under running cold water for 5 min, and treat with 98% formic acid (*see* Subheading 2.3.5, **item 14**) for a further 5 min. Wash the slides in running water for 5 min to remove excess formic acid.

Autoclaving Heat Retrieval

8. After removal of paraffin (*see* Subheading "Staining with Hematoxylin and Eosin (H&E)"), the slides are placed in an autoclave-resistant tub containing 1 liter of citrate buffer (*see* Subheading 2.3.5, **item 15**).
9. The tub is covered with aluminum foil and run in an autoclave at 121 °C for 20 min. Allow autoclave to return to low pressure before removing the tub. Place slides under running cold tap water.
10. Treat with 98% formic acid (*see* Subheading 2.3.5, **item 14**) for a further 5 min. Wash the slides in running water for 5 min to remove excess formic acid.

Automated Staining

11. The monoclonal anti-PrP antibody ICSM35 (*see* Subheading 2.3.5, **item 16**) (1 mg/ml stock) is used at a 1:1000 dilution in antibody diluent (*see* Subheading 2.3.5, **item 9**).
12. Automated staining is carried out on the BenchMark staining machine (*see* Subheading 2.3.5, **item 7**) from Ventana Medical Systems.
13. The slides are subjected to further pretreatment with Protease 3 (*see* Subheading 2.3.5, **item 17**) for 4 min and then 10 min with SuperBlock (*see* Subheading 2.3.5, **item 18**), a blocking agent. The slides are stained using the staining kit iView DAB (*see* Subheading 2.3.5, **item 10**) and counterstained using hematoxylin (*see* Subheading 2.3.5, **item 11**) and a bluing reagent (*see* Subheading 2.3.5, **item 12**).
14. Once the staining process is complete, the slides are washed in hot soapy water (diluted washing up liquid) and dehydrated through alcohol and xylene. They are mounted in a xylene-based mountant as described in Subheading "Staining with Hematoxylin and Eosin (H&E)" (*see* **Note 8**).

Manual Staining

15. Sections are treated to remove paraffin as far as 100% alcohol (*see* Subheading "Staining with Hematoxylin and Eosin (H&E)").
16. Block endogenous peroxidase activity on the sections by treatment with 2.5% (v/v) hydrogen peroxide in methanol (*see* Subheading 2.3.5, **item 19**) for 30 min.
17. Wash sections in running tap water for 5 min and then in purified water for 5 min, and transfer to an appropriate container for autoclaving. Autoclave at 121 °C for 20 min in Tris-buffered saline (TBS), pH 7.6 (*see* Subheading 2.3.5, **item 20**).

18. Cool slides in running tap water. Treat the slides in 98% formic acid (*see* Subheading 2.3.5, **item 14**) for 5 min, and wash in running tap water for 5–10 min.
19. Treat sections with 4 M guanidine thiocyanate (*see* Subheading 2.3.5, **item 21**) at 4 °C for 2 h and then wash in tap water and transfer to TBS (*see* Subheading 2.3.5, **item 20**).
20. Block nonspecific immunoglobulin staining with normal rabbit serum (*see* Subheading 2.3.5, **item 22**) diluted 1:10 in TBS for 30 min. Do not wash off.
21. Apply primary antibody ICSM35 (*see* Subheading 2.3.5, **item 16**) (1 mg/ml stock) at 1:1500 dilution in TBS containing 1:100 normal rabbit serum (*see* Subheading 2.3.5, **item 22**), overnight at 4 °C.
22. Wash in several changes of TBS.
23. Incubate in biotinylated rabbit anti-mouse immunoglobulins (*see* Subheading 2.3.5, **item 23**) 1:200 in TBS for 45 min.
24. Wash in several changes of TBS.
25. Incubate in AB complex (*see* Subheading 2.3.5, **item 24**) for 45 min.
26. Wash in several changes of TBS.
27. Develop in 3,3-diaminobenzidine tetrachloride (*see* Subheading 2.3.5, **item 25** (25 mg/100 ml of TBS) plus 30 μl of hydrogen peroxide (*see* Subheading 2.3.5, **item 19**) (added just before use) for 5–15 min. Check microscopically. Once chromogen has developed to satisfaction, wash slides in running tap water for 10 min.
28. Counterstain in Harris hematoxylin (*see* Subheading 2.3.5, **item 3**) for 3 min.
29. Differentiate in 1% acid alcohol (*see* Subheading 2.3.5, **item 4**) for 5 s.
30. Allow blue coloration to develop in tap water, 5 min.
31. Dehydrate, clear, and mount as described in Subheading "Staining with Hematoxylin and Eosin (H&E)" (*see* **Note 8**).

4 Notes

1. Whole brain homogenate can be analyzed by identical procedures; however, problems of high sample viscosity due to nucleic acid aggregation are often encountered. For processing of 20 μl of whole brain homogenate, preincubation with 0.5 μl of Benzonase for 10 min at 20 °C is recommended before further sample analysis by using appropriately adjusted volumes of subsequent reagents.

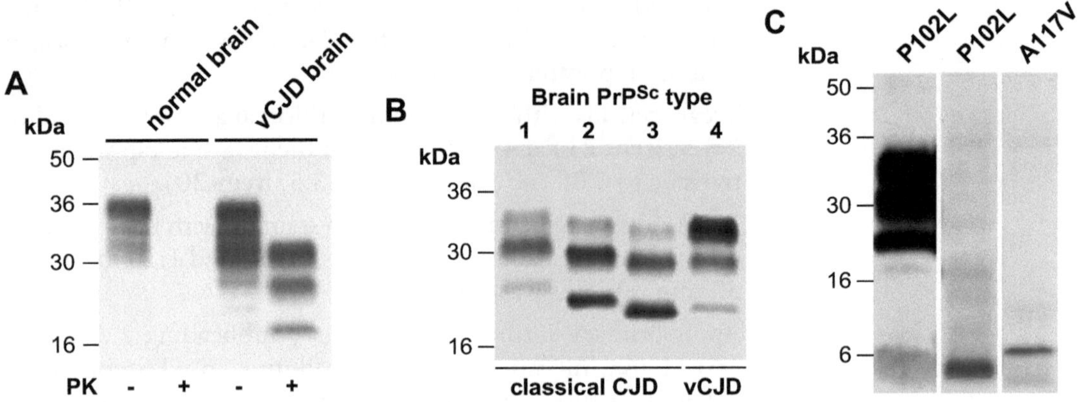

Fig. 4 Immunoblot analysis of human PrP. (**a**) Immunoblot analysis of normal human brain and vCJD brain homogenate before and after treatment with proteinase K (PK). PrP^{C} in both normal and vCJD brain is completely degraded by PK, whereas PrP^{Sc} present in vCJD brain shows resistance to proteolytic degradation leading to the generation of amino terminally truncated fragments of di-, mono-, and nonglycosylated PrP. (**b**) Immunoblot of PK-digested brain homogenate with monoclonal antibody 3F4 showing PrP^{Sc} types 1–4 in the human brain. Types 1–3 PrP^{Sc} are seen in the brain of classical forms of CJD (either sporadic or iatrogenic CJD), whereas type 4 PrP^{Sc} is uniquely seen in vCJD brain. Classification according to Hill et al. [9]. (**c**) Immunoblots of PK-digested brain homogenate from cases of inherited prion disease with *PRNP* mutations showing protease-resistant PrP fragments of ~6–8 kDa. The *PRNP* point mutation is designated above each immunoblot. Immunoblots were developed with anti-PrP monoclonal antibody 3F4

2. PrP^{Sc} is covalently indistinguishable from PrP^{C}, but it can be differentiated from PrP^{C} by its partial resistance to proteolysis and its marked insolubility in detergents [1, 77]. Under conditions in which PrP^{C} is completely degraded by the nonspecific protease, proteinase K, PrP^{Sc} in sporadic and acquired forms of human prion disease exists in an aggregated form with the C-terminal two thirds of the protein showing marked resistance to proteolysis, leading to the generation of amino terminally truncated fragments of di-, mono-, and nonglycosylated PrP [1, 77] (Fig. 4).
3. The procedures described here have been optimized for use with Novex® 16% acrylamide precast Tris-glycine gels. Variation in the resolution of the system may occur if other gel systems are used or if reagent compositions are varied from those listed here. Optimal resolution of PrP^{Sc} fragment size is achieved after electrophoresis for 80 min at 200 V. For improved separation of PrP glycoforms for densitometry analysis, electrophoresis is performed for 90 min at 200 V.
4. To date, we have identified four major types of human PrP^{Sc} associated with sporadic and acquired human prion diseases that can be differentiated by their fragment size on immunoblots after limited proteinase K digestion of brain homogenates [7, 9, 78, 79] (Fig. 4). These types can be further classified by

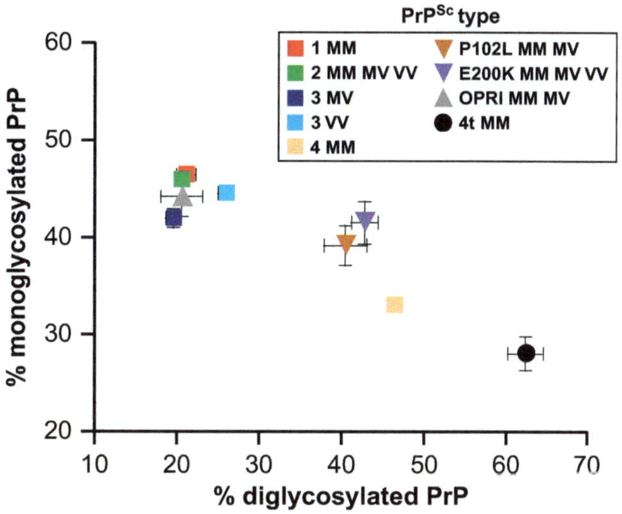

Fig. 5 PrP glycoform ratios in human prion disease. PK digestion of brain homogenate and analysis by enhanced chemifluorescence with anti-PrP monoclonal antibody 3F4 enables calculation of the proportions of di-, mono-, and nonglycosylated PrP. The plot shows the protease-resistant PrP glycoform ratio seen in classical CJD (PrP^{Sc} types 1–3), in vCJD (PrP^{Sc} type 4 in brain and type 4 t PrP^{Sc} in tonsil), and in cases of inherited prion disease. The key shows PrP^{Sc} type or mutation and *PRNP* codon 129 genotype (methionine [M] and valine [V]). Classification according to Hill et al. [9, 10]. Data points represent the mean relative proportions of di- and mono-PrP as percentage ± S.E.M. In some cases, the error bars were smaller than the symbols used

the ratio of the three PrP bands seen after protease digestion, corresponding to amino terminally truncated cleavage products generated from di-, mono-, or nonglycosylated PrP^{Sc} (Figs. 4 and 5). PrP^{Sc} types 1–3 are seen in classical (sporadic or iatrogenic) CJD brain, whereas type 4 PrP^{Sc} is uniquely seen in vCJD brain [7, 9, 78, 80]. An earlier classification of PrP^{Sc} types seen in classical CJD described only two banding patterns [81] with PrP^{Sc} types 1 and 2 that we describe corresponding with the type 1 pattern of Gambetti and colleagues and our type 3 fragment size corresponding to their type 2 pattern [8, 82]. Although type 4 PrP^{Sc} is readily distinguished from the PrP^{Sc} types seen in classical CJD by a predominance of the diglycosylated PrP glycoform, type 4 PrP^{Sc} also has a distinct proteolytic fragment size [9] (Fig. 4), although this is not recognized by the alternative classification, which designates type 4 PrP^{Sc} as type 2b [82]. Although proteinase K-resistant PrP fragments of ~21–30 kDa seen in inherited prion disease caused by *PRNP* P102L, D178N, and E200K mutations have molecular masses similar in size to those seen in classical CJD [10, 83–85], the glycoform ratio is distinct from PrP^{Sc}

fragments seen in both classical CJD [10, 83–85] and vCJD [10] (Fig. 5). Individuals with these mutations also propagate PrP^{Sc} with distinct fragment sizes [10, 83, 84]. The fragment sizes and glycoform ratios of PrP^{Sc} seen in 2-, 4-, and 6-OPRI cases are indistinguishable from those of PrP^{Sc} seen in classical CJD [10]. Importantly detection of PrP^{Sc} in the molecular mass range of ~21–30 kDa is by no means a consistent feature in inherited prion disease; and some cases, in particular those in which amyloid plaques are a prominent feature, show smaller protease-resistant PrP fragments of ~7–15 kDa derived from the central portion of PrP [10, 21, 83, 84, 86–88] (Fig. 4).

5. Sodium phosphotungstic acid precipitation facilitates highly efficient recovery and detection of PrP^{Sc} from human tissue homogenate, even when present at levels 10^4–10^5-fold lower than found in the brain [35, 89]. This procedure is now the preferred method for diagnostic analysis of tonsil in cases of suspected vCJD, and it should detect PrP^{Sc} in tonsil if levels reach 0.1% or above the maximum levels seen in necropsy vCJD tonsil [35, 90]. A distinctive PrP^{Sc} type, designated type 4 t, is seen in both antemortem and postmortem tonsil from patients with vCJD [35, 68] (*see* Figs. 5 and 7), including secondary vCJD infection resulting from blood transfusion [38]. Type 4 t PrP^{Sc} in tonsil differs in the proportions of the PrP glycoforms from type 4 PrP^{Sc} seen in vCJD brain [35, 68] (*see* Figs. 5 and 7), implying the superimposition of tissue- and strain-specific effects on PrP glycosylation [32, 68].

6. On H&E-stained sections, nuclei are stained deep blue, and the cytoplasm is stained pink. The cortex and subcortical white matter can be readily distinguished. In the cortex of a patient with prion disease, there may be variable degrees of spongiosis, accompanied by astroglial proliferation (Fig. 6). Neuronal loss also may be evident. Although synaptic PrP deposition is generally not recognizable on H&E sections, amyloid PrP plaques as seen in GSS and vCJD may be a prominent feature (Fig. 6). In the cerebellum, spongiosis is generally less evident; however, PrP plaques may be observed particularly in GSS.

7. Reactive astrocytes are readily visualized by GFAP immunohistochemistry. They are characterized by prominent processes (Fig. 6). In the white matter, there may be a diffuse fibrillary gliosis.

8. Abnormal PrP deposition can present with a multitude of intensities, shapes, and distributions. The *synaptic* pattern is characterized by a fine, dispersed distribution, and it is the predominant pattern of abnormal PrP staining seen in sporadic CJD (Fig. 6). In contrast, PrP amyloid plaques are a predominant feature in GSS, kuru, and vCJD. The histopathologic

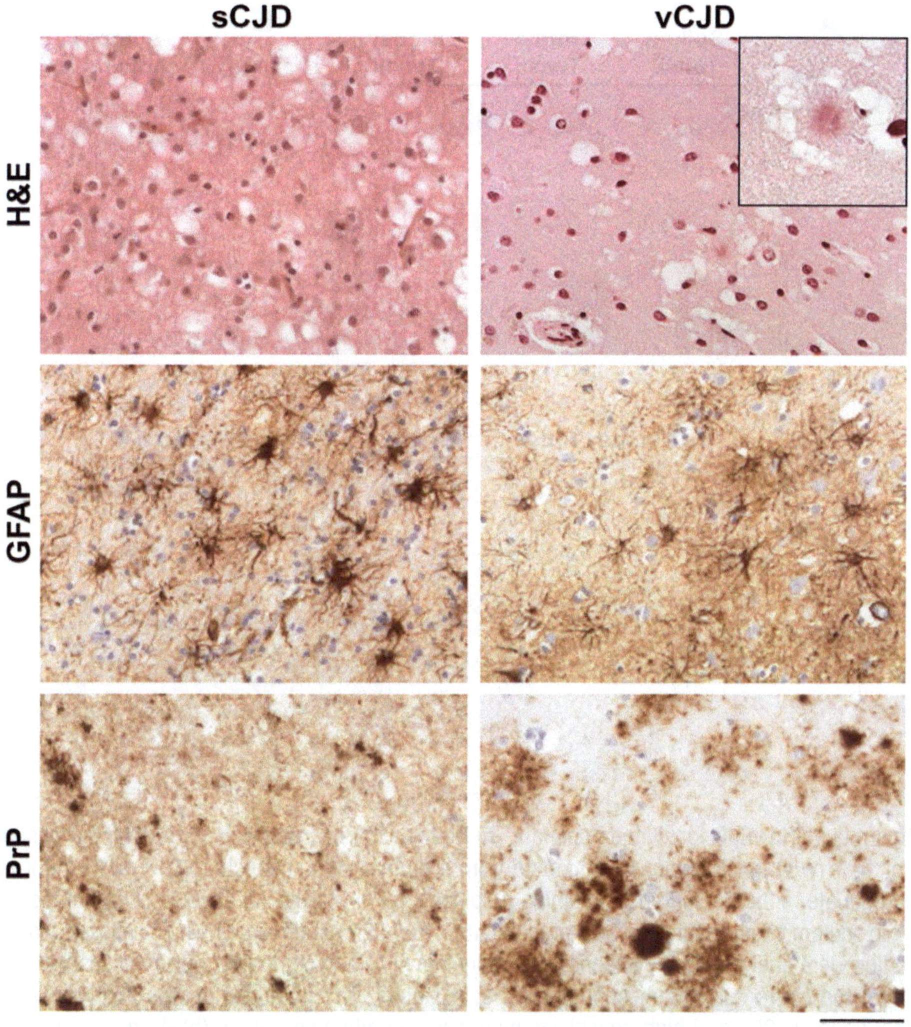

Fig. 6 Prion disease pathology. Brain sections from sCJD and vCJD show spongiform neurodegeneration after hematoxylin and eosin (H&E) staining, proliferation of reactive astrocytes after immunohistochemistry using anti-GFAP antibodies (GFAP), and abnormal PrP immunoreactivity after immunohistochemistry using anti-PrP monoclonal antibody ICSM35 (PrP). Abnormal PrP deposition in sCJD most commonly presents as diffuse, synaptic staining, whereas vCJD is distinguished by the presence of florid PrP plaques consisting of a round amyloid core surrounded by a ring of spongiform vacuoles. Bar = 100 μm. Inset, high-power magnification of a florid PrP plaque

features of vCJD are remarkably consistent and distinguish it from other human prion diseases with large numbers of PrP-positive amyloid plaques that differ in morphology from the plaques seen in kuru and GSS in that the surrounding tissue takes on a microvacuolated appearance, giving the plaques a florid appearance (Fig. 6). Abnormal PrP immunoreactivity in vCJD tonsil [38, 68, 90] and appendix [34, 91] is confined to lymphatic follicles with deposition mainly in dendritic cells (Fig. 7).

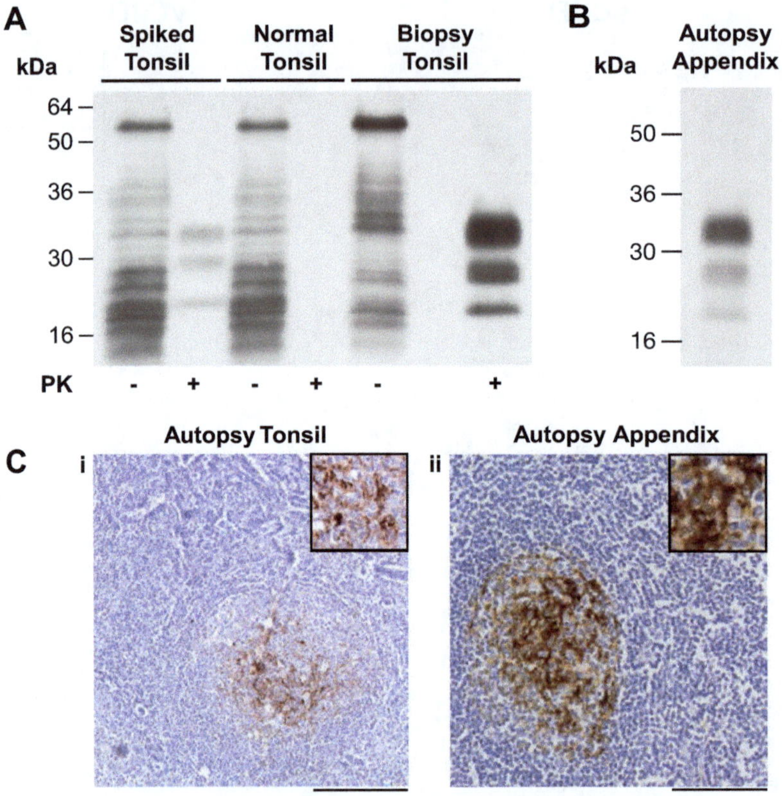

Fig. 7 Abnormal PrP in vCJD tonsil and appendix. (**a**) Diagnostic PrP^{Sc} analysis of tonsil biopsy tissue. Aliquots (0.5 ml) of 10% (w/v) tonsil biopsy homogenate from a patient with suspected vCJD or 10% normal human tonsil homogenate, either lacking or containing a spike of 50 nl of 10% (w/v) vCJD brain homogenate, were subjected to sodium phosphotungstic acid precipitation. Then, 20 μl aliquots of whole samples isolated before centrifugation were analyzed in the absence of PK digestion (−) and compared with PK digestion products (+) derived from the entire sodium phosphotungstic acid pellets. The immunoblot was analyzed with anti-PrP monoclonal antibody 3F4 and high-sensitivity enhanced chemiluminescence. (**b**) PK-digested sodium phosphotungstic acid pellet derived from 0.5 ml 10% (w/v) appendix homogenate from a patient with neuropathologically confirmed vCJD analyzed by high-sensitivity enhanced chemiluminescence using anti-PrP monoclonal antibody 3F4 (**c**). Immunohistochemical analysis of vCJD tonsil (**i**) and appendix (**ii**) obtained at autopsy. Abnormal PrP immunoreactivity is confined to lymphatic follicles with deposition mainly in dendritic cells. Anti-PrP monoclonal antibody ICSM 35. Bar = 160 μm (**c**, **i**) and 100 μm (**c**, **ii**). Insets, high-power magnification of PrP deposits

Acknowledgments

We especially thank all patients and their families for generously consenting to the use of human tissues in this research and the UK neuropathologists who have kindly helped in providing these tissues. The work was funded by the UK Medical Research Council and Department of Health (England) and was performed under approval from the Institute of Neurology/National Hospital for Neurology and Neurosurgery Local Research Ethics Committee and the code of practice specified in the Human Tissue Authority license held by UCL Institute of Neurology. Some of the work was undertaken at the University College London Hospital NHS Foundation Trust which received a proportion of funding from the Department of Health's NIHR Biomedical Research Centres funding Scheme. SB was also supported by the Department of Health's NIHR Biomedical Research Centres funding scheme. We are grateful to R. Young and R. Newton for preparation of the figures.

References

1. Prusiner SB (1998) Prions. Proc Natl Acad Sci U S A 95:13363–13383
2. Collinge J (2001) Prion diseases of humans and animals: their causes and molecular basis. Annu Rev Neurosci 24:519–550
3. Collinge J (2005) Molecular neurology of prion disease. J Neurol Neurosurg Psychiatry 76:906–919
4. Wadsworth JD, Collinge J (2011) Molecular pathology of human prion disease. Acta Neuropathol 121:69–77
5. Collinge J, Clarke A (2007) A general model of prion strains and their pathogenicity. Science 318:930–936
6. Telling GC, Parchi P, DeArmond SJ et al (1996) Evidence for the conformation of the pathologic isoform of the prion protein enciphering and propagating prion diversity. Science 274:2079–2082
7. Collinge J, Sidle KC, Meads J et al (1996) Molecular analysis of prion strain variation and the aetiology of 'new variant' CJD. Nature 383:685–690
8. Parchi P, Giese A, Capellari S et al (1999) Classification of sporadic Creutzfeldt-Jakob disease based on molecular and phenotypic analysis of 300 subjects. Ann Neurol 46:224–233
9. Hill AF, Joiner S, Wadsworth JD et al (2003) Molecular classification of sporadic Creutzfeldt-Jakob disease. Brain 126:1333–1346
10. Hill AF, Joiner S, Beck J et al (2006) Distinct glycoform ratios of protease resistant prion protein associated with *PRNP* point mutations. Brain 129:676–685
11. Safar JG (2012) Molecular pathogenesis of sporadic prion diseases in man. Prion 6:108–115
12. Puoti G, Bizzi A, Forloni G et al (2012) Sporadic human prion diseases: molecular insights and diagnosis. Lancet Neurol 11:618–628
13. Lloyd S, Onwuazor ON, Beck J et al (2001) Identification of multiple quantitative trait loci linked to prion disease incubation period in mice. Proc Natl Acad Sci U S A 98:6279–6283
14. Asante EA, Linehan J, Desbruslais M et al (2002) BSE prions propagate as either variant CJD-like or sporadic CJD-like prion strains in transgenic mice expressing human prion protein. EMBO J 21:6358–6366
15. Lloyd SE, Mead S, Collinge J (2013) Genetics of prion diseases. Curr Op Genet Dev 23:345–351
16. Collinge J (1997) Human prion diseases and bovine spongiform encephalopathy (BSE). Hum Mol Gen 6:1699–1705
17. Collinge J, Palmer MS (1997) Prion diseases. Oxford University Press, Oxford
18. Beck JA, Poulter M, Campbell TA et al (2010) PRNP allelic series from 19 years of prion protein gene sequencing at the MRC prion Unit. Hum Mutat 31:E1551–E1563
19. Wadsworth JD, Hill AF, Beck J et al (2003) Molecular and clinical classification of human prion disease. Br Med Bull 66:241–254
20. Collins SJ, Sanchez-Juan P, Masters CL et al (2006) Determinants of diagnostic investigation sensitivities across the clinical spectrum of sporadic Creutzfeldt-Jakob disease. Brain 129:2278–2287
21. Kovacs GG, Trabattoni G, Hainfellner JA et al (2002) Mutations of the prion protein gene phenotypic spectrum. J Neurol 249:1567–1582
22. Mead S (2006) Prion disease genetics. Eur J Hum Genet 14:273–281
23. Brown P, Preece MA, Will RG (1992) "Friendly fire" in medicine: hormones, homografts, and Creutzfeldt-Jakob disease. Lancet 340:24–27

24. Brown P, Preece M, Brandel JP et al (2000) Iatrogenic Creutzfeldt-Jakob disease at the millennium. Neurology 55:1075–1081
25. Brown P, Brandel JP, Sato T et al (2012) Iatrogenic Creutzfeldt-Jakob disease, final assessment. Emerg Infect Dis 18:901–907
26. Alpers MP (1987) In: Prusiner SB, MP MK (eds) Prions: novel infectious pathogens causing scrapie and Creutzfeldt-Jakob disease. Academic Press, San Diego, pp 451–465
27. Mead S, Stumpf MP, Whitfield J et al (2003) Balancing selection at the prion protein gene consistent with prehistoric kuru-like epidemics. Science 300:640–643
28. Collinge J, Whitfield J, McKintosh E et al (2006) Kuru in the 21st century—an acquired human prion disease with very long incubation periods. Lancet 367:2068–2074
29. Will RG, Ironside JW, Zeidler M et al (1996) A new variant of Creutzfeldt-Jakob disease in the UK. Lancet 347:921–925
30. Hill AF, Desbruslais M, Joiner S et al (1997) The same prion strain causes vCJD and BSE. Nature 389:448–450
31. Bruce ME, Will RG, Ironside JW et al (1997) Transmissions to mice indicate that 'new variant' CJD is caused by the BSE agent. Nature 389:498–501
32. Collinge J (1999) Variant Creutzfeldt-Jakob disease. Lancet 354:317–323
33. Wadsworth JD, Collinge J (2007) Update on human prion disease. Biochim Biophys Acta 1772:598–609
34. Gill ON, Spencer Y, Richard-Loendt A et al (2013) Prevalent abnormal prion protein in human appendixes after bovine spongiform encephalopathy epizootic: large scale survey. BMJ 347:f5675
35. Wadsworth JD, Joiner S, Hill AF et al (2001) Tissue distribution of protease resistant prion protein in variant CJD using a highly sensitive immuno-blotting assay. Lancet 358:171–180
36. Llewelyn CA, Hewitt PE, Knight RS et al (2004) Possible transmission of variant Creutzfeldt-Jakob disease by blood transfusion. Lancet 363:417–421
37. Peden AH, Head MW, Ritchie DL et al (2004) Preclinical vCJD after blood transfusion in a *PRNP* codon 129 heterozygous patient. Lancet 364:527–529
38. Wroe SJ, Pal S, Siddique D et al (2006) Clinical presentation and pre-mortem diagnosis of variant Creutzfeldt-Jakob disease associated with blood transfusion: a case report. Lancet 368:2061–2067
39. Wadsworth JD, Joiner S, Fox K et al (2007) Prion infectivity in variant Creutzfeldt-Jakob disease rectum. Gut 56:90–94
40. Collinge J, Palmer MS, Dryden AJ (1991) Genetic predisposition to iatrogenic Creutzfeldt-Jakob disease. Lancet 337:1441–1442
41. Palmer MS, Dryden AJ, Hughes JT et al (1991) Homozygous prion protein genotype predisposes to sporadic Creutzfeldt-Jakob disease. Nature 352:340–342
42. Windl O, Dempster M, Estibeiro JP et al (1996) Genetic basis of Creutzfeldt-Jakob disease in the United Kingdom: a systematic analysis of predisposing mutations and allelic variation in the *PRNP* gene. Hum Genet 98:259–264
43. Lee HS, Brown P, Cervenáková L et al (2001) Increased susceptibility to Kuru of carriers of the *PRNP* 129 methionine/methionine genotype. J Infect Dis 183:192–196
44. Mead S, Poulter M, Uphill J et al (2009) Genetic risk factors for variant Creutzfeldt-Jakob disease: a genome-wide association study. Lancet Neurol 8:57–66
45. Wadsworth JD, Asante EA, Desbruslais M et al (2004) Human prion protein with valine 129 prevents expression of variant CJD phenotype. Science 306:1793–1796
46. Wadsworth JD, Asante EA, Collinge J (2010) Contribution of transgenic models to understanding human prion disease. Neuropathol Appl Neurobiol 36:576–597
47. Mok T, Jaunmuktane Z, Joiner S et al (2017) Variant Creutzfeldt-Jakob disease in a patient with heterozygosity at *PRNP* codon 129. N Engl J Med 376:292–294
48. Collinge J, Harding AE, Owen F et al (1989) Diagnosis of Gerstmann-Straussler syndrome in familial dementia with prion protein gene analysis. Lancet 2:15–17
49. Collinge J, Owen F, Poulter M et al (1990) Prion dementia without characteristic pathology. Lancet 336:7–9
50. Collinge J, Brown J, Hardy J et al (1992) Inherited prion disease with 144 base pair gene insertion: II: clinical and pathological features. Brain 115:687–710
51. Mallucci G, Campbell TA, Dickinson A et al (1999) Inherited prion disease with an alanine to valine mutation at codon 117 in the prion protein gene. Brain 122:1823–1837
52. Mead S, Poulter M, Beck J et al (2006) Inherited prion disease with six octapeptide repeat insertional mutation—molecular analysis of phenotypic heterogeneity. Brain 129:2297–2317

53. World Health Organisation (2003) WHO manual for surveillance of human transmissible spongiform encephalopathies http://www.who.int/bloodproducts/TSE-manual2003.pdf
54. Collinge J, Poulter M, Davis MB et al (1991) Presymptomatic detection or exclusion of prion protein gene defects in families with inherited prion diseases. Am J Hum Genet 49:1351–1354
55. Owen J, Beck J, Campbell T et al (2014) Predictive testing for inherited prion disease: report of 22 years experience. *Eur J Hum Genet* 22(12):1351–1356. doi:10.1038/ejhg.2014.42
56. Mead S, Gandhi S, Beck J et al (2013) A novel prion disease associated with diarrhea and autonomic neuropathy. N Engl J Med 369:1904–1914
57. Zerr I, Kallenberg K, Summers DM et al (2009) Updated clinical diagnostic criteria for sporadic Creutzfeldt-Jakob disease. Brain 132:2659–2668
58. Hyare H, Thornton J, Stevens J et al (2010) High-b-value diffusion MR imaging and basal nuclei apparent diffusion coefficient measurements in variant and sporadic Creutzfeldt-Jakob disease. AJNR Am J Neuroradiol 31:521–526
59. Carswell C, Thompson A, Lukic A et al (2012) MRI findings are often missed in the diagnosis of Creutzfeldt-Jakob disease. BMC Neurol 12:153
60. McGuire LI, Peden AH, Orru CD et al (2012) Real time quaking-induced conversion analysis of cerebrospinal fluid in sporadic Creutzfeldt-Jakob disease. Ann Neurol 72:278–285
61. Atarashi R, Satoh K, Sano K et al (2011) Ultrasensitive human prion detection in cerebrospinal fluid by real-time quaking-induced conversion. Nat Med 17:175–178
62. Budka H, Aguzzi A, Brown P et al (1995) Neuropathological diagnostic criteria for Creutzfeldt-Jakob disease (CJD) and other human spongiform encephalopathies (prion diseases). Brain Pathol 5:459–466
63. Budka H (2003) Neuropathology of prion diseases. Br Med Bull 66:121–130
64. Brandner S (2013) Prion diseases. In: Ellison D, Love S, Chimelli L, Harding BN, Lowe JS, Vinters HV, Brandner S, Yong WH (eds) Neuropathology: a reference text of CNS pathology, 3rd edn. Elsevier, Mosby, pp 659–676
65. Brandner S, Whitfield J, Boone K et al (2008) Central and peripheral pathology of kuru: pathological analysis of a recent case and comparison with other forms of human prion disease. Philos Trans R Soc Lond Ser B Biol Sci 363:3755–3763
66. Hainfellner JA, Brantner-Inthaler S, Cervenáková L et al (1995) The original Gerstmann-Straussler-Scheinker family of Austria: divergent clinicopathological phenotypes but constant PrP genotype. Brain Pathol 5:201–211
67. Ironside JW, Head MW (2004) Neuropathology and molecular biology of variant Creutzfeldt-Jakob disease. Curr Top Microbiol Immunol 284:133–159
68. Hill AF, Butterworth RJ, Joiner S et al (1999) Investigation of variant Creutzfeldt-Jakob disease and other human prion diseases with tonsil biopsy samples. Lancet 353:183–189
69. Glatzel M, Abela E, Maissen M et al (2003) Extraneural pathologic prion protein in sporadic Creutzfeldt-Jakob disease. N Engl J Med 349:1812–1820
70. Hilton DA, Sutak J, Smith ME et al (2004) Specificity of lymphoreticular accumulation of prion protein for variant Creutzfeldt-Jakob disease. J Clin Pathol 57:300–302
71. Head MW, Ritchie D, Smith N et al (2004) Peripheral tissue involvement in sporadic, iatrogenic, and variant Creutzfeldt-Jakob disease: an immunohistochemical, quantitative, and biochemical study. Am J Pathol 164:143–153
72. Joiner S, Linehan J, Brandner S et al (2005) High levels of disease related prion protein in the ileum in variant Creutzfeldt-Jakob disease. Gut 54:1506–1508
73. Mead S, Wadsworth JD, Porter MC et al (2014) Variant Creutzfeldt-Jakob disease with extremely low lymphoreticular deposition of prion protein. JAMA Neurol 71:340–343
74. Wadsworth JD, Powell C, Beck JA et al (2008) Molecular diagnosis of human prion disease. Methods Mol Biol 459:197–227
75. Advisory Committee on Dangerous Pathogens' Transmissible Spongiform Encephalopathy (ACDP TSE) Risk Management Subgroup (2012) Prevention of CJD and vCJD. UK Department of Health. https://www.gov.uk/government/publications/guidance-from-the-acdp-tse-risk-management-subgroup-formerly-tse-working-group.
76. Safar J, Wille H, Itri V et al (1998) Eight prion strains have PrP^{Sc} molecules with different conformations. Nat Med 4:1157–1165
77. Riesner D (2003) Biochemistry and structure of PrP(C) and PrP(Sc). Br Med Bull 66:21–33
78. Wadsworth JD, Hill AF, Joiner S et al (1999) Strain-specific prion-protein conformation determined by metal ions. Nat Cell Biol 1:55–59

79. Wadsworth JD, Joiner S, Linehan JM et al (2008) Kuru prions and sporadic Creutzfeldt-Jakob disease prions have equivalent transmission properties in transgenic and wild-type mice. Proc Natl Acad Sci U S A 105:3885–3890
80. Wadsworth JD, Joiner S, Linehan JM et al (2008) Review. The origin of the prion agent of kuru: molecular and biological strain typing. Philos Trans R Soc Lond Ser B Biol Sci 363:3747–3753
81. Parchi P, Castellani R, Capellari S et al (1996) Molecular basis of phenotypic variability in sporadic Creutzfeldt-Jakob disease. Ann Neurol 39:767–778
82. Parchi P, Capellari S, Chen SG et al (1997) Typing prion isoforms. Nature 386:232–233
83. Piccardo P, Dlouhy SR, Lievens PMJ et al (1998) Phenotypic variability of Gerstmann-Straussler-Scheinker disease is associated with prion protein heterogeneity. J Neuropathol Exp Neurol 57:979–988
84. Parchi P, Chen SG, Brown P et al (1998) Different patterns of truncated prion protein fragments correlate with distinct phenotypes in P102L Gerstmann-Sträussler-Scheinker disease. Proc Natl Acad Sci U S A 95:8322–8327
85. Furukawa H, Doh-ura K, Kikuchi H et al (1998) A comparative study of abnormal prion protein isoforms between Gerstmann-Sträussler-Scheinker syndrome and Creutzfeldt-Jakob disease. J Neurol Sci 158:71–75
86. Piccardo P, Liepnieks JJ, William A et al (2001) Prion proteins with different conformations accumulate in Gerstmann-Straussler-Scheinker disease caused by A117V and F198S mutations. Am J Pathol 158:2201–2207
87. Tagliavini F, Lievens PMJ, Tranchant C et al (2001) A 7-kDa prion protein (PrP) fragment, an integral component of the PrP region required for infectivity, is the major amyloid protein in Gerstmann-Straussler-Scheinker disease A117V. J Biol Chem 276:6009–6015
88. Wadsworth JD, Joiner S, Linehan J et al (2006) Phenotypic heterogeneity in inherited prion disease (P102L) is associated with differential propagation of protease-resistant wild-type and mutant prion protein. Brain 129:1557–1569
89. Safar JG, Geschwind MD, Deering C et al (2005) Diagnosis of human prion disease. Proc Natl Acad Sci U S A 102:3501–3506
90. Frosh A, Smith LC, Jackson CJ et al (2004) Analysis of 2000 consecutive UK tonsillectomy specimens for disease-related prion protein. Lancet 364:1260–1262
91. Joiner S, Linehan J, Brandner S et al (2002) Irregular presence of abnormal prion protein in appendix in variant Creutzfeldt-Jakob disease. J Neurol Neurosurg Psychiatry 73:597–598

Chapter 23

Molecular Subtyping of PrP^{res} in Human Sporadic CJD Brain Tissue

G.M. Klug, V. Lewis, and S.J. Collins

Abstract

Across the spectrum of sporadic human prion diseases (also known as transmissible spongiform encephalopathies: TSE), there is considerable phenotypic diversity. Cumulative scientific evidence supports that prions, the infectious agents of prion diseases, are constituted predominantly, if not exclusively, by misfolded, typically protease-resistant, disease-associated isoforms of the prion protein (PrP^{res}). Consequently, tissue deposition of PrP^{res} is considered a hallmark of prion disease pathology, and this can be visualized by Western blotting after tissue homogenization and treatment with proteinases, particularly proteinase K (PK). Indeed, Western blot profiles of PrP^{res} are utilized as one marker of different prion strains, with such strains thought to contribute to at least part of the phenotypic variation observed in sporadic human prion disease. Typically, Western blotting of PrP^{res} demonstrates three bands of different electrophoretic mobility, depicting the di-glycosylated, mono-glycosylated and unglycosylated species although further subclassification and the delineation of novel sporadic disease subtypes, such as variably protease-sensitive prionopathy, has contributed greater complexity. Nevertheless, it is the mobility of the unglycosylated PrP^{res} band, the relative abundance of the two glycosylated bands or overall profile of the banding post-PK, in combination with the prion protein gene (*PRNP*) codon 129 genotype that allows the categorisation of molecular subtypes of sporadic human prion disease. These subtypes appear to correlate with distinct clinico-pathological profiles of sporadic Creutzfeldt-Jakob disease.

Key words Glycotyping, Prion protein, Transmissible spongiform encephalopathy (TSE), Western blotting

1 Introduction

The aetiology of human prion diseases includes those with a genetic basis (genetic Creutzfeldt-Jakob disease (gCJD), fatal familial insomnia and Gerstmann-Sträussler-Scheinker syndrome), those acquired through accidental horizontal transmission (iatrogenic CJD, Kuru and variant CJD) and for the majority of human CJD cases (85–90%) and those with no known attributable cause (sporadic CJD; sCJD) [1]. Sporadic CJD is a heterogeneous group with a number of phenotypic subtypes. Such phenotypic subtypes demonstrate correlates with the patient's molecular PrP^{res} type, in

Victoria A. Lawson (ed.), *Prions: Methods and Protocols*, Methods in Molecular Biology, vol. 1658,
DOI 10.1007/978-1-4939-7244-9_23, © Springer Science+Business Media LLC 2017

combination with the *PRNP* genotype at the polymorphic codon 129 site.

PrP^{res} is defined as the protease-resistant, disease-associated conformer of the normal prion protein (PrP^{C}). The altered secondary structure with much higher β-sheet content of PrP^{res} confers poorer solubility and a relative protease resistance of the C-terminal residues of the protein, which allows for its recognition in biochemical assays, with digestion of the N-terminal amino acids up to alternate residues resulting in residual proteins with distinct sizes [2]. The distinct patterns of PrP^{res}, as visualized on Western blots, contributes to defining the PrP^{res} molecular types. Specifically, the relative electrophoretic mobility of the unglycosylated PrP^{res} band after PK digestion in combination with the relative abundance of the di- and mono-glycosylated isoforms can be used to differentiate between different PrP^{res} types. There is strong evidence that these different PrP^{res} types visualized by Western blot, in part, represent different prion strains [2–4], and there has been congruence with molecular subtype reporting in various geographical regions, suggesting not environmental but endogenous determinants of prion molecular subtypes and strains [3]. Different prion strains are thought to be largely enciphered by the conformations of PrP^{res} molecules, with the precise folding into slightly different conformations correlating with different prion strains.

The recent delineation of variably proteinase-sensitive prionopathy (VPSPr) has further expanded the spectrum of sporadic human prion disease, the nosology of molecular subtypes and our understanding of the pathobiology of PrP^{res} [5]. VPSPr is thought to represent approximately 3% of sCJD, with some phenotypic differences occurring across the three codon 129 (MM, MV and VV) types, albeit sharing less typical clinical features such as relatively longer survival and infrequency of characteristic findings on clinical investigations correlating with the altered protease resistance and often relatively lower abundance of PrP^{res} [5, 6]. To optimally identify this molecular subtype of PrP^{res}, it is necessary to utilize an antibody distinct from the standard 3F4 antibody, (epitope spanning residues 109–112 of PrP^{res}) which is widely employed for routine diagnostic molecular subtyping. The 1E4 antibody, which recognizes an epitope from residues 97–108 of PrP^{res}, can be utilized as described by Zou et al. [5] and characteristically demonstrates a "ladder-like" banding pattern comprised of five principle bands of approximately 26, 23, 20, 17 and 7 kDa, with the 7 kDa band generally predominant, especially in VV and MV VPSPr. In this chapter, the routine practices for PrP^{res} molecular subtyping are described.

2 Materials

2.1 Preparation of Brain Homogenate and PK Digestion

1. Sterile scalpels and screw-capped tubes and caps.
2. Phosphate-buffered solution (PBS): prepare 10× stock from 100 mM $Na_2HPO_4.2H_2O$, 31 mM KH_2PO_4 and 1.23 M NaCl:
   (a) Adjust to pH 7.4 with HCl if necessary.
   (b) Autoclave before storage at room temperature (RT).
   (c) Prepare working stock by dilution in sterile, distilled water (1 part 10× PBS: 9 parts distilled water).
3. Hybaid Ribolyser™ (BioRad, Hercules, CA, USA) or dounce tissue grinder, size 21 (Kimble/Kontes, NJ, USA).
4. Zirconium beads (1.0 mm: Biospec Products, Bartlesville, OK).
5. Proteinase K (PK; Life Technologies, Carlsbad, CA, USA): 10 mg/mL in distilled water.
   (a) Store at −20 °C.
   (b) Prepare working stocks of 1 mg/mL in distilled water.
6. Pefabloc (Roche Molecular Biochemicals, Mannheim, Germany): prepare working stock of 100 mM in distilled water.
   (a) Store at −20 °C.
7. 2× Sample Buffer: 125 mM Tris, pH 6.8, 4% (w/v) sodium dodecyl sulphate (SDS), 6% (v/v) 2-mercaptoethanol, 20% (v/v) glycerol and 0.005% (w/v) bromophenol blue.

2.2 SDS-Polyacrylamide Gel Electrophoresis (PAGE)

1. Novex 16% precast Tris–glycine gels—1.0 mm, 10-well (Life Technologies).
2. Precast gel/Sure-Lock system (Life Technologies).
3. 10× running buffer: 250 mM Tris, 1.92 M glycine, 1% (w/v) SDS.
   (a) Store at room temperature.
   (b) Prepare 1× working stock by dilution in distilled water (1 part 10× running buffer: 9 parts distilled water).
4. Benchmark pre-stained protein ladder (Life Technologies).

2.3 Western Blotting

1. 10× transfer buffer: 250 mM Tris, 1.92 M glycine.
   (a) Store at RT.
   (b) Prepare working stock by the dilution of 100 mL 10× transfer buffer, 20% (v/v) methanol in distilled water.
   (c) Store at 4 °C until ready for use.

2. Immobilon-P transfer membrane (polyvinylidene difluoride; PVDF) (Millipore, Billerica, MA, USA).
3. Phosphate-buffered saline with Tween-20 (PBS-T): prepare working stock by diluting 100 mL 10× PBS with 900 mL distilled water, 0.05% (v/v) Tween-20.
   (a) Store at RT.
4. Block buffer: 5% (w/v) nonfat dry milk powder in PBS-T.
5. Primary antibody: 3F4 monoclonal mouse antibody or 1E4 monoclonal mouse antibody (Cell Sciences, Canton, MA).
6. Secondary antibody: Sheep anti-mouse horseradish peroxidase-conjugated whole antibody (GE Healthcare, Little Chalfont, Buckinghamshire, UK).
7. Enhanced chemiluminescence (ECLTM) prime Western blotting detection reagents (GE Healthcare).
8. Biomax light film (Sigma-Aldrich, St Louis, MO).
9. LAS 3000 Imager (Fujifilm Life Science, Minato-ku, Tokyo, Japan).

3 Methods

All steps up to and including loading of samples onto the SDS-PAGE gel are carried out in a Biohazard safety cabinet Class II for operator protection and minimization of sample cross-contamination.

3.1 Brain Sampling and Homogenization

1. Frozen (stored at −80 °C) brain tissue is partially thawed for a minimum of 48 h at -20 °C. Using a long-handled sterile scalpel, remove a small sample of cerebral cortex and place in an appropriately labelled screw-cap tube, seal and weigh.
2. Add 500 mg of zirconium beads to a second sterile screw-capped tube. Add preweighed tissue to prepared tube and the appropriate volume of sterile PBS to give a 10% (w/v) brain homogenate preparation.
3. Sealed samples are then transferred to a Hybaid Ribolyser™ and homogenised with a 45-s pulse at maximum speed (6.5 m/s). A repeat homogenization may be required to achieve a uniform homogenate. If this is the case, a 10-min incubation at 4 °C is necessary between homogenization steps to avoid the sample overheating. Store homogenate at −80 °C until ready for analysis.

3.2 PK Digestion

1. Add 1 μL 1 mg/mL PK to sterile screw-capped tube, and then add 9 μL 10% (w/v) brain homogenate (test samples and known Type 1, 2, 3 or 4 controls) giving a final concentration

of 100 μg/mL PK. For suspected VPSPr brain samples, lower final concentrations of 25 or 50 μg/mL PK may provide more ideal protein banding for analysis.

2. Seal and incubate samples for 1 h at 37 °C with no agitation.
3. Add 100 mM Pefabloc stock to a final concentration of 10 mM to inhibit enzyme activity.
4. Immediately, add the appropriate volume of 2× sample buffer, and boil the sample for 10 min. Spin briefly to condense sample 1000 × *g*, 10 s in Eppendorf centrifuge (Eppendorf, Hamburg, Germany).

3.3 SDS-PAGE

The following instructions are to be used in conjunction with the SureLock system (Life Technologies).

1. Open 16% Tris–glycine gel packet, decant buffer solution, remove comb and adhesive strip, and rinse lanes gently with distilled water. Assemble Safe-Lock gel tank and precast gel, using buffer dam as appropriate.
2. Prepare a 1× running buffer solution by mixing 100 mL of 10× running buffer with 900 mL of distilled water. Fill both the inner and outer chambers of the Sure-Lock tank with 1× running buffer, up to the level of approximately 5–10 mm above the wells.
3. Load appropriate volume of each test sample and controls to wells (*see* **Notes 1** and **2**). This will usually vary between 5 μL and 20 μL of PK-digested brain tissue, depending on the relative amounts of PrP^{res} in the samples. The aim is to visually balance the amounts of PrP^{res} for optimal resolution and comparison of the unglycosylated bands (*see* **Note 3**).
4. Connect the gel tank to a power supply, and run the gel at 125 V for approximately 2 h (or 70 V for 2 h for VPSPr analysis) to allow for adequate resolution.

3.4 Western Blotting

1. Immediately before the end of the SDS-PAGE, cut a gelsized piece of PVDF transfer membrane filter. Pre-wet the membrane in 100% methanol for 1–2 s, rinse in distilled water and then equilibrate in precooled (4 °C) 1× transfer buffer 10–15 min.
2. Upon completion of PAGE, prepare a transfer "sandwich" in a tray containing precooled (4 °C) 1× transfer buffer (*see* **Note 4**). Add transfer cassette to tray, and immerse cassette sponges in transfer buffer. Add one sponge and two pieces pre-cut filter paper to the cassette.
3. Using the spatula tool provided with the Sure-Lock system, open the gel cassette, and carefully excise the stacking gel/wells

(approximately top 1.5 cm of gel) and the lower 0.5 cm of the gel.

4. Carefully load the gel slab to the sandwich, and overlay the pre-wet PVDF. Use the roller tool provided with the Sure-Lock system, to gently remove air bubbles between the gel slab and membrane. Add two pieces of pre-cut filter paper and the second prewet sponge. Close cassette and add to transfer tank.
5. Add ice-block and magnetic stirrer to ensure buffer circulation. Fill the tank with precooled (4 °C) 1× transfer buffer.
6. Connect to power supply and transfer proteins from gel to membrane at 0.38 Amperes for 60 min.
7. After transfer, turn off power supply and disassemble the transfer cassette. Carefully remove the membrane from the cassette and any excess membrane can be trimmed and discarded.

3.5 Immunoprobing

1. Add PVDF membrane to an incubation tray with 20-mL block buffer, and incubate at room temperature for 1 h with agitation on rocking or orbital shaking platform.
2. After incubation, the block buffer is discarded. Add the primary antibody.
   - 3F4 prepared in block buffer at a 1:5000 dilution and incubate overnight at 4 °C with agitation (standard analysis).
   - 1E4 prepared in block buffer at a 1:500 dilution and incubate for 2 h at RT (for VPSPr analysis).
3. Discard primary antibody and wash membrane with PBS-T; three brief rinses then 10-min wash; two brief rinses, then 3-min wash; and one brief rinse, then 3-min wash. Typically, the volume per wash is 15–30 mL.
4. During the final wash, dilute the secondary antibody.
   - 1:15,000 dilution of sheep anti-mouse horseradish peroxidase-conjugated whole antibody in block buffer (standard analysis).
   - 1:3000 dilution of sheep anti-mouse peroxidase-conjugated whole antibody in block buffer (for VPSPr analysis).
5. After discarding the last wash, the secondary antibody preparation is added to the membrane and incubated at room temperature RT for 1 h with agitation.
6. After the secondary antibody solution is discarded, the membrane is washed with PBS-T as in **step 3**.
7. During the wash step, the ECL Prime reagents are equilibrated to room temperature. During the final wash, ECL Prime reagents A and B are combined according to manufacturer's guidelines.

8. Using tweezers, the membrane is removed and blotted using Kim-Wipes to remove excess wash buffer and then applied to a clean and dry acetate sheet. To avoid drying of the membrane, the combined ECL reagents are added directly to the membrane and agitated as necessary to ensure even coverage of entire membrane during the incubation period. Incubate for 5 min at room temperature RT.
9. The membrane is transferred to a second appropriately sized acetate sheet and then carefully overlaid with an acetate top sheet, avoiding air bubbles forming between the membranes and sheeting. Excess ECL fluid can be blotted with Kimwipes.
10. The chemiluminescence can then be visualized by either an X-ray film or digital imaging, for example, using a LAS imager. The acetate membrane-PVDF "sandwich" is transferred to either
    - An X-ray film cassette is placed in a dark room under photographic safe lighting. Biomax light film is then trimmed to an appropriate size and then exposed to the acetate membrane-PVDF "sandwich" for a sufficient period of time to enable a clear comparison of the glycotypes. Routine exposure times range from 30 s to 10 min. Often a shorter exposure period provides the best clarity of images for distinguishing and classifying the glycotypes (*see* **Note 5**). Develop and fix film as per manufacturer's instructions.
    - LAS 3000 imager tray for imaging. Using the LAS 3000 software, routine exposures are typically 30 s to 20 min using the standard or high sensitivity settings.

4 Notes

1. To ensure optimal transfer and resolution of molecular subtypes, avoid using the first and last lanes of the 16% polyacrylamide gels. It is preferable to use these lanes for molecular weight markers or 1× sample buffer.
2. Often samples will need to be re-tested to ensure accurate typing of PrP^{res}. When this is necessary, it is useful to adjust the SDS-PAGE loading pattern so that the sample is positioned between the most relevant control glycotypes.
3. The concentrations of PrP^{res} vary between brain samples. It is preferable to have approximately equivalent protein loadings for the controls and test samples. To achieve this, PK-digested samples can be diluted using 1× sample buffer and then reanalysed by loading equivalent sample volumes for PAGE.
4. 1× Transfer buffer can be reused four to five times without a loss of transfer efficiency, and it should be stored at 4 °C.

5. It is useful to prepare at least one short and one long exposure blot. Short exposures provide visualization of greater detail of the glycotype patterns, which is especially important when a brain sample may comprise of more than one PrP^{res} molecular subtype. Longer exposures may be necessary for the visualization of unexpectedly lower concentrations of PrP^{res}.

Acknowledgements

The Australian National Creutzfeldt-Jakob Disease Registry (ANCJDR) is funded by the Commonwealth Department of Health. The authors thank the families and clinicians for their ongoing support in the surveillance of CJD in Australia. Original T1–T4 standards were supplied to the ANCJDR by the National Institute for Biological Standards and Control (NIBSC) as part of the World Health Organisation Collaborative Nomenclature Study (2001–2002). Since then Australian cases of T1-3 glycotypes have been determined and are used as control samples for future glycotyping gels.

References

1. Collins S, Boyd A, Lee JS et al (2002) Creutzfeldt-Jakob disease in Australia 1970–1999. Neurology 59:1365–1715
2. Hill AF, Joiner S, Wadsworth JD et al (2003) Molecular classification of sporadic Creutzfeldt-Jakob disease. Brain 126:1333–1346
3. Lewis V, Hill AF, Klug GM et al (2005) Australian sporadic CJD analysis supports endogenous determinants of molecular-clinical profiles. Neurology 65:113–118
4. Parchi P, Giese A, Capellari S et al (1999) Classification of sporadic Creutzfeldt-Jakob disease based on molecular and phenotypic analysis of 300 subjects. Ann Neurol 46:224–233
5. Zou WQ, Puoti G, Xiao X et al (2010) Variably protease-sensitive prionopathy: a new sporadic disease of the prion protein. Ann Neurol 68:162–172
6. Gambetti P, Dong Z, Yuan J et al (2008) A novel human disease with abnormal prion protein sensitive to protease. Ann Neurol 63:677–708

Part VI

Methods for Investigation of Prion-Like Disease

Chapter 24

Intercellular Prion-Like Conversion and Transmission of Cu/Zn Superoxide Dismutase (SOD1) in Cell Culture

Leslie I. Grad, Edward Pokrishevsky, and Neil R. Cashman

Abstract

The prion hypothesis has extended to the fatal motor neuron disease, amyotrophic lateral sclerosis (ALS), as a means to explain the spatiotemporal spread of pathology from one or more focal points through the neuroaxis. About 20% of inheritable cases of ALS are due to mutation in the gene encoding the Cu/Zn superoxide dismutase (SOD1), causing the protein to misfold and form neurotoxic aggregates. Mutant SOD1 has been shown to impart its misfold onto natively folded wild-type SOD1 in living cells. Furthermore, misfolded wild-type SOD1 can itself induce further rounds of propagated SOD1 misfolding. Finally, this prion-like mechanism of propagated SOD1 misfolding can be transmitted from cell to cell in human cell culture. Here, we describe a protocol for the induction of wild-type SOD1 misfolding inside living cells and its subsequent transmission from cell to cell in a prion-like fashion.

Key words Cu/Zn superoxide dismutase, Propagated protein misfolding, Intercellular transmission, Immunoprecipitation, Immunoblotting, Cell culture, Transfection, Conditioned medium

1 Introduction

Amyotrophic lateral sclerosis (ALS) is a fatal neuromuscular disorder characterized by degeneration of the upper and lower motor neurons causing progressive paralysis and spasticity [1]. ALS is categorized as a proteinopathy; spinal cord histology of ALS patients often reveals abnormal accumulations of ubiquitinated proteinaceous inclusions in motor neurons and neural accessory cells. These large aggregates are thought to be the result of the aggregation of nonnatively folded proteins, triggered to misfold by a multitude of factors, including mutation, oxidation, aberrant posttranslational modification, dysfunctional proteostasis, and seeded aggregation. Prion-like mechanisms have been proposed and identified for certain proteins involved in ALS, including SOD1, a small soluble and ubiquitously expressed free radical scavenging enzyme that normally exists as a protease-resistant homodimer [2]. In its native state, SOD1 is a highly stable protein;

Victoria A. Lawson (ed.), *Prions: Methods and Protocols*, Methods in Molecular Biology, vol. 1658,
DOI 10.1007/978-1-4939-7244-9_24, © Springer Science+Business Media LLC 2017

however, it is prone to destabilization and subsequent aggregation when aberrantly oxidized or mutated. Recent experimental evidence supports that a protein-to-protein conversion process observed for both mutant and overexpressed wild-type SOD1 ($SOD1^{WT}$) can be self-sustaining [3]. Induced misfolding of SOD1 can persist even in the absence of the original misfolded "seed" in cell culture [3, 4], suggesting that newly misfolded SOD1 can act as template for subsequent cycles of misfolding. The observation that $SOD1^{WT}$ is capable of facilitating intracellular propagated SOD1 misfolding and that misfolded $SOD1^{WT}$ can be detected in non-SOD1 cases of sporadic ALS [5–8] further implicates its misfolding as a possible common pathogenic mechanism of ALS regardless of the presence of mutations in SOD1. Interestingly, the prion-like paradigm of propagated SOD1 misfolding may explain the clinical observations of ALS disease progression, which suggests that motor neuron injury begins at one, or possibly more [9], focal points followed by progressive and contiguous spread of disease through the neuroaxis, i.e., pathology appears to spread without skipping neuroanatomical regions [10].

We have developed a cell-based misfolded human SOD1 conversion-transmission model system that demonstrates the intermolecular conversion and propagated transmission of misfolded $SOD1^{WT}$ [3, 8]. Human mesenchymal or neuroblastoma cells can be transiently transfected with expression plasmids carrying human mutant SOD1 genes that produce obligate misfolded SOD1 protein, which can induce misfolding in endogenous natively-folded SOD1. This induced misfolded $SOD1^{WT}$ has been shown to be secreted by cells into their culture medium [3] and can be subsequently collected and used as template, or "seed," to treat untransformed target cells in separate plates where further rounds of endogenous SOD1 misfolding can occur. The amount of misfolded $SOD1^{WT}$ generated by prion-like propagation can be determined via immunoprecipitation under non-denaturing conditions using conformation-specific antibodies [3, 8, 11] that specifically detect regions of SOD1 that are only exposed upon pathological misfolding and normally buried within the protein under native conditions.

2 Materials

All commercial reagents were stored and used in accordance with the manufacturer's instructions. All buffers and reagents prepared in-house were made using distilled water purified by the Milli-Q Synthesis A10 water purification system (Millipore, Billerica, MA, USA).

2.1 Tissue Culture Components

1. 100 mm × 20 mm cell culture dishes, treated polystyrene (Corning Inc., Corning, NY, USA).
2. Disposable serological pipets, polystyrene. Assorted volumes (5, 10, 25 mL).
3. Dulbecco's Modified Eagle's medium (DMEM) (Gibco by Life Technologies, Grand Island, NY, USA) supplemented with 10% (vol/vol) fetal calf serum, 100 U/mL penicillin, 100 μg/mL streptomycin, and 2 mM L-glutamine. Warm medium to 37 °C.
4. HEK293FT (human embryonic kidney) or SH-SY5Y (human neuroblastoma) cell lines (*see* **Note 1**).
5. DH Autoflow CO_2 Air-Jacketed Incubator (NuAire, Plymouth, MN, USA). Set to 37 °C; 5% CO_2.

2.2 Mammalian Cell Transfection Components

1. Disposable serological pipets, polystyrene, sterile. Assorted volumes (5, 10, 25 mL).
2. 15 mL polystyrene conical centrifugation tubes.
3. Opti-MEM® I Reduced Serum Medium, (Gibco by Life Technologies, Grand Island, NY, USA).
4. Lipofectamine™ LTX with PLUS™ Reagent (Thermo Fisher Scientific, Waltham, MA, USA).
5. Plasmid DNA at a concentration of 1 mg/mL (*see* **Note 2**). Store at 4 °C.

2.3 Immunoprecipitation Components

1. Plastic cell scraper.
2. Phosphate-buffered saline (PBS), pH 7.4 (1×), without calcium chloride or magnesium chloride (Gibco by Life Technologies, Grand Island, NY, USA). Store at 4 °C.
3. 1.5 mL microtubes, polypropylene.
4. 0.65 microtubes, polypropylene.
5. Lysis buffer: PBS, pH 7.4, 0.5% Triton X-100, 0.5% sodium deoxycholate. Add 9 mL PBS to a 10 mL conical plastic tube. Add 0.5 mL 10% Triton X-100 (*see* **Note 3**) and 0.5 mL 10% sodium deoxycholate (*see* **Note 4**). Mix gently by vortexing and store at 4 °C until ready for use.
6. 50× protease inhibitor cocktail: add 1.0 mL of PBS to a 1.5 mL microtube. Add one tablet of cOmplete, EDTA-free Protease Inhibitor Cocktail (Roche, Basel, Switzerland), and vortex until tablet dissolves (*see* **Note 5**). Store at 4 °C for immediate use and −20 °C for long-term storage.
7. Antibody-coupled M-280 Tosylactivated Dynabeads® (Thermo Fisher Scientific, Waltham, MA, USA; *see* **Note 6**).
8. DynaMag™ magnetic rack (Thermo Fisher Scientific, Waltham, MA, USA) for 1.5 mL microtubes.

9. RIPA buffer: 150 mM NaCl, 50 mM Tris–HCl, pH 8.0, 1% NP-40 (Nonidet P-40), 0.5% deoxycholate, 0.1% SDS. Dissolve 146.1 g NaCl in water to a total volume of 500 mL to make 5 M stock. Dissolve 121.1 g Tris into 900 mL. Mix and adjust pH to 8.8 with HCl (*see* **Note 7**). Make up to 1 L with water for 1.0 M stock. For RIPA buffer, add 3 mL 5 M NaCl, 5 mL Tris–HCl, pH 8.8, 1 mL NP-40 (Nonidet), 5 mL 10% sodium deoxycholate, 1 mL 10% sodium dodecyl sulfate (SDS) into 85 mL water in a glass reagent bottle. Mix thoroughly and store at 4 °C.
10. Sample buffer (5×): to a 15 mL plastic conical tube, add 0.5 g SDS, 2.5 mL glycerol, 2 mL 0.5 M Tris–HCl, pH 6.8, 250 μL β-mercaptoethanol (Sigma-Aldrich, St. Louis, MO, USA), 0.0025 g bromophenol blue, and add water to a final volume of 5 mL. Mix gently, but thoroughly. Store at room temperature for short-term, −20 °C for long-term storage (*see* **Note 8**).

3 Methods

Carry out all procedures at room temperature unless otherwise specified. All cell culturing and cell transfection were performed using aseptic technique in a class II type A2 biological safety cabinet.

3.1 Cell Transfection with Mutant SOD1 DNA Constructs

1. One day prior to transfection, plate cells in 8 mL of complete DMEM so that cells will be 50–70% confluent at the time of transfection, and incubate at 37 °C in incubator (*see* **Note 1**).
2. For transfection of cells in 100 mm plates, prepare the following in a 15 mL plastic conical tube: dilute 12.0 μg of plasmid DNA (*see* **Note 9**) in 2.4 mL Opti-MEM® I Reduced Serum Medium, and mix thoroughly.
3. For transfection of SH-SY5Y, add 12.0 μL PLUS™ Reagent to diluted DNA. Mix gently and incubate at room temperature for 5 min. This step is not necessary for HEK293FT cells.
4. Add 30 μL Lipofectamine™ LTX to diluted DNA. Mix thoroughly and incubate for 30–45 min at room temperature.
5. Remove pre-plated cells from incubator (*see* **Note 10**). Add mixture to plates and mix gently by rocking plate back and forth several times.
6. Place plates with cells and transfection reagent back into incubator for 48 h. Replace medium with fresh DMEM after 24 h (*see* **Note 11**).

3.2 Passaging of Conditioned Medium to Untransformed Cells

1. One day prior to incubating untransformed cells with conditioned medium, plate fresh cells so that they will be 70–80% confluent the next day (*see* **Note 1**).
2. Following 48 h from time of transfection, remove plates containing transfected cells from incubator, and remove medium

to fresh 15 mL plastic conical tubes for each transfection plate. Collect transfected cells to test for transfection efficiency (*see* **Note 12**).

3. Centrifuge conditioned media at 1000 × *g* to remove cellular debris (*see* **Note 13**). Remove 8 mL of supernatant to a fresh 15 mL plastic conical tube. Add 2 mL of fresh complete DMEM to replenish nutrients and maintain correct pH. This tube represents the conditioned medium. Remove old media from untransfected recipient cells, and slowly add conditioned media so as not to dislodge cells from plate. Incubate for 20–24 h in incubator at 37 °C. During this time, secreted misfolded SOD1 from the medium will enter recipient cells and induce misfolding of endogenous SOD1 (*see* **Note 14**). Multiple cycles of conditioned medium generation and recipient cell incubation can be performed as required (Fig. 1).

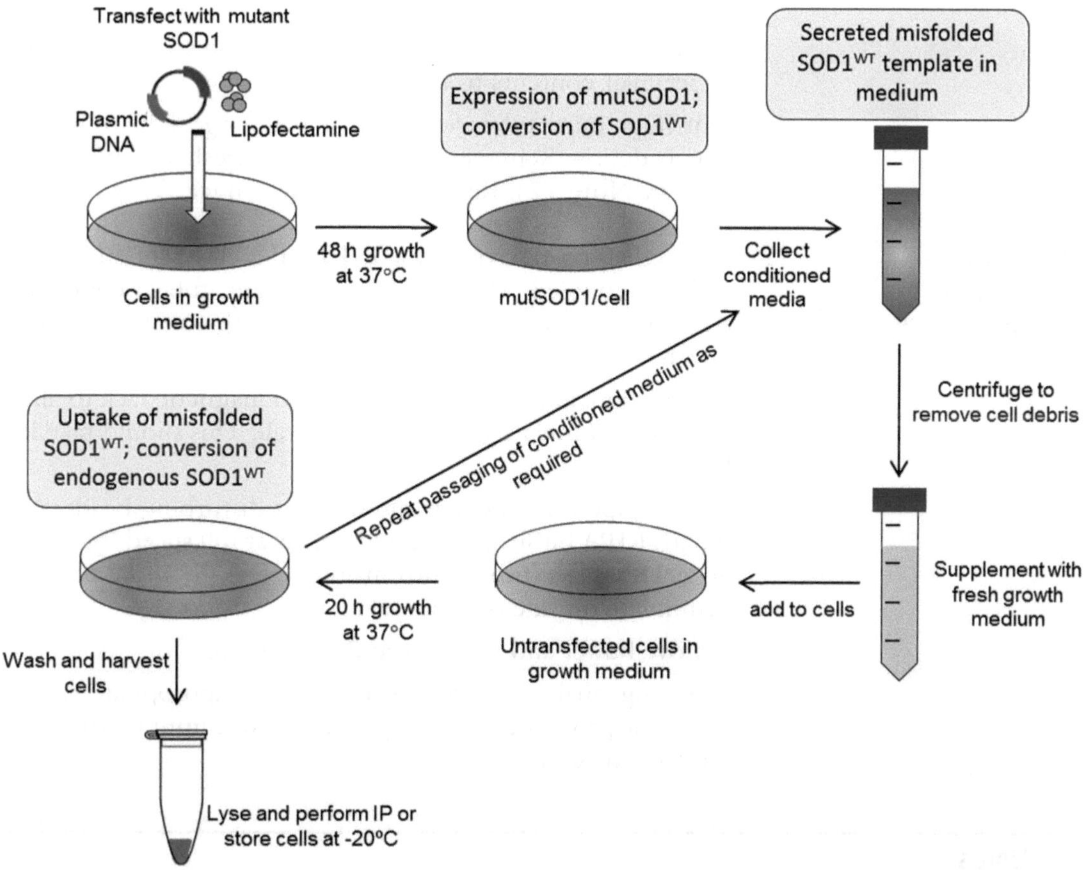

Fig. 1 Diagram of general work flow for cell-to-cell propagation and transmission of misfolded SOD1 in human cell culture. Conditioned medium from transfected cells expressing mutant SOD1 template is collected, clarified, and placed onto untransfected recipient cells. During the subsequent incubation, misfolded SOD1 template enters new cells and induces misfolding of endogenous SOD1 substrate [8]. This process can be repeated over multiple passages as new misfolded SOD1 template is continuously being generated and secreted into the medium

3.3 Cell Collection and Lysis

1. Remove plates containing cells incubated with conditioned media from incubator. Remove media and wash twice with 2 mL of cold PBS. Collect cells in 1 mL cold PBS. For moderately adhesive cells (i.e., HEK293FT), gently pipette PBS against plate to remove cells from plate surface using a P1000 micropipettor. For strongly adhesive cells (i.e., SH-SY5Y), use a plastic cell scraper to loosen cells from plate surface. Once cells have been removed from plate surface, gently tip plate to pool cells and remove to a fresh 1.5 mL microtube.
2. Centrifuge cells for 5 min at 1000 × *g*. Remove supernatant and add 1 mL cold PBS. Gently flick tube to resuspend pellet. Repeat step once more. This will wash remainder of medium from cells.
3. Upon final centrifugation, remove supernatant (*see* **Note 15**). Add ice-cold 500 μL lysis buffer and 10 μL 50× protease inhibitor cocktail, and resuspend pellet by gentle flicking, and then incubate tubes on ice for 2 min (*see* **Note 16**). Cell membranes will preferentially lyse during this time, releasing intracellular contents to the supernatant.
4. Centrifuge microtubes for 5 min at 1000 × *g*, 4 °C to remove cellular debris. Remove supernatant to a fresh 1.5 mL microtube (*see* **Note 17**). Keep lysate samples on ice.

3.4 Immunoprecipitation of Misfolded SOD1

1. In a fresh 0.65 microtube, add 100 μL cell lysate and 10 μL antibody-coupled Dynabeads®. Incubate tubes containing lysate and beads at room temperature with constant rotation for 3 h.
2. Following incubation, place tubes onto magnetic rack to allow beads to accumulate on side of tube wall. This should take less than 15 s.
3. Using a pipette, remove lysate without disturbing beads. Add 150 μL RIPA buffer to tube and vortex at full speed for 2 s (*see* **Note 18**). Place tubes onto magnetic rack to allow beads to accumulate on side of tube wall.
4. Remove buffer and repeat **step 3** twice more.
5. Following final wash (*see* **Note 19**), add appropriate sample buffer for protein gel electrophoresis and immunoblot detection (*see* **Note 20**).

4 Notes

1. We typically store stocks of low-passage cell lines in complete DMEM (*see* **step 3** of Subheading 2.1) supplemented with an additional 20% fetal bovine serum and 10% DMSO, stored at −80 °C or in a liquid nitrogen containment unit. In advance of

experiments involving cell culture, HEK293FT or SH-SY5Y cells are rapidly thawed from frozen stock in a 37 °C water bath, added to 15 mL pre-warmed complete DMEM, and centrifuged at 1000 × *g* for 5 min. Supernatant is then removed and cells are resuspended in 10 mL complete DMEM, plated at high density to optimize recovery, and incubated at 37 °C to ensure proper surface adherence and optimal growth. Cells are subcultured as needed prior to an experiment to ensure desired confluence. In general, the doubling time for HEK293FT cells (~33 h) is much faster than for SH-SY5Y cells (48–55 h); therefore, appropriate time should be given for cell passaging prior to experiment. Cells are discarded approximately 12 passages after thawing.

2. Purification of plasmid DNA from bacterial culture is typically performed using the PureLink® HiPure Plasmid DNA Maxiprep (Invitrogen by Life Technologies, Carlsbad, CA) according to manufacturer's instructions. Following elution of DNA from column with purified water, we quantify plasmid preparations using the NanoDrop 2000 Spectrophotometer (Thermo Fisher Scientific, Wilmington, DE). Prior to transfections, plasmid DNA preparations are diluted to a concentration of 1 mg/mL and stored at 4 °C for short-term use or −20 °C for long-term storage.
3. A 100 mL stock solution of 10% Triton X-100 is made by diluting 10 mL Triton™ X-100 Surfactant (EMD Chemical Inc., Gibbstown, NJ) in 90 mL purified water in a glass bottle. Mix solution by gentle swirling or inversion so as not to cause frothing. Store at room temperature.
4. A 100 mL stock solution of 10% sodium deoxycholate is made by dissolving 10 g sodium deoxycholate powder (Sigma-Aldrich, St. Louis, MO) in purified water to 100 mL in a glass bottle. Mix solution by gentle swirling or inversion so as not to cause frothing. Store at room temperature.
5. Tablets dissolve more quickly when they are broken up to increase surface area. Typically, we place a tablet onto a sheet of weigh paper and cut it using a razor blade until tablet is chopped and pulverized into powder. Tablet powder is then poured carefully into a 1.5 mL microtube and dissolved in 1 mL PBS. Vortexing will ensure powder goes into solution quickly and completely. Resulting solution is a 50× stock of protease inhibitor cocktail. Stock can be stored according to manufacturer's instructions. Short-term storage is typically at 4 °C for up to a week.
6. Magnetic M-280 Tosylactivated Dynabeads® are prepared for antibody coupling according to manufacturer's instructions. We typically prepare 500 μL of antibody-coupled bead slurry

at a time and store at 4 °C. We normally couple 50 μg of rabbit polyclonal antibody and corresponding isotype control antibody and 80 μg of mouse monoclonal antibody and corresponding isotype control antibody per bead preparation.

7. Initially, concentrated HCl (12 N) is used to narrow the gap from starting pH to the desired pH. As you approach the desired pH, more diluted concentrations of HCl are used so as not to dip below the desired pH.
8. SDS precipitates at 4 °C. We therefore warm aliquots of sample buffer to room temperature prior to use.
9. Transfected DNA constructs represent the template used to induce misfolding in native $SOD1^{WT}$. In order to differentiate misfolded SOD1 template from natively-folded $SOD1^{WT}$ substrate, we utilize SOD1 mutants whose translation products migrate faster than $SOD1^{WT}$ via standard SDS-PAGE. These constructs include the full-length mutant $SOD1^{G85R}$ and the truncated mutant $SOD1^{G127X}$ subcloned in place of eGFP in the mammalian expression vector, pFUGW [3]. The latter construct is utilized more than the former as an inducing agent since it does not carry the C-terminal epitopes recognized by our misfolded SOD1-specific antibodies [3]; therefore, any misfolded SOD1 detected must be that of full-length $SOD1^{WT}$ and can be unambiguously visualized using other methodologies such as immunohistochemistry or immunofluorescence. Additionally, other constructs used to induce misfolding of $SOD1^{WT}$ include pathological forms of TDP-43 and FUS [7].
10. Prior to adding transfection mixture, plates with cells are examined briefly under a transmitted light microscope to ensure cells are at the correct confluency, are generally healthy, and show good adherence to plate surface. A high number of floating cells, floating particulate, or morphologically abnormal cells may indicate culture contamination.
11. Cell culture medium is changed after 24 h in order to remove any excess transfection reagent present in the medium. Also, it is important that the culture medium remains nutrient rich and at the correct pH in order to optimize cell health and the exogenous protein expression from transfected constructs. We use DMEM that contains phenol red, a pH-sensitive dye that is red at physiological pH 7.4 but turns yellow when the media is acidic. Growth media on plates should be monitored both after 24 h and after 48 h to ensure pH is near normal. Media that turns pink-orange or orange-yellow indicates a decrease in pH, which may cause unintended cell stress, confounding any downstream results.

12. The amount of induced SOD^{WT} misfolding is often dependent upon the efficiency of mutant SOD1 expression plasmid transfection. In order to confirm successful expression of mutant SOD1 protein, transfected cells can be collected following removal of conditioned medium. Cells can be lysed as described in **steps 1–4** of Subheading 3.3. 1–2 μL of cell lysate can be included in subsequent SDS-PAGE experiments and immunoblotted for SOD1. The relative amount of mutant SOD1 expression can then be determined. Successful mutant SOD1 expression correlates with efficient production of misfolded SOD^{WT} in cells that can subsequently be secreted into the growth medium.
13. We have discovered that the SOD1-associated transmitting agent can be isolated via high-speed centrifugation [8], indicating a relatively large transmitting particle composed of protein aggregates, extracellular vesicles, or both. As an additional step, conditioned medium can be harvested from cells and ultracentrifuged for 1 h at 100,000 × *g*. Following centrifugation, there is rarely a visible pellet, but the medium near the bottom of tube will appear slightly darker and denser. Carefully remove the supernatant from the top down using a 5 mL serological pipet, leaving approximately the last 0.5 mL of medium at the bottom of tube. This represents the fraction of medium containing high molecular weight particles. To this fraction, an additional 9.5 mL of fresh complete DMEM is added. This can then be added to untransfected target cells as stated in **step 3** of Subheading 3.2.
14. The cell type of transfected cells used to generate the conditioned medium does not have to be identical to the target cells that provide new substrate for propagated SOD1 misfolding. Due to the relatively higher amenability of HEK293FT cells to transfection and subsequent transgenic expression and the closer physiological relevance of the SH-SY5Y neuroblastoma cell line to mammalian neurons, we often use the former to generate template for conditioned medium and the latter as target cells to provide endogenous SOD1 substrate for propagated misfolding.
15. At this point, the procedure can be stopped and the cell pellets can be stored at −20 °C until needed.
16. It is important not to vortex the tubes or resuspend the cell pellet through vigorous shaking in order to preferentially lyse the plasma membrane of cells and not the nucleus. Nuclear lysis will result in genomic DNA contamination causing the lysate to become viscous and unusable for immunoprecipitation. Gentle flicking and inversion of the tube should be done until the cell pellet is dislodged from the bottom of the tube and broken up into the buffer.

17. Remove as much supernatant as possible without disturbing the cell debris pellet. This may result in leaving up to 50 μL of cell lysate in the tube, which is acceptable. It is important to have a cell lysate that is uncontaminated with cell debris particulate, which may confound antibody-protein interactions during immunoprecipitation.
18. It is crucial that all samples receive the same amount of vortexing. Too much vortexing may result in the unintentional dissociation of target protein from antibodies, whereas too little vortexing may result in elevated levels of nonspecific protein binding. Initial immunoprecipitation results are often used as a litmus test to determine the relative affinity of the antibodies for specific and nonspecific protein binding, allowing one to gauge how much washing is needed to optimize the signal to background ratio.
19. Once wash buffer is removed from tubes, wait 1 minute to allow residual buffer to pool at bottom of tube, then remove with a pipettor. Also, use pipettor to remove any droplets of buffer stuck to the underside of the tube cap. The presence of excess buffer with the beads increases the total volume of the sample once sample buffer is added to beads to elute protein (and antibody) prior to gel electrophoresis. This will unintentionally dilute your protein sample.
20. Although any electrophoresis system and immunoblotting protocol can be used to assess SOD1 immunoprecipitation, we typically load samples onto Novex® 4–20% Tris–Glycine Gels (Life Technologies, Carlsbad, CA) and electrophorese at 135 V until dye front has reached the bottom of the gel (~1.5 h). Immunoblotting is as previously described [3].

Acknowledgments

This work was supported by donations from the Allen T. Lambert Neural Research Fund and the Temerty Family Foundation, as well as by grants from PrioNet Canada and the Canadian Institutes of Health Research (CIHR).

References

1. Cleveland DW, Rothstein JD (2001) From Charcot to Lou Gehrig: deciphering selective motor neuron death in ALS. Nat Rev Neurosci 2:806–819
2. Ratovitski T, Corson LB, Strain J et al (1999) Variation in the biochemical/biophysical properties of mutant superoxide dismutase 1 enzymes and the rate of disease progression in familial amyotrophic lateral sclerosis kindreds. Hum Mol Genet 8:1451–1460
3. Grad LI, Guest WC, Yanai A et al (2011) Intermolecular transmission of superoxide dismutase 1 misfolding in living cells. Proc Natl Acad Sci U S A 108:16398–16403
4. Munch C, O'Brien J, Bertolotti A (2011) Prion-like propagation of mutant superoxide

dismutase-1 misfolding in neuronal cells. Proc Natl Acad Sci U S A 108:3548–3553
5. Bosco DA, Morfini G, Karabacal NM et al (2010) Wild-type and mutant SOD1 share an aberrant conformation and a common pathogenic pathway in ALS. Nat Neurosci 13:1396–1403
6. Forsberg K, Jonsson PA, Andersen PM et al (2010) Novel antibodies reveal inclusions containing non-native SOD1 in sporadic ALS patients. PLoS One 5:e11552
7. Pokrishevsky E, Grad LI, Yousefi M et al (2012) Aberrant localization of FUS and TDP43 is associated with misfolding of SOD1 in amyotrophic lateral sclerosis. PLoS One 7: e35050
8. Grad LI, Yerbury JJ, Turner BJ et al (2014) Intercellular propagated misfolding of wild-type Cu/Zn superoxide dismutase occurs via exosome-dependent and -independent mechanisms. Proc Natl Acad Sci U S A 111:3620–3625
9. Sekiguchi T, Kanouchi T, Shibuya K et al (2014) Spreading of amyotrophic lateral sclerosis lesions–multifocal hits and local propagation? J Neurol Neurosurg Psychiatry 85:85–91
10. Ravits JM, La Spada AR (2009) ALS motor phenotype heterogeneity, focality, and spread: deconstructing motor neuron degeneration. Neurology 73:805–811
11. Vande Velde C, Miller TM, Cashman NR, Cleveland DW (2008) Selective association of misfolded ALS-linked mutant SOD1 with the cytoplasmic face of mitochondria. Proc Natl Acad Sci U S A 105:4022–4027

INDEX

Victoria A. Lawson (ed.), *Prions: Methods and Protocols*, Methods in Molecular Biology, vol. 1658, DOI 10.1007/978-1-4939-7244-9, © Springer Science+Business Media LLC 2017

H

I

K

L

M

N

P

MIX
Papier aus verantwortungsvollen Quellen
Paper from responsible sources
FSC® C105338

If you have any concerns about our products,
you can contact us on
ProductSafety@springernature.com

In case Publisher is established outside the EU,
the EU authorized representative is:
Springer Nature Customer Service Center GmbH
Europaplatz 3, 69115 Heidelberg, Germany

Printed by Libri Plureos GmbH
in Hamburg, Germany